T0074774

Convergence of Blockchain Technology and E-Business

Green Engineering and Technology: Concepts and Applications

Series Editors: Brujo Kishore Mishra, GIET University, India and Raghvendra Kumar, LNCT College, India

Environment is an important issue these days for the whole world. Different strategies and technologies are used to save the environment. Technology is the application of knowledge to practical requirements. Green technologies encompass various aspects of technology which help us reduce the human impact on the environment and creates ways of sustainable development. Social equability, this book series will enlighten the green technology in different ways, aspects, and methods. This technology helps people to understand the use of different resources to fulfill needs and demands. Some points will be discussed as the combination of involuntary approaches, government incentives, and a comprehensive regulatory framework will encourage the diffusion of green technology, least developed countries and developing states of small island requires unique support and measure to promote the green technologies.

Handbook of Research for Green Engineering in Smart Cities
Edited by Kanta Prasad Sharma, Abdel-Rahman Alzoubaidi, Shashank Awasthi, Ved Prakash Mishra

Green Internet of Things for Smart Cities
Concepts, Implications, and Challenges
Edited by Surjeet Dalal, Vivek Jaglan, and Dac-Nhuong Le

Green Materials and Advanced Manufacturing Technology
Concepts and Applications
Edited by C. Samson Jerold Samuel, M. Suresh, Arunseeralan Balakrishnan, and S. Gnansekaran

Cognitive Computing Using Green Technologies
Modeling Techniques and Applications
Edited by Asis Kumar Tripathy, Chiranji Lal Chowdhary, Mahasweta Sarkar, and Sanjaya Kumar Panda

Handbook of Green Engineering Technologies for Sustainable Smart Cities
Edited by K. Saravanan and G. Sakthinathan

Green Engineering and Technology
Innovations, Design, and Architectural Implementation
Edited by Om Prakash Jena, Alok Ranjan Tripathy, and Zdzislaw Polkowski

Machine Learning and Analytics in Healthcare Systems
Principles and Applications
Edited by Himani Bansal, Balamurugan Balusamy, T. Poongodi, and Firoz Khan KP

Convergence of Blockchain Technology and E-Business:
Concepts, Applications, and Case Studies
Edited by D. Sumathi, T. Poongodi, Balamurugan Balusamy, Bansal Himani and Firoz Khan K P

For more information about this series, please visit: www.routledge.com/Green-Engineering-and-Technology-Concepts-and-Applications/book-series/CRCGETCA

Convergence of Blockchain Technology and E-Business

Concepts, Applications, and Case Studies

Edited by
D. Sumathi, T. Poongodi, Balamurugan Balusamy,
Bansal Himani, and Firoz Khan K P

CRC Press is an imprint of the
Taylor & Francis Group, an **informa** business

First edition published 2022
by CRC Press
6000 Broken Sound Parkway NW, Suite 300, Boca Raton, FL 33487–2742

and by CRC Press
2 Park Square, Milton Park, Abingdon, Oxon, OX14 4RN

CRC Press is an imprint of Taylor & Francis Group, LLC

Library of Congress Cataloging-in-Publication Data

Names: Sumathi, D., editor.
Title: Convergence of blockchain technology and E-business : concepts, applications, and case studies / edited by D. Sumathi, T. Poongodi, Balusamy Balamurugan, Bansal Himani, and Firoz Khan K P.
Description: First edition. | Boca Raton : CRC Press, 2021. | Series: Green engineering and technology : concepts and applications | Includes bibliographical references and index.
Identifiers: LCCN 2021004131 (print) | LCCN 2021004132 (ebook) | ISBN 9780367498146 (hbk) | ISBN 9781003048107 (ebk)
Subjects: LCSH: Business—Data processing. | Electronic commerce. | Blockchains (Databases)—Industrial applications.
Classification: LCC HF5548.2 .C6388 2021 (print) | LCC HF5548.2 (ebook) | DDC 658/.0557—dc23
LC record available at https://lccn.loc.gov/2021004131
LC ebook record available at https://lccn.loc.gov/2021004132

ISBN: 978-0-367-49814-6 (hbk)
ISBN: 978-0-367-49919-8 (pbk)
ISBN: 978-1-003-04810-7 (ebk)

Typeset in Times
by Apex CoVantage, LLC

Contents

Preface

The current wave in this digital world is the digital transformation that involves various technologies like blockchain. It has a significant impending approach to expand business processes and to construct new business models. The adoption of blockchain in various industries has gained momentum in several domains because of its features like trust, distributed ledger, security, and digital identities, etc. Several consensus protocols and methods have been devised by various start-ups, organizations that are associated with the blockchain standards, and universities so that the workload could be disseminated in terms of computation and storage and thereby facilitate a high throughput in the transaction and a tremendous increase in scalability. This book has been designed to address all these facts by providing information in the form of chapters. The objective is to attract all professionals, students, and research people from various domains through the sharing of perspectives from several authors. It is organized as follows:

Chapter 1 starts with the basic concepts of blockchain, the architecture of the blockchain and also illustrates in detail the blockchain escrow service.

Chapter 2 presents the consensus mechanisms and explores the behavior of the distributed ledger in blockchain transactions.

Chapter 3 analyzes the security in each layer of the blockchain and describes several attacks and countermeasures.

Chapter 4 discusses the intervention of blockchain in international trading, ways of boosting the economy, and diversified business models.

Chapter 5 deals with the implication of blockchain in the Internet of Things and presents several case studies that cover various domains like agriculture, logistics, and the pharmacy industry.

Chapter 6 provides more information about blockchain-based health care applications.

Chapter 7 covers the solutions for the integration of blockchain in various fields like the management of water, energy, and waste.

Chapter 8 emphasizes the benefits of the involvement of blockchain in supply chain management.

Chapter 9 discusses the process of blockchain interoperability and its intervention in the health care domain.

Chapter 10 presents a holistic approach to the integration of cybersecurity and blockchain, distributed denial-of-service (DDoS) attacks, and other factors associated with the data.

Finally, Chapter 11 highlights the issues and challenges that deal with the blockchain-based online education and learning environment.

Editors

Balusamy Balamurugan has served up to the position of Associate Professor in his 14 years of experience with VIT University, Vellore. He has completed his bachelors, masters and Ph.D. degrees from premier institutions. His passion is teaching, and he adapts different design thinking principles while delivering his lectures. He has contributed to 30 edited books on various technologies and visited more than 15 countries for his technical discourse. He has several top-notch conferences in his resume and has published over 150 articles in peer reviewed quality journals, conferences and book chapters combined. He serves on the advisory committee for several start-ups and forums and does consultancy work for industry on Industrial IoT. He has presented more than 175 talks in various events and symposium. He is currently working as a professor at Galgotias University, teaching students and doing research on blockchain and IoT.

Bansal Himani has over 14 years of wide experience in academics and the IT industry. She is currently working as Assistant Professor in Jaypee Institute of Information Technology, Noida, India and possesses many reputed certifications such as UGC National Eligibility Test (NET), IBM Certified Academic Associate DB2 9 Database and Application Fundamentals, Google Analytics Platform Principles by Google Analytics Academy, E-Commerce Analytics by Google Analytics Academy and RSA (Rational Seed Academy) and SAP-ERP professional. Her general research interests include machine learning and data analytics, cloud computing, business analytics, data mining and information retrieval. She has filed 4 patents and has around 40 publications including edited books, an authored book, international journals and conferences of high repute. She has served as section editor, guest editor, convener and session chair for various exemplary Journals and Conferences such as SCPE, NGCT, IndiaCom, CSI Digital Life, IJAIP, JGIM, ICACCI, ICCCA, etc. and has reviewed many research papers. She is a Life Member of various professional societies such as CSI, ISTE, CSTA and IAENG and is an active member of IEEE and ACM. Recently, IEEE has conferred her with Senior Membership.

Firoz Khan K P was born in Kerala, India, in 1974. He received his B.Sc. degree in Electronics from the Bharatiyaar University, Coimbatore, India, in 1991 and master's degree in information technology from University of Southern Queensland, Australia, and another master's degree in information network and computer security (with Honors) from New York Institute of Technology, Abu Dhabi, UAE, in 2006 and 2016, respectively. He is currently working toward his Ph.D. in computer science from the British University in Dubai, UAE. In 2001, he joined the Higher Colleges of Technology in the Computer Information Science Department as a teaching technician and subsequently became a faculty member in 2005. He is currently holding the position of a lecturer, with security and networking being his primary areas of teaching. His current research fields include computer security, machine learning, deep learning and computer networking.

T. Poongodi is an associate professor in the School of Computing Science and Engineering, Galgotias University, Delhi—NCR, India. She has completed her Ph.D. in information technology (information and communication engineering) from Anna University, Tamil Nadu, India. She is a pioneer researcher in the areas of Big Data, wireless ad-hoc networks, Internet of Things (IoT), network security and blockchain technology. She has published more than 50 papers in various international journals, national/international conferences, and book chapters in CRC Press, IGI global, Springer, Elsevier, Wiley, DeGruyter and edited books for CRC, IET, Wiley, Springer and Apple Academic Press.

D. Sumathi is an Associate Professor Grade 1-SCOPE at VIT-AP University, Andhra Pradesh. She received the B.E computer science and engineering degree from Bharathiar University in 1994 and M.E computer science and engineering degree from Sathyabama University, Chennai in 2006. She completed her doctorate degree in Anna University, Chennai. She has 18 years' overall experience of which 6 years were in industry and 12 years in the teaching field. Her research interests include cloud computing, network security, data mining, machine learning, natural language processing and theoretical foundations of computer science. She has published papers in international journals and conferences. She has organized many international conferences and also acted as Technical Chair and tutorial presenter. She is a life member of ISTE. She has published chapters in books under Springer, De Gruyter, Wiley and IGI Global publications.

Contributors

A.M. Abirami
Thiagarajar College of Engineering
Tamil Nadu, India

M. Vivek Anand
Galgotias University
Greater Noida, India

Balusamy Balamurugan
Galgotias University
Delhi-NCR, India

Jeyamala Chandrasekaran
Thiagarajar College of Engineering
Tamil Nadu, India

R. Roopa Chandrika
Malla Reddy College of Engineering and Technology
Hyderabad, Telangana, India

Savita Dahiya
Galgotias University
Uttar Pradesh, India

N.S. Gowri Ganesh
Malla Reddy College of Engineering and Technology
Hyderabad, Telangana, India

Vaishali Gupta
Galgotias University
Uttar Pradesh, India

Sathya Karunanithi
Coimbatore Institute of Technology
Coimbatore, India

Hari Kishan Kondaveeti
VIT-AP University
Andhra Pradesh, India

Annapurani Kumarappan
SRM Institute of Science and Technology
Tamil Nadu, India.

M.R. Manu
Faculty, Ministry of Education UAE
Abudhabi, UAE

Divya Menon
Ministry of Education, UAE
Fujairah, UAE

A. Mummoorthy
Malla Reddy College of Engineering and Technology
Hyderabad, Telangana, India,

Namya Musthafa
Royal College of Engineering and Tech
Chiramanangad, Akkikavu, India

S H. Shah Newaz
Universiti Teknologi Brunei (UTB)
Brunei Darussalam

J. Prabhudas
VIT-AP University
Andhra Pradesh, India

S. Prithi
Rajalakshmi Engineering College
Chennai, India

Pradeep Reddy CH
VIT-AP University
Andhra Pradesh, India.

Karthikeyan Saminathan
VIT-AP University
Andhra Pradesh, India

Prabha Selvaraj
VIT-AP University
Andhra Pradesh, India

Nidhi Sengar
Maharaja Agrasen Institute of Technology
Delhi, India

Yogesh Sharma
Maharaja Agrasen Institute of Technology
Delhi, India

S. Sindhu
Jyothi Engineering College
Kerala, India

Apeksha Singh
Galgotias University
Greater Noida, India

Kiran Singh
Galgotias University
Uttar Pradesh, India

Abhilasha Sisodia
Galgotias University
Greater Noida, India

S. Suganthi
Cauvery College for Women
Tamil Nadu, India

D. Sumathi
VIT-AP University
Andhra Pradesh, India

Annie Uthra
SRM Institute of Science and Technology
Tamil Nadu, India.

Soumya Varma
Ministry of Education
Abu Dhabi, UAE

S. Vijayalakshmi
Galgotias University
Uttar Pradesh, India

1 Blockchain and Bitcoin Scripts

S. Prithi, Prabha Selvaraj,
S. Suganthi and D. Sumathi

CONTENTS

1.1 INTRODUCTION

Blockchain technology is an innovation that enables the distribution of digitized information and was originally designed for the bitcoin cryptocurrency. The authors Don and Alex Tapscott of "Blockchain Revolution" defined blockchain as a scrupulous digital ledger of economic transactions that is programmed to record not only financial transactions but everything that has a value. As a basic definition, blockchain is a chain of blocks, in which block represents digital information and chain denotes the data stored in a public database. Blockchains are tamper-proof digital ledgers carried out in a distributed manner without a centralized database and authority/organization. They allow a community of end users to perform transactions in a ledger which is shared within that association. Under the regular operation of the blockchain network, once the transactions are published they cannot be modified. In 2008, the idea of blockchains was integrated with other computing technologies and concepts to devise modern cryptocurrencies in which electronic cash is protected by means of cryptographic mechanisms in place of a central repository (Nakamoto, 2008). In 2009, with the establishment of the bitcoin network, blockchain became extensively in use. The electronic cash, represented as digital information, is transferred in a distributed system. The bitcoin users digitally sign and transfer their rights to another user and the bitcoin blockchain keeps the record of this transfer public so that the other participants can independently confirm the validation of the transactions. Thus, with cryptographic mechanisms, the blockchain is robust to attempts and the ledger can also follow up the forging of transactions.

Blockchain technology has empowered the advancement of cryptocurrency system such as Bitcoin, Litecoin, Peercoin, Namecoin and Ethereum. Blockchain permits members who don't know each other to perform business safely. It verifies the identification of the participants, authenticates the transactions and guarantees that everyone performs by its rules. For instance, once the technology is fully developed and incorporated with supplementary technologies like artificial intelligence (AI) and internet of things (IoT), self-contained agents acting as the driver could negotiate an insurance tariff with multiple car insurance companies instantly using data from sensors. David Furlonger, Distinguished Vice President Analyst, Gartner says that "Blockchain technologies provide a set of abilities that deliver new business, economic and societal paradigms." According to an estimation done by Gartner, blockchain will yield $3.1 trillion in new operation value by 2030, and the organizations should start exploring the new technologies. He also predicted that by 2025, 50% of people who are using a smartphone but don't have a bank account will utilize the mobile-accessible cryptocurrency account.

1.2 CATEGORIES OF BLOCKCHAIN

Blockchain can primarily be categorized into two types, private blockchain and public blockchain. Conversely, there are numerous variations, like Consortium/Federated and Hybrid blockchain. Each blockchain comprises a cluster of nodes operating on a peer-to-peer (P2P) network-based system. Each node contains a copy of the shared ledger that will be updated periodically. Every node can initiate or receive transactions, authenticate transactions and create blocks.

1.2.1 Public Blockchain

A public blockchain is a non-limitative variant where every peer can have a copy of the ledger. It is a permissionless distributed ledger technique where anybody can join and perform transactions. This implies that anybody can have accessibility to public blockchain if they have access to the internet. Bitcoin public blockchain was the first public blockchain that was launched to the public. It allowed everyone who had access to the internet to do transactions in a decentralized manner. The authentication of transactions was done through some agreements such as Proof-of-Work (PoW), Proof-of-Stake (PoS), etc. Some examples are Bitcoin, Litecoin and Ethereum.

1.2.1.1 Advantages

Some of the advantages of public blockchain are:

- Anybody can participate in public blockchain.
- The entire community could be trusted.
- Each and every person feels motivated toward the enhancement of the public network.
- Based on the count of the participating nodes, the public blockchains are protected.
- Transparency is brought to the entire network as the data are accessible for authentication purposes.

1.2.1.2 Disadvantages

Public blockchain has a few disadvantages. They are:

- **Transaction Speed:** To complete a transaction, blockchain takes a few minutes to hours. For example, bitcoin can handle only 7 transactions per second when compared with 24,000 transactions per second by VISA. This is because of the time taken to solve mathematical problems to complete the transaction.
- **Scalability:** As the number of nodes is increased, the performance of the network becomes slow. For instance, bitcoin is working on moving the transactions off-chain to quicken the bitcoin network and to make it more scalable.
- **Choice of agreement:** The choice of agreement method also degrades the performance of the network. For instance, bitcoin uses PoW which absorbs lot of energy.

1.2.2 Private Blockchain

A private blockchain works in a restrictive/closed environment. It is also known as permissioned blockchain, which is under the control of a system. For internal use-cases, the private corporation or organization makes a wonderful usage of private blockchains. The private blockchain could be used effectively and permits only certain participants to access the blockchain network. One of the differences of private blockchain from public blockchain is its accessibility. The next major difference is that only one authority keeps control over the network so the transactions are not done in a decentralized manner. Some examples are Multichain, Hyperledger Fabric, Corda, Hyperledger Sawtooth, etc.

1.2.2.1 Advantages

Some of the advantages of private blockchain are:

- They are relatively fast because there are fewer participants when compared with public blockchain. The network takes less time to attain consensus resulting in faster transactions.
- They are more scalable because only a few blocks are verified to validate transactions. So even if the network grows, the performance of the network, such as speed and efficiency, will not be affected. The essential factor is the centralized system that takes control over the network.

1.2.2.2 Disadvantages

The disadvantages of private blockchain are:

- Private blockchains are not decentralized, which contradicts the core doctrine of distributed ledger technology.
- Because it is a centralized network, it becomes quite difficult to attain trust within the private blockchain.
- Finally, there is a chance to lose security as only few nodes are utilized by the private network.

1.2.3 Consortium/Federated Blockchain

A consortium or federated blockchain is a creative approach to solving the requirements of the corporation where there is a demand for the features of both private and public blockchains. In a federated blockchain, some features of the corporation are made public whereas others remain private. One or more organizations manage the consortium blockchain so there is no one single authority of centralized results. To guarantee suitable functional capability, the consortium contains a validator node that performs two functions, authenticate transactions and also initiate or receive transactions. It provides all the features of a private blockchain such as transparency, efficiency, privacy and power. Some examples of consortium blockchain are Energy Web Foundation, R3 Corda, Marco Polo and IBM Food Trust.

1.2.3.1 Advantages

Some of the advantages are:

- It provides better customization capabilities and control over provisions.
- Federated blockchain is more secure and provides better expandability.
- Consortium blockchain is more efficient than public blockchain networks.
- It operates upon clearly defined structures of governance.
- It provides access controls.

1.2.3.2 Disadvantages

- Despite the fact that it is secure, the entire network can be compromised because of the members' integrity.
- It is less transparent.
- Standards and restraint have a significant influence on network functional capability.

1.2.4 Hybrid Blockchain

Hybrid blockchain is the final type of blockchain and is a combination of private and public blockchain. It does have instances of uses in a corporation that neither desires to organize a private blockchain nor public blockchain and needs to deploy the best of both. Some examples of hybrid blockchain are Dragonchain and XinFin's Blockchain.

1.2.4.1 Advantages

- Need not consider everything public as it operates in a closed ecosystem.
- Based on the requirements, the rules can be modified.
- Hybrid blockchain networks are invulnerable to 51% of attacks.
- Even though it connects with a public network, it provides privacy as well as scalability.

1.2.4.2 Disadvantages

- Not fully transparent.
- It is a challenge to upgrade to the hybrid blockchain network.
- There is no provocation to contribute or participate in the network.

1.3 COMPONENTS OF BLOCKCHAIN

The several components of blockchain technology besides its dependence on cryptographic primitives such as cryptographic hash functions, transactions, digital signatures, asymmetric key cryptography and distributed systems are discussed in this section. Conversely, each component is described in simple terms, and they are used as building blocks to comprehend the complex system.

1.3.1 Cryptographic Hash Functions

Hashing is a technique to apply a cryptographic hash function to input data, which for input of any size like text or image estimates a unique result known as the message digest. The users can in turn use the input data, apply hash function and derive back the same result. The smallest deviation in the input, for instance even a deviation in a single bit, results in a unique output message digest.

Cryptographic hash function possesses three main security properties:

1. **Preimage Resistant:** This property implies that if some output value is given, it is infeasible to compute the correct input value. For instance, given a message digest, find x such that hash(x) = message digest.

2. **Second Preimage Resistant:** This property implies that if the input value is given, it is infeasible to compute a second input that yields the same output. For instance, given x, find y such that hash(x) = hash(y). The only method available is to search the input space exhaustively, nevertheless this is computationally infeasible to do with any chance of success.
3. **Collision Resistant:** This property implies that it is infeasible to obtain two inputs that hash to the message digest. For instance, find an x and y such that hash(x) = hash(y).

The various tasks carried out by cryptographic hash function inside a blockchain network are

- Derivation of address
- Unique identifier creation
- Block data security
- Block header security

Secure Hash Algorithm (SHA) is the most common cryptographic hash function used in the implementation of blockchain; it has an output size of 25 bits so it is also called SHA-256. This algorithm is supported by most computer hardware because of its computation speed. SHA-256 contains an output of 32 bytes, which is usually represented as a 64-character hexadecimal string. The SHA-25 algorithm has approximately $2^{256} \approx 10^{77}$ possible message digests. There are an infinite number of possible input values and a finite number of possible output message digest values, it is highly infeasible to have a collision where hash(x) = hash(y). To identify whether a collision in SHA-256 has occurred, one needs to execute the algorithm, on average, about 2128 times; therefore, SHA-256 is considered collision resistant. The algorithm for SHA-256 is stated in the Federal Information Processing Standard (FIPS PUB 180–184). The specifications of FIPS for all NIST hashing algorithms are specified on the NIST Secure Hashing website.

1.3.2 Transactions

The interactions between parties are referred to as transactions. In terms of cryptocurrencies, the transfer of cryptocurrency among blockchain network users are represented as a transaction. In the blockchain network, each block holds zero or more transactions. It is crucial to sustain the security of the blockchain for an uninterrupted provision of new blocks in the implementation of the blockchain network. The user transmits information to the blockchain network. The transmitted information includes the identification of the sender, the public key of the sender, a digital signature and the transaction inputs and outputs.

The information that is needed for the transaction of a single cryptocurrency transaction are the input and output. In some cases, more information is also required.

- **Inputs:** The list of digital assets that are to be transferred are known as inputs. A transaction refers to the origin of the digital asset, which can be a former transaction given to the sender or new digital assets. The value can neither be attached nor detached from the existing digital assets. The sender should produce proof to access the referenced inputs by signing the transactions digitally.

- **Outputs:** The ledger of the recipients of the digital assets in addition to the digital assets they obtain represent the outputs. Each output indicates the new owner's identification, the amount of digital assets transmitted to the new owner and the set of criteria that should be met by the new owners to disburse the value.

It is very crucial to determine the validity and trustworthiness of the transaction in the blockchain network. The validity of a transaction guarantees that the protocol standards as well as the authorized data format of a transaction are met in the implementation of the blockchain network. The trustworthiness of a transaction determines that the sender of the digital assets had access to those digital assets. Transactions are normally signed digitally by the sender's associated private key, known as asymmetric-key cryptography, and verified in any given time using the corresponding public key.

1.3.3 Asymmetric Key Cryptography

The public key cryptography or asymmetric key cryptography are commonly used by blockchain technology, which consists of a private key and public key bonded with each other. The private key should retain its confidentiality in order to sustain its cryptographic protection, whereas the public key can be available to the public without diminishing the security process. Asymmetric key cryptography allows a relationship of mutual trust among users who do not trust each other by granting a procedure to validate the authenticity and integrity of transactions while permitting the transaction to remain public. To achieve this, the transactions are signed digitally. This is obtained by encrypting the transaction by a private key and decrypting it using a public key. The public key has unrestricted access; the signer of the transaction can gain access to a private key by encrypting the transaction with a private key. The limitation of asymmetric key cryptography is slow computation. The process of asymmetric key cryptography could be explained as follows.

- The transactions are digitally signed by private keys.
- The address is derived by a public key.
- The signatures are verified by public keys generated with private keys.
- It provides the capability of authenticating that the user who transfers value to another user is in control of the private key capable of signing the transaction.

1.3.4 Derivation of Addresses

Addresses are alphanumeric strings of characters that are extracted from the public key of the blockchain network user by means of cryptographic hash function in addition to other data such as checksum and version number. The to and from end points in a transaction use addresses that are shorter than the public key and are not secret.

The address is generated by creating a public key, then cryptographic hash function is applied to the public key and finally the hash function is converted to text. A different method is used to derive an address for each implementation of blockchain networks.

1.3.5 Ledgers

The collection of transactions is called a ledger. Ledgers are stored digitally in a large repository that is possessed and operated by a centralized, trusted third party who is the owner of the ledger as a representative of the user's community. Blockchain technology facilitates both distributed ownership and distributed physical architecture. Some of the major aspects of a ledger are as follows:

- Centrally owned ledgers could be lost or devastated; there should be trust among the users that the backup of the system is done properly by the owner.
- Centrally owned ledgers could be on a homogeneous network, meaning all software, hardware and network infrastructure might be the same. Because of this feature, there might be a reduction in the resiliency of the overall system, because an attack on one part of the network will work everywhere.
- The transactions on a centrally owned ledger are not made transparently and might be invalid; there exists a trust among users that the owner is validating each received transaction.
- There might not be a complete transaction list on a centrally owned ledger; therefore, a trust should prevail among the users that the owner is including all valid transactions that have been received.
- There might be an alteration in the transaction data on a centrally owned ledger; therefore, a trust should prevail among the users that the owner is not altering past transactions.

1.3.6 Blocks

The transactions are submitted by the blockchain users through software such as smartphone/desktop applications, digital wallets, web services, etc. The transactions are sent by the software to the node or nodes across the blockchain network. The nodes could be publishing nodes or non-publishing full nodes. The transactions that are submitted are broadcast to other nodes in the network. When a publishing node publishes a block to the network, the transactions are added to the blockchain. The block comprises a block data and block header where the block header holds the metadata for that particular block and the block data holds a list of authentic and validated transactions that were submitted to the blockchain network. The full nodes verify the authenticity and validity of all the transactions in a published block and if it contains invalid transaction then it will be rejected.

1.4 ARCHITECTURE OF BLOCKCHAIN

The aspects of the blockchain architecture consist of the platform, nodes, transactions, implementation security and the process of adding new blocks to the chain.

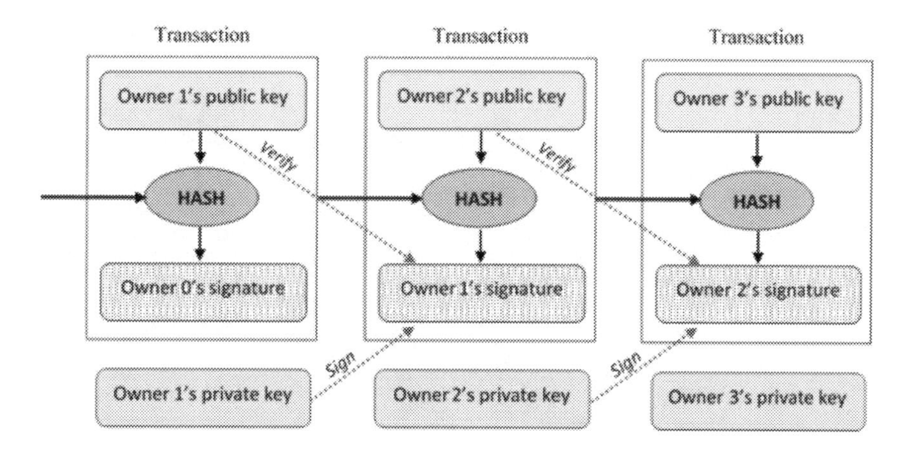

FIGURE 1.1 Architecture of blockchain.

The blockchain architecture is certainly complicated, but in simplified form it is depicted in Figure 1.1.

Once the blocks are connected with each other along with their corresponding hash codes, the entire blockchain ecosystem becomes a repository. Each time a blockchain transaction flag is raised, a blockchain agreement/consensus requirements are accomplished to update the details in the blockchain. The member nodes in the blockchain network adhere to a blockchain agreement/consensus protocol to approve the ledger content. The cryptographic hashes and digital signatures are also approved to confirm the integrity of the transactions. Once the integrity of the transaction is authenticated, the transactions on the blockchain are judged as successful and irrevocable.

1.5 THE BLOCKCHAIN AS A DECENTRALIZED SECURITY FRAMEWORK

The technology has undergone many changes in its development over the past decade to enhance its characteristics and provide more efficient, reliable and secure data. The evolution of the Internet started the digital era, which empowered data at various levels ranging from online transactions to business transactions. The whole information and money transaction framework experiences a secure middle person that maintains the consistency in the data even though the transaction occurs in multiple records. This authorized party is flawed if any fraudulent activities are found in updating information or during delays in conveyance or misrepresentation (Puthal et al., 2017). Blockchain is the only solution for the fraudulent data problems suggested by Satoshi Nakamoto when he presented the standard for decentralization of currency formally known as "Bitcoin" (Grinberg, 2012; Barber et al., 2012; Nakamoto, 2008). A shared distributed record is accountable for the transaction in Bitcoin and is responsible for data storage and processing along the network in a secure manner. Redundancy is avoided by synchronizing the database by accessing the single record and maintaining the consistency of the database transactions. The risk due to redundant data is avoided by synchronization of the data. Bitcoin uses the concept of public key infrastructure (PKI) to secure data from allowing unauthorized users to control the system. To manage simultaneous transactions, they are grouped into similar transactions in a block and generate the hash value and time stamp to retrieve the data. An agreement mechanism is proposed for handling the malicious users performing transactions while updating in the shared ledger. The update process involves an agreement algorithm called Proof-of Work that helps to find the hash value in less than a minute and thus avoid the conflicts in a secure way. The retrieved hash value is the target value to find the unique key to unblock in 10 minutes of time. The procedure to find the target value by utilizing the available resources is called mining, and the nodes involved in the process are called miners. The features of the blockchain are shared, distributed, consensus-based updating, time stamped, sequential, reduced third-party interference, independent, digital ledger that is secured cryptographically and irreversibly with an auditable, immutable, transparent log and decentralized P2P communication (Li et al., 2017; Zheng et al., 2017).

1.5.1 Working Model

The important components of the network setup, blockchain phases, functionalities and their vital role are explained in this section. The functionalities comprise its collaboration with the data and the communication among the nodes involved in the transaction using the agreement mechanism.

1.5.1.1 Background on Blockchain Technologies

Bitcoin started to emerge in 2009, as many architectures were designed to satisfy various technical, commercial and legal transaction preferences. Considering the existing complex design of blockchain architectural enhancement, it would be neither detailed nor inclusive to provide the complete design of the existing blockchain technologies implemented till date. Hence there are detailed reports describing the various principles that include decentralization, robustness, transparency, privacy and security. The distributed model of the network setup needs unauthorized users to find consensus. Considering blockchain, consensus is framed depending upon the protocols or by considering the logs of transactions. The consensus mechanism for the decentralized network manages the updates in the shared ledger by exchanging the work to other nodes that could manage their computation independently. The sharing of responsibility among the nodes means that they no longer exists in the network lagging of failure because of a single node.

Records are subject to audit by a group of users that may constitute some number. In case of transparent blockchain, public users in the network with an Internet data connection have high priority to access the shared ledger. Hence the records can be accessed by everyone and easily tracked. The Internet users have the rights and the resources to manage the ledger. They also have the privilege to combine their independent rights whenever needed. In terms of security, the data in blockchain are shared and subject to replication in other records that cannot be reversed and are also difficult to fake when creating cryptographic value in hash functions.

Security is ensured in blockchain because the data transactions can be performed by users having the private key. The key that generates the respective signature for the corresponding block in the blockchain during transaction. The generated signature helps to validate the transaction with the authorized user and prevent malicious users entering into the system to access or update the data. The principle behind the working of blockchain is the non-repudiation of the records. It also follows the irreversibility process. Blockchain provides high security and immutability, i.e., no other intruders can access the ledger once the data is updated. They are tamper-resistant. In the context of blockchain, immutability is protected or maintained by the utilization of hashes (a kind of a numerical functions that transforms any sort of information into a unique mark of fixed size, that information called a hash. On the off chance that the information changes even eventually, the hash changes in the same way) and frequently of blocks. Each block incorporates the earlier generated hash value as a private key that helps to make a blockchain. The quality of a given blockchain's immutability is virtual and associated with how hard the transaction logs are to change. It is a challenging task for an individual or any group of users to modify the ledger, except if these users have maximum control over the system. For public proof-of-work blockchains, for example, bitcoin, the immutability nature is identified with the expense of actualizing "51% assault." For private blockchains, the block adding system will in general be somewhat different when compared with public blockchain, and as opposed to depending on high cost PoW, the blockchain is just legitimate and acknowledged when blocks are recognized by an authorized user. This implies that, to reproduce the chain, one needs to obtain the private keys from the neighbor block-adders. A total conversation of intrusion to the immutability of transaction logs suggested by Barber et al. (2012) "Bitter to Better." Although blockchains are actually immutable, from an administration point of view, this changelessness is rarely completely figured it out. There are numerous examples where community choices made the bitcoin users return bitcoin blocks. The difference between Ethereum and Ethereum exemplary and later among Bitcoin and Bitcoin Cash and Bitcoin Gold are not simply episodic proof: they are pure pointers of the significance that administering bodies—regardless of informal that closes having on the data ultimately kept in the blockchain. Data automation and storage repository are incorporated as the fundamental characteristics of blockchain. Without the requirement for manual intervention and communication, validation and verification, or assertion, the product is composed so that parallel transactions are not written in the blockchain permanently. Any contention that occurs is accommodated and each legitimate transaction is added just a single time (no twofold updating). In addition, computerization respects to improvement and organization of smart lawful agreements (or savvy contract codes, see Clack, Bakshi, and Braine, "Shrewd Contract Templates" (Clack et al., 2016) with result contingent upon calculations that are self-executable, self-enforceable, self-irrefutable and self-compelled. The repository or the database accessible on blockchain networks can be utilized for the storage capacity and randomly transacting data structures. The information stored in the database can have some size constraints in order to eliminate the "blockchain bloat" problem (Cawrey, 2014). For instance, metadata can be utilized to generate meta-coins: second-layer frameworks that affect the features like reliability and portability of the fundamental coin utilized distinctly as "fuel." Any transaction in the subsequent layer gives the transaction of the other preceding basic network. On the other hand, the storage capacity of extra information can happen "off-chain" through a private cloud on the customer's side or on open (P2P or outsider) database. Some blockchains, such as Ethereum, likewise permit the database of information as a variable of user agreements or as a user agreement log occasion.

1.5.2 Taxonomy of Blockchain

The research in blockchain has been carried out and gives an opportunity to blend thoughts and imagination, yet it can likewise bring about discontinuity in the field and duplication of endeavors. One arrangement is to build up a normalized framework to plan the field and advance by innovative work activities. In any case, regarding blockchain programming design configuration, little has been proposed (Xu et al., 2017), so far, and the issue of reliably designing enormous, complex blockchain frameworks remains unsolved to a great extent. An approach used for this issue suggests a component-based blockchain that starts from coarse-grained connector–segment investigation. The system compartmentalizes the blockchain connectors parts and builds up the connections between them in a categorized way. An approach called reverse-engineering that reveals the blockchains and categorizes them into principle (coarse-grained) modules is also used (Meguerditchian, 2017). Every primary component is then divided into more (fine-grained) sub segments (where fundamental). For every one of these sub (and additionally sub-sub) parts, various designs (models) are distinguished and looked at. By inferring the logical connection between (principle, sub, or sub-sub) parts, the examination helps to explain the usual operation of the blockchains and assists with building up the conceptual blockchain plan and design.

1.5.3 Security and Privacy

The advancement and new executions of blockchain frameworks bring insecurity and issues, both technical and operational, related to security and protection. Accordingly, we bunch together security and confidentiality as the two are related with similar issues. Additionally, ISO TC 307 is being implemented as a Working Group on "Security and Privacy" (www.iso.org/committee/6266604.html). The security of blockchain frameworks is a huge concern. Cryptocurrencies, one of the broadly implemented blockchain frameworks, have experienced cyberattacks, which became conceivable as a result of the mismanagement of delicate information and the imperfect plan and blueprint of the systems (Lin & Liao, 2017). Without going into the elaborate differentiation between "risks," "threats," "attack surfaces," and "vulnerabilities," the security of blockchain frameworks for the most part involves: (i) data mismanagement (variation, deletion, destruction, disclosure, etc.); (ii) implementation vulnerabilities (counting crypto mechanisms' deployed vulnerabilities, run-time loss of data, and so on); (iii) cryptographic framework mismanagement (using algorithms that are weak, key discovery, and so on); and (iv) user rights mismanagement. For an ongoing overall survey explicitly specifying the security and authentication process of bitcoin and its interrelated concepts, see Conti (et al.) "A Survey on Security and Privacy Issues of Bitcoin" (2017), (Conti et al., 2017). These security standards apply to any ICT framework containing or handling PII, including blockchain frameworks.

1.5.4 Security and Protection

For information by data encryption, a concept called cryptographic primitives is proposed. These cryptographic primitives (list of algorithms for cryptographic mechanisms) are utilized to guarantee authentication, confidentiality and requests of events to occur. For instance, Bitcoin blockchain utilizes the ECDSA signature method for validation and integrity and the SHA-2 hash work for the sequence of events to occur. Hash values are additionally utilized as a process of a PoW agreement framework. We distinguish two different formats for data encryption. The first is SHA-2. SHA represents Secure Hash Algorithm. In its two definitions, SHA-256 and SHA-512, SHA (initially created by the National Security Agency of the United States of America) is the most generally applied for hashing value in storage mechanisms, likely using the stemming concept used in bitcoin (Crosby et al., 2016; Harvey, 2016). When applied to hash value transactions, it requires a snippet of data from the end user, i.e., the public key for the approval to occur (Meiklejohn & Orlandi, 2015). The second format is ZK-SNARKS. The Zero-Knowledge Succinct Non-Interactive Argument of Knowledge is a more up to date innovation in which no information must be given to approve a particular hash (Ben-Sasson et al., 2014), as the hash itself fills in as confirmation. Along these lines, both the hashed message and the encoded one are adequate confirmation of presence. This approach anonymizes the personal data to a much greater degree.

1.6 BITCOIN TECHNOLOGY

Bitcoin is a mechanism to make electronic payments between the willing parties directly over a communication channel in the form of cryptocurrency or digital money without a trusted third party. It is the first application of the disruptive blockchain technology introduced by Satoshi Nakamoto and is based on the concept of distributed consensus-based validation. Blockchains were initially developed for the purpose of storing transactions in a shared ledger but later, when combined with other technologies such as cryptography and other computing concepts, were used for cryptocurrencies. There were many solutions proposed for cryptocurrencies, but they necessitated a third party and could not overcome the problem of double spending in which the same bitcoin was used for more than one transaction. In 2008, a model of blockchain was established in a white paper and was posted to a cryptographic mailing list under the pseudonym Satoshi Nakamoto. This may be a single person or a group of developers working on the blockchain, which is not known yet. The system described was a P2P, decentralized cash system called bitcoin. In 2009, the bitcoin source code was released as open-source software.

On January 3, 2009, Satoshi Nakamoto mined the first block (genesis block). On January 12, 2009, Hal Finney received 10 bitcoins from Satoshi Nakamoto and was the first recipient in the world's bitcoin transactions. With the introduction of cryptocurrencies, blockchains have become more prominent nowadays. Bitcoin uses distributed ledger technology (DLT) in which digital data is stored in a secure way and distributed across all nodes connected in a P2P network. It uses a P2P distributed time stamp server network in which all transactions are irreversibly time stamped, which stands as a computational proof of the chronological order of the transactions (Nakamoto, 2008). If the transactions are irreversible and publicly announced to a system of participants, it would stand as a proof and would protect buyers and sellers from potential fraud by augmenting certain security mechanisms. The participants are aware of all the transactions and the system works fine as long as the majority of the nodes are honest when compared with attacker nodes. Thus, bitcoin overcomes the problem of double-spending and does not require a third-party intervention, thereby reducing the transaction costs and eliminating privacy issues.

1.6.1 Bitcoin Basics

1.6.1.1 Transactions

A transaction takes place when two different parties interact and the data is transmitted over the bitcoin network. The transfer of funds from the source is called the transaction input and the value transferred to a destination is called the transaction output. The data includes the transaction identifier, sender's public key, digital signatures and transaction inputs and outputs. Details regarding the receiver's identifier and the amount of data to be transferred are also included. A transaction block is the number of transactions received over a particular period of time. The number of transactions stored in a block depends upon the size of the transaction and the block. If a user wants to send some bitcoins to another user in the network, he can publicly announce the transaction and the correctness of the transaction is verified by the users on the network. The transactions are not encrypted and so all the technical details regarding the transaction can be viewed publicly by all the participants on the network. The ownership can be determined by the payee by verifying the chain of signatures. An electronic coin is a chain of digital signatures. The coin is transferred from one owner to the next when he digitally signs a hash of previous transactions by using his private key, and the next owner verifies it by using the public key. Figure 1.2 shows the transaction structure of bitcoin blockchain. Each transaction output can be used only once thus avoiding the problem of double spending. The output of the transaction is called the unspent transaction output (UTXO), which is the basis of bitcoin transaction (www.oreilly.com/library/view/mastering-bitcoin). The output is called unspent transaction output (UTXO) if it was not referenced before; if it was referenced, it is called spent transaction output (STXO) (Vujičić et al., 2018) UTXO are chunks of bitcoin currency that are locked to a particular user. When a user receives a bitcoin, it is recorded in the bitcoin network in the form of UTXOs, and thus a user's bitcoin is spread among hundreds of transactions and blocks. The user's balance is calculated with a wallet application that aggregates all UTXOs in the bitcoin network specific to that user.

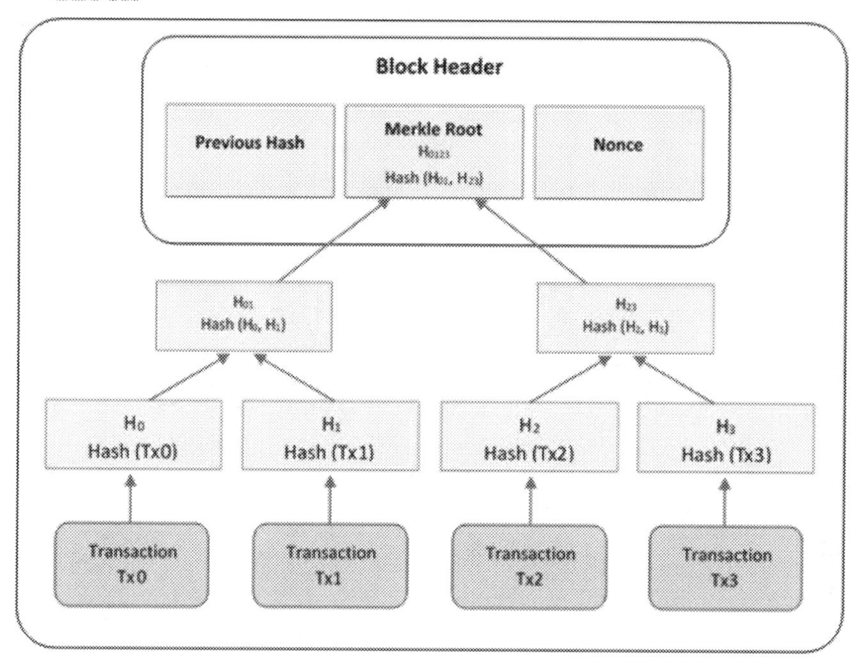

FIGURE 1.2 Transaction structure of the bitcoin blockchain.

The transaction output consists of two parts:

- Bitcoin amount—which is denominated in the smallest bitcoin unit (Satoshi)
- Locking script—which locks the amount by specifying the conditions for spending the amount

A transaction input is a pointer having reference to the transaction hash and the sequence number that contains the UTXO for spending. It also contains unlocking scripts that are usually a signature that proves the ownership of the coin and specifies conditions for spending the amount.

The true validation in the bitcoin network depends on the locking script and the unlocking script. A locking script usually contains a public key or bitcoin address and the unlocking script in the transaction input mostly contains a digital signature generated from the private key of the user. Usually, the transactions are validated by every bitcoin client by concatenating the unlocking script and the locking script and executing them in sequence. As this method is also vulnerable to malformed scripts, the scripts are executed separately by transferring the stack between executions. With this method, the unlocking script is executed first by the stack execution engine. If it is executed without errors, then the locking script is executed by copying the main stack contents. If this is also executed correctly, then the unlocking script has met the constraints of the locking script. Hence, the input is valid for spending the UTXO and is recorded in the blockchain.

The bitcoin miners are rewarded with a small amount as transaction fees for securing the network. The transaction fees are automatically calculated by the wallet application and the fees depend on the size of the transaction and not the value that is transferred.

1.6.1.2 Ledger

Bitcoin uses a distributed ledger database that stores all bitcoin transactions and ownership details of the nodes in the network. The transactions are made public and are visible to everyone on the network. They are synchronized and shared among multiple nodes situated across different locations. The transactions are unalterable once they are stored. The copies of the transactions are present in multiple nodes and so altering or making changes to it is difficult as it requires alteration of all copies of the transaction spread across multiple nodes. The bitcoin ledger can be represented as a state transition system in which each state depicts the ownership details of all existing bitcoins and the state transition function depicts the transactions (Vujičić et al., 2018). The output of this function provides a new state after the transition.

1.6.1.3 Blocks

Blocks are like pages in the ledger having one or more transactions. They are records storing unalterable digital information related to transactions that are connected together to form a blockchain, and any new block can be appended to the chain of blocks. Once a block reaches an approved number of transactions, a new block is formed. Each block contains a block header containing metadata about the block and block data storing a set of transactions and other related data. Each block header except the very first block, known as the genesis block, contains a link to the immediately previous block's header. The link is usually a reference that is the hash value of the previous block or the parent block. Thus, a change of data in a block would change its hash value and thereby be reflected in the subsequent blocks.

The Merkle binary tree is a data structure used to hash the transactions within a block. The Merkle tree has many leaf nodes, and the parent node is the hash of its children. Each transaction is hashed and paired with the hash of another transaction and hashed together. This process is done for all transactions until a single hash value is obtained, which is the root hash. The root hash is stored in the block header and a transaction can be verified by using this root hash without the necessity to download the whole blockchain. Figure 1.3 shows a bitcoin block with hashed transactions in

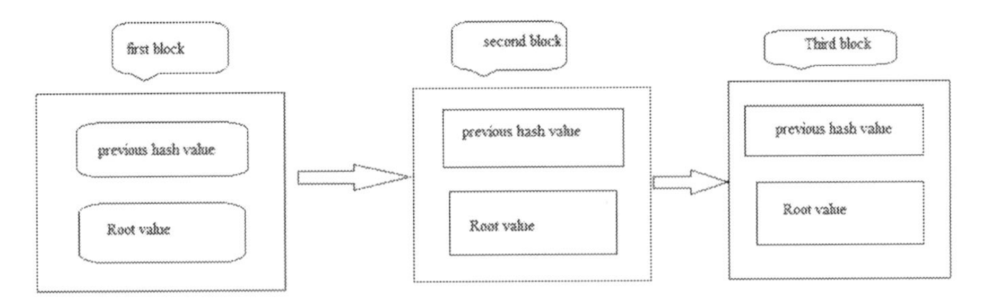

FIGURE 1.3 A bitcoin block with hashed transactions in a Merkle tree.

a Merkle tree. In a large data set, it supports the efficient verification process of the complete data available in the data set. Also, it is essential for long-term maintenance and liberation of storage space to store blockchains on the nodes.

1.6.1.4 Time stamp Server

In bitcoin, double-spending is a problem in which the same bitcoin is used for different transactions with different users. There should be a system that ensures the payee that the payer did not sign any previous transactions with some other users. To overcome this problem without a trusted third party, the transactions are announced publicly to all users in the network and the earliest transaction is the one that is taken into account whereas later transactions with the same bitcoin are ignored. Also, there should be majority of the participants who are aware and agree a single history of transactions which stands as a proof for the payee that it was his transaction that came first. In bitcoins this is implemented by using a time-stamped server. The block or the set of transactions is time stamped by the server by producing a hash of the transactions in the block and publicly announcing it. The time stamp stands as a proof for the existence of the transactions at that time and also for the order of the transactions to arrive at a particular hash value. The hashes are linked in chain with each time stamp including the hashes of the previous time stamp and the time-stamped server checks whether the time stamp of the block is greater than the previous block. With this system the transactions can be traced back to any point in its history.

1.6.1.5 Proof-of-Work

The PoW system is used to implement the distributed time stamped server in a P2P network. PoW, which is used to enhance the immutability of the blockchain, is a solution to a difficult mathematical puzzle that depends on the data in the candidate block (Fullmer et al., 2018). In this, all nodes compete to publish a block by performing a lot of calculations to prove that they are legitimate. This is done by frequently changing the block header's hash value. The node that has computed a hash value less than or equal to some target value is selected for publishing the block. The selected node is rewarded for its work and all other nodes validate it and finally the transaction block is added to the chain of blocks. This system is similar to that of Adam Back's Hashcash, and is based on SHA-256, in which a value is found that when hashed begins with a particular number of zero bits.

In bitcoins, PoW is implemented by incrementing the nonce for every calculation, which is a 4-byte field in the block header usually starting with 0. This is done by expending a lot of CPU effort to find a value that gives the required number of zero bits to the block's hash. The work required is exponential to the number of zero bits in the nonce, which changes the hash output of the block contents. If the PoW is satisfied, the block cannot be changed and is added to the chain of blocks. Also, it is witnessed and validated by the majority of the nodes in the network. If a malicious attacker changes the block, it would require changing all the blocks as it would yield invalid hashes for the following blocks. The longest chain is considered to be the correct one, and thus an attacker has to overcome the majority of the honest nodes, which requires heavy computations.

1.6.1.6 Bitcoin Network

The bitcoin network starts by broadcasting new transactions to all nodes in the network. The transactions are gathered into a block by each node that works on finding the PoW for its block. After which, the block is broadcast in the network and the remaining nodes accept the block after validating the transactions. The accepted block is added to the chain of blocks by creating the next block and adding the hash of the accepted block to the new block as its previous hash.

1.6.1.7 Bitcoin Script

Bitcoin scripts are simple stack-based scripting language used for creating transactions. The scripts are implemented using programming language in which the bitcoin core or the bitcoin software is written. Many languages such as java, python, etc. are used nowadays for writing bitcoin software, but it was originally implemented using C++. The data structure of the bitcoin script can be described as a linear stack-like structure that follows Last In, First Out (LIFO) queue. Bitcoin scripts are intentionally made to be Turing incomplete, which is the inability to run malformed scripts. The malformed scripts may be written intentionally by the hackers or may be done unintentionally by mistake. Running such complex codes may lead to infinite looping, which would cause network slowdown and difficulty in freeing up resources. This may lead to a crash of the whole system. Also, there are certain limitations imposed on the script types used, which are processed by the reference client. The function isStandard() is used for encoding these limitations by which standard transactions are defined and they may change in the future.

1.6.2 Data Insertion in Bitcoins

Bitcoin, which uses decentralized distributed ledger technology, is not only used for digital currency exchange but also for publishing information on a global scale. The data stored in bitcoin is immutable and cannot be retracted and is always available as long as the bitcoin exists. Large amounts of data can be inserted by content insertion services by using low-level data insertion methods that insert small chunks of data (Matzutt et al., 2018). In the following, we have enumerated data insertion through standard transaction methods, nonstandard transaction methods and intended data insertion methods. Data can be inserted in several ways in the bitcoin blockchain and the efficiency depends on the data quantity to be stored, cost incurred and the user's priority (Sward et al., 2018).

1.6.2.1 Data Insertion through Standard Transactions

The standard transaction types used for data insertion are as follows:

1. *Pay-to-public-key-hash (P2PKH)*

Most of the transactions on the bitcoin network are P2PKH, which is more commonly used in today's transactions (https://komodoplatform.com/p2pkh-pay-to-pubkey-hash/). In this type of transaction, the output is encumbered by the locking script using the bitcoin address. The bitcoin address is a unique identifier that consists of a sequence of alphanumeric characters and serves as a virtual location for sending and receiving bitcoins on the network. It is a hash of the recipient public key and exists until the transaction is over.

The public key hashing mechanism used in this type of transaction provides an additional layer of security in which public keys are not made visible to other users in the network. The public key hash is a sequence of alphanumeric characters that is shorter than the public key. Hence it is easy to handle and also plays an effective role in error detection. The public key hash that creates P2PKH transactions is generated by hashing the public key twice by using algorithms, and the resultant public key hash is converted into a bitcoin address. The encoded public key hashes are very effective in detecting errors as the blockchain software will not create transactions for the public key hash if the address executed is invalid.

2. *Pay-to-public-key (P2PK)*

In this transaction method, the locking script contains the public key itself. The unlocking script contains a simple signature for unlocking this type of output. The bitcoin addresses are shorter and easy to use.

3. *Multi-signature*

In this type of script, if N public keys are recorded, a threshold of at least M private keys or signatures are needed for validation. The transactions are very large as the script contains a long public key. They are very efficient but difficult to use.

4. *Pay-to-script hash (P2SH)*

In this transaction method, the complexity of the locking script is overcome by using digital fingerprints. Digital fingerprints are cryptographic hashes that are unique for each file in which the algorithm gives the same output only when the exact same input is given. In this type, in addition to the unlocking script, the transaction spending UTXO should contain a script that should match with the hash.

These standard transaction methods can be misused to insert data by changing the values of the output scripts and the input scripts. In output scripts, data can be inserted by replacing public keys and script hash values with arbitrary data. Although large amounts of data can be stored, it incurs more cost in the way of transaction fees and burning bitcoins for altering the valid identifier of the receiver.

1.6.2.2 Data Insertion through Nonstandard Methods

Transactions which do not fit under the defined transaction template can carry arbitrary data through their input and output scripts. Transactions using nonstandard output scripts can carry data chunks at low cost but are insufficient and mostly ignored by the miners. However, nonstandard input scripts can carry large amounts of data as long as the semantics is not changed and matches with the output script.

1.6.2.3 Intended Data Insertion Methods

1. *Data output (OP-RETURN)*

OP-RETURN is a standard script that allows data and metadata to be stored for each transaction in a small amount. This creates provably unspendable UTXO and hence avoids the tracking done by the miners. OP-RETURN can currently store only 80 bytes per transaction. There can be only one OP-RETURN output among many outputs in a transaction. Multiple transactions are needed to use more than one OP-RETURN for storing data.

2. *Coin-base transaction*

In bitcoin, the new currency for incentivizing miners is introduced by the coin-base transaction and exactly one coin-base transaction is contained in each block. The coin-base transaction is a way of storing data on the blockchain, and it can be used to store up to 100 bytes of data. The general users of the bitcoin cannot use this field, and it is available only to the miners. The arbitrary data is inserted by the miners in the form of ASCII-encoded strings and may contain the mining pool name or any other messages.

1.7 BITCOIN SCRIPTS

Digital signatures and validation are done with private and public key pairs. Bitcoin script is used for the implementation. The nitty-gritty factor is that the input is given as scripts. Transaction authentication could be done to check whether it is in line with the earlier blocks of the chain appropriately so that the input script of a new transaction is combined with the previous transaction's output script.

1.7.1 Bitcoin Scripting Language

The objective of developing the bitcoin script was to have characteristics like key cryptographic functions, simplicity and compactness. The predominant characteristics are

- **Not-Turing Complete**: This refers to the incapability of the language to execute methods using recursion and loops.
- **Limited size**: It is small in size. 256 instructions are bundled as one instruction. It is denoted as one byte.
- **Stack-based:** Every instruction is executed once and linearly. Therefore, the implementation of loops is not used.

1.7.2 Working of a Bitcoin Script

Bitcoin script is defined as a group of instructions that are logged with the corresponding transaction that defines the procedure for the receiver of the funds to access it. Let us discuss an example in P2PKH, or "Pay to Pubkey Hash" payment. Assume a person 'X' sends a bitcoin to another person 'Y' with a P2PKH transaction. Person 'Y' must produce a private key/public key pair before the commencement of the transaction. The public key of 'Y' must be converted to a pubkey hash and then it is changed into an address. 'Y' sends the address to 'X', and then the payment is sent by 'X' to the address of 'Y'. The digital signature is issued from the private key so that it gets matched with the public key of 'Y' through which the control of funds could be obtained by 'Y'. Suppose 'Y' maintains the private key in a protected way, then 'Y' takes control of the funds both in receiving and spending. Pubkey hashes and addresses are used for communication between 'X' and 'Y'. Hence, an additional layer of secured encryption among the private key and public is provided.

For example, the long string in the middle could be given as an instance of a hashed public key.

<sig> <pubKey> OP_DUP OP_HASH160

<371c20fb2e9899338ce5e99908e64fd30b789313> OP_EQUALVERIFY OP_CHECKSIG

The first two in the script are used to denote the signature and the public key. Signature verification is done with the help of the public key. The public key is pushed on to the stack with the help of OP DUP instruction. Using the OP HASH160, the top value is popped and it is used to compute the cryptographic hash. After the computation, the resultant value is pushed into the stack.

1.7.3 Applications of Bitcoin Scripts

Escrow transactions: It is found to be useful under various scenarios (http://learningspot.altervista.org/applications-of-bitcoin-scripts/) such as

- When a person 'X' wants to buy an item from another person 'Y'
- To pay for the item, a person might use bitcoins, and based on the payment, the person could send the item
- Person 'X' doesn't like to pay the amount until the items are received and as well as person 'Y' is not willing to send the item until the payment is received.

To overcome these types of issues,

Person 'X' constructs a MULTISIG transaction. The main objective of the transaction is that it needs more than person to put a signature so that the coins are redeemed. For example, consider there are three persons namely 'X','Y', and 'Z'. In this example, 'Z' is identified as a judge who participates when there is any disagreement between 'X' and 'Y'. Person 'X' puts a signature in a transaction and redeems a few coins that the person possesses. These coins are found to be in escrow among the members involved in the transaction. Among these members, any two could set down the location that the coin should go to. Currently, person 'Y' transfers the item and puts the signature in the transaction that results in releasing the money. At the other end, if person 'X' is satisfied with the received item, then that person could release the money to 'Y' by signing in the transaction. Hence, there is no need for person 'Z' to be involved in this transaction. If the process takes a topsy-turvy turn then there is the need for the intervention of person 'Z'. This happens if person 'X' is not satisfied with the item and asks for the money or the other person 'Y' does not wish to put the signature in the transaction. The role of the person 'Z' is to find out the real reasons and solve this issue. The solution is that there is a need for signatures so that the money will be transferred to anyone. As it has been observed that all bitcoin transactions are irrevocable, there is a necessity to depend on the escrow service if the two parties are not showing trust toward each other. Therefore, a mediator is considered as a third party that can decide which corresponding party could withdraw funds.

This section provides you information about active and optimistic protocols as per the suggestions from Goldfeder et al. (2017). If any dispute occurs between the two parties, then the role of the mediator must come into play. There are two protocols on which the mediator should abide.

1. Lively deposit: The protocol gets its lively nature when the two parties agree to deposit money.
2. Lively withdrawal: The two parties agree to withdraw money even if there is no dispute. At this time also, the mediator must participate actively.
3. Optimistic protocol: In this, the mediator would not be active during the transactions.

1.8 ESCROW PROTOCOLS—SECURITY

The parties must depend on the mediator to decide in the case of a dispute. Hence, the fundamental characteristic of the mediator is that the parties must rely on them to make an impartial decision. There are possibilities of external and internal malevolent mediators. The feature of an external malicious mediator is that there is no control over any of the parties whereas the internal malicious mediator can control. Hence, it is inferred that there is a security breach in the case of an internal malicious mediator. The objective of the mediator is to resolve an issue that occurs between the two parties. The quality of the mediator is to allocate the funds to the respective party when that party agrees. The mediator has the chance of being controlled by the internal antagonist and also any one of the parties. This results in directing the funds to the respective party by going in the party's favor. Hence, there is a need to frame the security for the mediators based on the idea of an external attacker.

4. Secure: The escrow protocol is considered secured if there is no chance of transferring the amount that has been put in the escrow to a random address. This action does not require the support of both parties.

1.9 ESCROW PROTOCOLS—PRIVACY

To perform escrow transactions, three principles of privacy must be followed. They are

5. Dispute-hiding: In this case, there is no chance for the external observer to determine whether the dispute has to be resolved by the mediator.

6. External-hiding: In this case, the external observer could not identify the particular transaction on the blockchain that constitutes the components of that respective protocol.
7. Internal-hiding: Protocol execution might not be known by the mediator itself.

This section describes the protocols that have been adopted by the mediators and the properties of these protocols.

1.9.1 Escrow through Uninterrupted Payment

Transfer of money is done through the mediator. If there is any dispute, then it is the responsibility of the mediator to settle the money to the respective person honestly. The privacy is augmented by sending the funds to the blinded address. Through this method, in spite of not using the intermediator's address, it is made as not active on deposit. Hence, the two parties involved in this transaction execute this scheme by producing the randomness in such a way that both parties are satisfied in the proper execution of the algorithm. The escrowed funds could be transferred by providing the random number to the intermediator. It is combined with the secret key x to sign over the public key 'y' to generate the payout transaction.

Features:

There are several limitations, such as insecure, not ensured and it does not have internal hiding. Additionally, this scheme is lacking in active-on-deposit and simplicity. It preserves privacy to a certain extent because it has the features of external hiding and dispute hiding.

1.9.2 Escrow through MultiSig

As per the name, the fund is transferred to 2-of-3 multisig address. Here, the control is with one of the parties and another control is taken by the intermediator. If there is no dispute, then it is possible for the parties to cooperate in generating the payout transaction. But, if a dispute arises, then the intermediator is involved to cooperate with one of the parties so that the funds would be redeemed to any of them.

Features:

- Secure
- Optimistic
- Internal hiding
- Dispute detection is made possible with the analysis of the transaction graph.

1.9.3 Escrow through Threshold Signatures

The privacy could be improved still with the deployment of the threshold address. This works like a joint account in banking. The three parties involved in the transaction are able to construct a single key address that is made regular so that any of the two parties could spend the money in a cooperative manner. Through this scheme, it is not possible to determine the party that took part in the transaction.

Features

- Dispute hiding
- Secure
- External hiding
- Prone to denial of service
- New key generation is done every time and thereby it is the responsibility of the transaction parties and intermediator to monitor it.

1.9.4 Escrow through Bond

The intermediator might fail to release the funds from the escrow. This kind of service must be done by the inclusion of an incentive system so that the intermediator might be penalized for not sending the funds. Therefore, the intermediator is required to provide a surety bond to the parties that are involved in the transaction. Prevention of denial of service attacks by making the third party put funds in bond exists in other perspectives (Jakobsson, 1995). To generate a value, bitcoin scripting language is used. In this scheme, SHA(256) is applied on ‘x’ to get ‘y’. This behaves as the condition for spending the fund. It is extensively used in the generation of automatic cross-chain swap protocols and offline micropayment channels (Poon & Dryja, 2016).

1.9.5 Blockchain Escrow Services

This section describes the several blockchain escrow services. They are:

MarketPay: With this, economy sharing is made possible. In particular, it is deployed for end-to-end payment providers, flexible escrow contract rules and for construction of a built-in second ledger so that this scheme provides transparency and security.

IBC Escrow Service: This service makes use of smart contract as the third party to do transfers. A time limit has been fixed for the buyer so that the transactions are done through the escrow platform. It is a licensed escrow provider that adopts the policies of Know Your Customer (KYC) and Anti-Money Laundering (AML).

Themis: This is used to provide a decentralized escrow service for resolving the issues that are raised during the use of digital currencies as the platform. It is mainly deployed in P2P payments, Over-the-Counter market trading and more.

REFERENCES

Barber, S., Boyen, X., Shi, E., and Uzun, E. “Bitter to Better—How to Make Bitcoin a Better Currency.” In A. Keromytis (Ed.), *International Conference on Financial Cryptography and Data Security*. Berlin: Springer, 399–414, 2012. https://doi.org/10.1007/978-3-642-32946-3_29.

Ben-Sasson, E., Chiesa, A., Tromer, E., and Virza, M. "Scalable Zero Knowledge Via Cycles of Elliptic Curves." In J. A. Garay and R. Gennaro (Eds.), *Advances in Cryptology—CRYPTO 2014*. Berlin: Springer, 276–294, 2014. https://doi.org/10.1007/978-3-662-44381-1_16.

Cawrey, D. "Why New Forms of Spam Could Bloat Bitcoin's Block Chain." *CoinDesk*, 2014 (accessed 13 January 2017). www.coindesk.com/new-forms-spam-bloat-bitcoins-block-chain/.

Clack, C. D., Bakshi, V. A., and Braine, L. "Smart Contract Templates: Essential Requirements and Design Options." *arXiv*, 2016 (accessed 29 October 2018). https://arxiv.org/abs/1612.04496.

Conti, M., Lal, C., Ruj, S., and Sandeep, K. E. "A Survey on Security and Privacy Issues of Bitcoin." *arXiv*, 2017 (accessed 29 October 2018). https://arxiv.org/abs/1706.00916.

Crosby, M., Pattanayak, P., Verma, S., and Kalyanaraman, V. "Blockchain Technology: Beyond Bitcoin." *Applied Innovation Institute*, 2016 (accessed 29 October 2018). www.appliedinnovationinstitute.org/blockchain-technology-beyond-bitcoin/.

Fullmer, D., and Stephen Morse, A. "Analysis of Difficulty Control in Bitcoin and Proof-of-Work Blockchains." *IEEE*, pp. 5988–5992, 2018.

Goldfeder, S., Bonneau, J., Gennaro, R., and Narayanan, A. "Escrow Protocols for Cryptocurrencies: How to Buy Physical Goods Using Bitcoin." In A. Kiayias (Ed.), *Financial Cryptography and Data Security*. FC 2017. Lecture Notes in Computer Science, Vol. 10322. Cham: Springer, 2017. https://doi.org/10.1007/978-3-319-70972-7_18.

Grinberg, R. "Bitcoin: An Innovative Alternative Digital Currency." *Hastings Science & Technology Law Journal*, Vol. 4, pp. 159–208, 2012.

Harvey, C. R. "Cryptofinance." SSRN, 2016 (accessed 29 October 2018). https://ssrn.com/abstract=2438299.

Jakobsson, M. "Ripping Coins for a Fair Exchange." In *Advances in Cryptology Eurocrypt95*, Vol. 921. Berlin, Heidelberg: Springer, 220–230, 1995.

Li, X., Jiang, P., Chen, T., Luo, X., and Wen, Q. "A Survey on the Security of Blockchain Systems." *Future Generation Computer Systems*, 2017. https://doi.org/10.1016/j.future.2017.08.020.

Lin, I.-C., and Liao, T.-C. "A Survey of Blockchain Security Issues and Challenges." *International Journal of Network Security*, Vol. 19, No. 5, pp. 653–659, 2017. https://dx.doi.org/10.6633/IJNS.201709.19(5).01.

Matzutt, R., Hiller, J., Henze, M., Ziegeldorf, J. H., Müllmann, D., Hohlfeld, O., and Wehrle, K. "A Quantitative Analysis of the Impact of Arbitrary Blockchain Content on Bitcoin." In *22nd International Conference, FC 2018, Nieuwpoort, Curaçao*. Netherlands: Springer, 420–438, 2018.

Meguerditchian, V. "Roadmap for Blockchain Standards Report—March 2017." *Standards Australia*, 2017 (accessed 29 October 2018). www.standards.org.au/getmedia/ad5d74db-8da9-4685-b171-90142ee0a2e1/Roadmap_for_Blockchain_Standards_report.pdf.aspx.

Meiklejohn, S., and Orlandi, C. "Privacy-Enhancing Overlays in Bitcoin." In M. Brenner, N. Christin, B. Johnson, and K. Rohloff (Eds.), *Financial Cryptography and Data Security, FC 2015 International Workshops, BITCOIN, WAHC, and Wearable, San Juan, Puerto Rico, January 30, 2015, Revised Selected Papers*. Berlin: Springer, 127–141, 2015. https://doi.org/10.1007/978-3-662-48051-9_10.

Nakamoto, S. "Bitcoin: A Peer-to-Peer Electronic Cash System." 2008 (accessed 11 November 2017). https://bitcoin.org/bitcoin.pdf.

National Institute of Standards and Technology (NIST). Secure Hashing website. https://csrc.nist.gov/projects/hash-functions.

National Institute of Standards and Technology (NIST), Secure Hash Standard (SHS), Federal Information Processing Standards (FIPS) Publication 180–184, August 2015. https://doi.org/10.6028/NIST.FIPS.180-4.

Poon, J., and Dryja, T. *The Bitcoin Lightning Network: Scalable Off-Chain Instant Payments*. Tech. rep, 2016.

Puthal, D., Mohanty, S., Nanda, P., and Choppali, U. "Building Security Perimeters to Protect Network Systems against Cyber Threats." *IEEE Consumer Electronics Magazine*, Vol. 6, No. 4, pp. 24–27, 2017.

Sward, A., Vecna, I., and Stonedahl, F. "Data Insertion in Bitcoin's Blockchain." *Ledger*, Vol. 3, 2018.

Vujičič, D., Jagodić, D., and Ranđić, S. "Blockchain Technology, Bitcoin, and Ethereum: A Brief Overview." In *17th International Symposium INFOTEH-JAHORINA (INFOTEH)*. IEEE, 2018.

Xu, X., et al. "A Taxonomy of Blockchain-Based Systems for Architecture Design." In *2017 IEEE International Conference on Software Architecture*, 243–252, 2017. https://doi.org/10.1109/ICSA.2017.33; www.iso.org/committee/6266604.html.

Zheng, Z., Xie, S., Dai, H., Chen, X., and Wang, H. "An Overview of Blockchain Technology: Architecture, Consensus, and Future Trends." In *Proceedings of the IEEE International Congress on Big Data*, 557–564, 2017.

2 An Intuitive Approach behind Distributed Ledger Technology

An Elaborative Glance of Blockchain, Consensus Mechanisms, and Their Applications

J. Prabhudas and Pradeep Reddy CH

CONTENTS

2.1 INTRODUCTION

Blockchain is an innovation that empowered the traditional way of organizing transactional data securely among 'n' number of participants connected across diversified areas. To elaborate, it is a revolutionary technology that lets the computers or individuals or nodes organized as a network (P2P) record a value of a transaction via a global ledger. In general, every unique node in the network is distributed and decentralized, meaning that no middle-authority holds control over transactions. Its emergence is marked as outstanding because of the characteristics it possesses such as reliable transfer of data, transactional security, system fault tolerance, data integrity maintenance, and synchronization among nodes.

The underlying working concept of blockchain was initiated by Stuart Haber and W. Scott Stornetta (Haber and Stornetta, 1990), (Narayanan et al., 2016) in the year 1991 to make the documents in use tamperproof. They introduced a method that marks the documents with a time stamp whenever the concerned document is altered so that the originality is maintained as well as making it impossible for an intruder to update it. Further, in 1992, the conceptual improvement was made to an existing system (Haber and Stornetta, 1990) design by the researchers Haber, Stornetta, and Dave Bayer (Bayer, Haber and Stornetta, 1993) who introduced the "Merkle trees" data structure that yielded better efficiency. The discovered technique (Bayer, Haber and Stornetta, 1993) is capable of incorporating multiple time stamped documents into one group (Narayanan et al., 2016). Eventually, the present in depth blockchain systems were built with the conceptual idea described in (Bayer, Haber and Stornetta, 1993) and with "Merkle trees" as one of the data structures (Yang et al., 2019).

However, the prior idea in (Haber and Stornetta, 1990), (Bayer, Haber and Stornetta, 1993) failed to draw much attention from industrialists until it was systematically introduced by the authors "Satoshi Nakamoto" (Nakamoto et al., 2008). Significantly, the proposed blockchain framework in (Nakamoto et al., 2008) became a breakthrough to enormous hype in information technology. Blockchain is an empowering technology with distributed database documents or public ledger of transactions around the network of computers or nodes. Without any centralized authority, the nodes in the network must agree on a common point via consensus so that transactions are passed as granted. The general data scheme of blockchain is depicted in Figure 2.1, the nodes of the network (holds distributed ledger, common to every node) are organized chronologically with links to each other.

FIGURE 2.1 Linear organization of data blocks in (global ledger) blockchain.

As seen in Figure 2.1, the records of data are organized linearly via a listing connected with each block. Each typical block of data is encrypted using a cryptographic algorithm and is given a unique data identifier. Note that the explained blocks of data (in Figure 2.1) are maintained for every node in the network. This means redundant copies of blocks exist with every participant node in the network, making the system difficult to update unofficially (transactional immutability). To make a transactional entry in the ledger, one has to broadcast the transaction to the members. The entry is noted only if the members of the group validate the transaction. If the transaction is valid compared with the record of data available, the entry is made into the block with consensus agreement by the rest of the members. The new block of transactions is linked with the existing block using an encrypted unique identifier (hash in the case of bitcoin), in this way every block is connected (attached) with the former block and the same is available for every node in the network; a more detailed description is provided in Section 1.3.

Primarily, the basic idea of blockchain technology is generalized as an innovative technology that is composed of secure cryptography with distributed technology as shown in Figure 2.2.

Nakamoto et al. (2008) presented a framework that portrays the Internet in a new dimension in which a network of systems collaborate to maintain a ledger or database securely by merging the idea of cryptography with distributed systems and decentralization. As mentioned earlier, the database or ledger contains a list of blocks linked with each other with a set of transactions (as shown in Figure 2.1), which together with all linked blocks (the same will be available at every node in the network) constitute the blockchain. Note that a block is only added to the existing blockchain if the rest of the nodes agree with the consensus agreement; details about consensus are demonstrated in Section 2.2. Having a shared ledger among nodes in the network (decentralization criteria) make the system efficient and not prone to manipulation (transaction immutability) of the entry in the ledger.

Consequently, the system is extensively secured by having a common global ledger communicated across the network. Because all the blocks of data are arranged with time stamps in a linear linking, intruders can't make a change. To tamper with an existing value of a particular block requires the intruder to manipulate the associated block also, which is nearly impossible. Because for every change that is made, for the transaction to be successful or the ledger to be updated, the nodes of the network must validate it by agreeing in consensus, details of which are depicted in Figure 2.3.

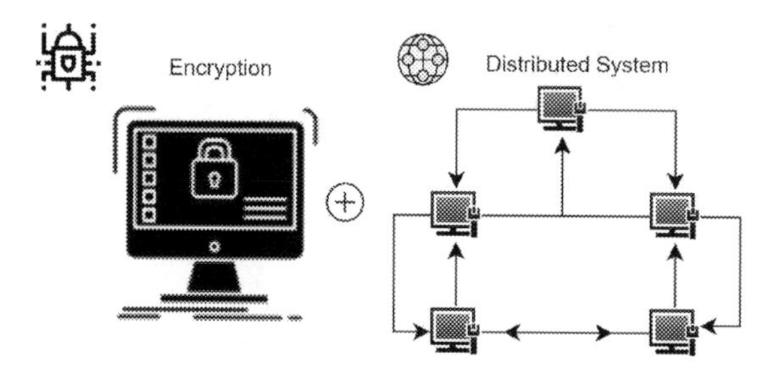

FIGURE 2.2 Composition of cryptography and distributed systems.

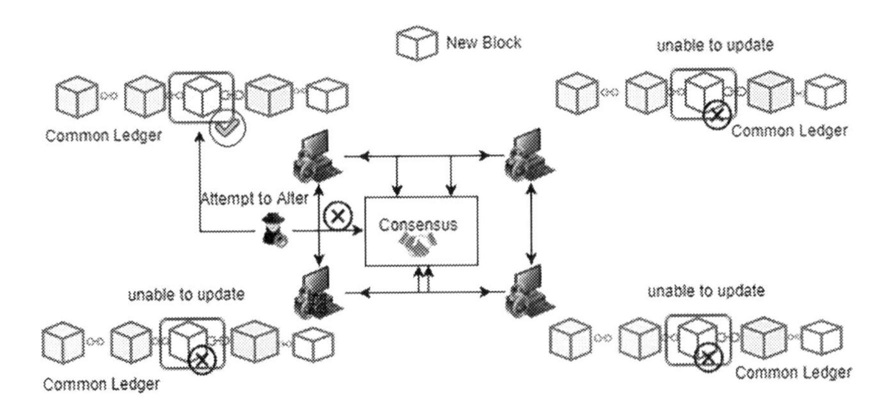

FIGURE 2.3 Blockchain as a basic tamperproof model.

From Figure 2.3, it is clear that the central actor to the system is a Ledger. The underlying mechanism of blockchain revolves around the distributed ledger synchronizing with the nodes of the network. Blockchain can act as a distributed ledger but a distributed ledger cannot be a blockchain, the details of distributed ledger technology are described in Section 2.1.1.

2.1.1 Distributed Ledger Technology (DLT)

DLT can be described as a registry or ledger or database that is distributed across different parties to record transactions. The idea of sharing a database started in 1978 by the researchers in (Davenport, 1978) to make the system efficient by eliminating central authority. Generally, centralized servers are more vulnerable to failure or get attacked by the intruders. To eliminate the risk of failure, researchers in (Davenport, 1978) introduced a distributed database shared among peer systems to carry out tasks (validate, process transactions) without a central server. All the participation nodes in the distributed network maintain an identical ledger (transparency) equipped with a synchronization facility that reflects the transactional details performed by one node to the ledgers of the remaining nodes in a matter of seconds.

Recent centralized systems expanded to distributed systems with high availability features to overcome the point of failure. These systems are capable of providing fault tolerance and maintaining the integrity of the data by synchronizing the central server (primary) with the rest of the secondary servers. By introducing a virtual IP configuration with hot standby to the servers, if any point of failure occurs to a server, the network will not be disturbed and the failure is limited to the current system. In case of primary server failure, a secondary server from the rest will take the initiative to support the application and, once the failed system is back to its job, it will be synced automatically and the data is maintained consistently.

However, it is possible that the configured nodes in the centralized distributed systems can be tampered with at any point in time. Despite the advantage of opting for the high availability of centralized distributed systems, it still has a central authority to authenticate transactions (making the system inefficient compared with DLT). The conceptual view of the centralized sever compared with the distributed ledger is shown in Figure 2.4.

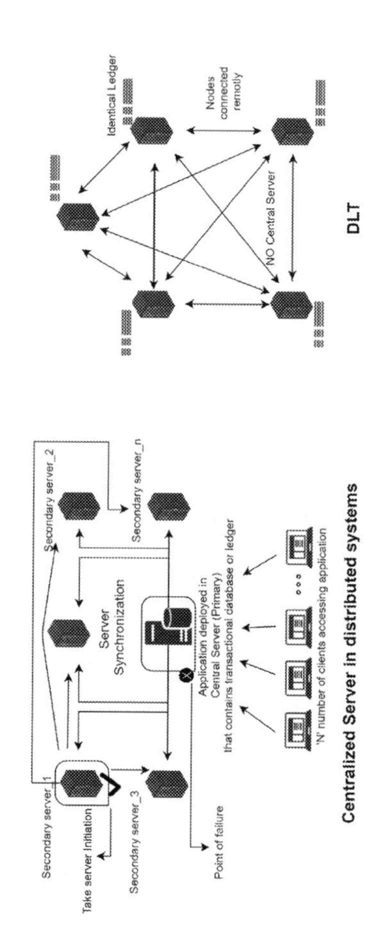

FIGURE 2.4 Conceptual view of the centralized sever compared with the distributed ledger.

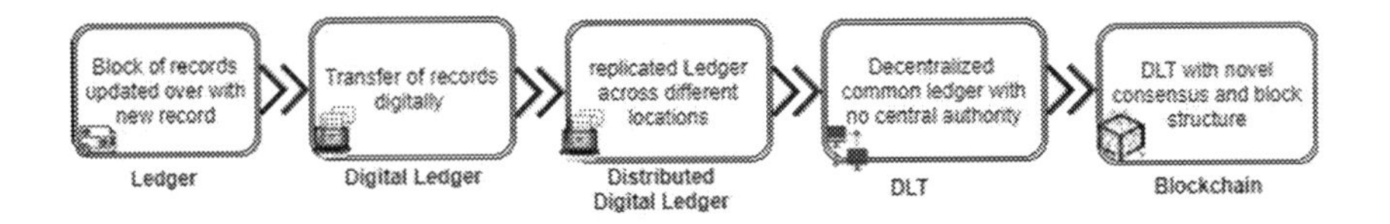

FIGURE 2.5 Evolution of ledger to blockchain.

The limitations of centralized servers are covered with DLT by having a shared distributed database or ledger that records transactions of any value with no central authority. Organizations use DLT to authenticate various types of transactions; the transactions are recorded into the ledger only when the network with active nodes reaches consensus. The evolution of the ledger from the digital ledger to the distributed digital ledger and DLT, Blockchain is demonstrated in Figure 2.5.

All details in the ledger are time stamped and given a unique identifier (encryption key) so that each node in the network is facilitated with an audit of records. The conceptual idea of DLT is applied to blockchain and the only difference that distinguishes blockchain from DLT is the adaptation of consensus and the structure of the data block.

2.1.2 Decentralized Blockchain Architecture

As mentioned earlier, blockchain is composed of cryptography (set of protocols that enable the network to run securely) and distributed environment that lets the peer nodes of a network share a common ledger without a central authority (Tschorsch and Scheuermann, 2016); details are shown in Figure 2.3 and Figure 2.4. The conceptual idea behind the technology is drawn from DLT with some changes. All the components behind the technology work for reliable data transfer, synchronization among nodes, maintenance of the global ledger, and obtaining node consensus in the network. Protocols pave a major significant component that constructs consensus amid the nodes. The transactional workflow of blockchain is indicated in Figure 2.6.

Figure 2.6 shows that in a network of participating nodes, if someone (say ‘a’) in the network initiates a transaction to node ‘b’, succeeding steps will occur.

- Node ‘a’ will broadcast the transaction via an application to the rest of the network.
- The broadcast transaction appears in an unconfirmed group from which the nodes of network capture it to validate.
- Participant nodes (usually referred to as Miners) will try to capture it and make their copy of the block with the broadcast transaction.
- The nodes of the network will check the eligibility of the transaction and validate the transaction.
- The network will legitimate the details of the transaction; on successful legitimacy, the blockchain is updated with the new one. The details of it are specified in Section 2.1.3.

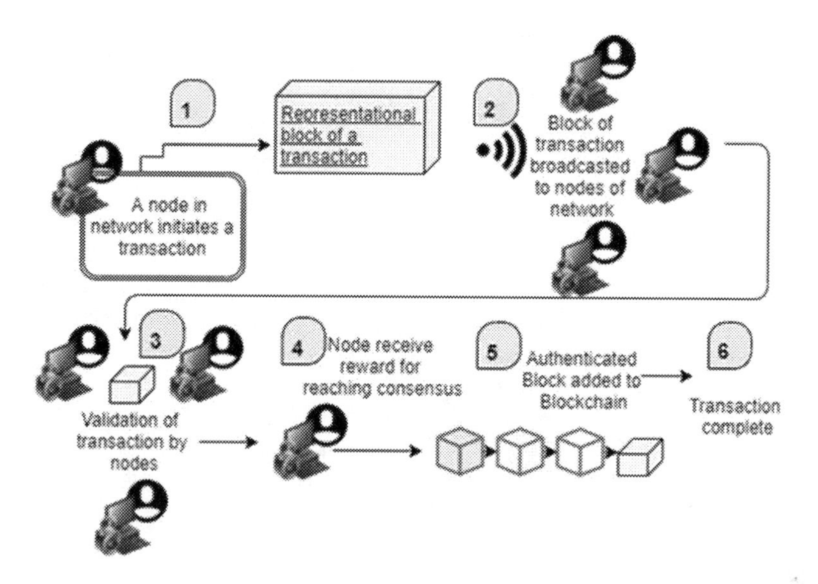

FIGURE 2.6 Transactional workflow of blockchain.

Architectural components of blockchain are as follows:

- Participant Nodes of the Network: This refers to the peer computer or node or participant user connected to a network remotely (distributed environment). The framework of blockchain is designed to collaborate with the operating system that lets the user connect to a trusted network remotely. The network with participating nodes are connected in lightweight. The lightweight node will track the header of a block (only to verify if a transaction is included inside a block) in its local storage, whereas the complete node will extract an entire replicated block of data (Belotti et al., 2019), (Wang et al., 2019). Lightweight nodes are operated by complete nodes.
- Details of Transaction: The technical components in blockchain will circulate around the transaction. Transactions are performed by individuals that require a logical asset transfer or exchange. Signed transactions are simply a record of static information, containing a value associated with a destination address. It is processed by the nodes through a pair of keys (public and private) shared by a node to the remaining nodes of the network. The transaction record becomes immutable only after validation by miners (Kroll, Davey and Felten, 2013) by being sealed into a block and linked with the existing blockchain. Each transaction sealed inside a block is assigned a unique identifier. To preserve the context of the transaction, it is protected with a cryptographic signature (SHA-256 in the case of bitcoin). A typical signed transaction usually contains a unique identifier of an existing block, a time stamp of it, and an alterable nonce. The details of it are clarified in Section 2.1.3.

- Shared Ledger: This is referred to as a registry or database file that is common to every user. Generally, the entries of a ledger or registry or database file are time stamped sequentially on every usage so that the entries are immutable. The Ledger is filled with valid transactions by nodes, and each transaction in the ledger represents communication among the nodes. The entry into the ledger is made after all the nodes agree with it.
- Use of Encryption: The validity of the actual information inside a ledger is protected by the implemented encryption technique and hashing. Using hashing will turn any arbitrary input into a fixed output. Note that it is nearly impossible to revive any input from the output. For every input string of data, the output generated by a hash function is unique. Each node produces a pair consisting of a private and a public key; the public key is distributed to every node whereas the private key is confidential. The private key is utilized for decryption purposes and is generated by digital signature encryption function. The public key is utilized for the verification process. On successful signing and sharing of the transaction, miners in the network validate it via the private key and broadcast it. The members of a network verify the correctness of a shared transaction signature with data provided using the public key. Several cryptographic techniques were used in this process and the details are shown clearly by the researchers in (Yang et al., 2019).
- Miners in the network: Every participant or node in a network serves as either a miner or a verifier. In general, every node will possess a miner with their computational expenditure that is used to process a transaction using encryption algorithms as mentioned previously in the use of encryption. A digitally signed transaction, once transmitted by a node, must be verified by the miners and every node in a network should agree with the correctness of the transaction via consensus protocols. Verification of a transaction by miners involves complex computational loads to solve using cryptographic algorithms. The verified transaction by a miner is then broadcast to the network so that the nodes will verify the block. Note that a block is only reflected or added to the blockchain if the majority (at least 51 percent) of the nodes in the network certifies its validity by verification of the block using a shared public key. Different signed transactions from the unconfirmed pool can be mined by different miners in which the common ledgers differ from miner to miner. To make the ledger consistent, different consensus mechanisms were implemented by the researchers in the field (Chaudhry and Yousaf, 2018). On successful block entry to the blockchain, a miner is awarded the price allocated with the transaction. Consequently, a miner trying to associate an invalid transaction (rejected by the majority of nodes) could cost a suitable amount of loss in terms of computational power and electricity (Sturm et al., 2018).
- The agreement followed (Consensus): Conventionally, the centralized authority scheme that manages the confidential transactions is more prone

to being attacked by intruders. Because the blockchain doesn't rely on a centralized authority, in a distributed workplace the nodes of the network should trust each other. The trusting scheme is facilitated by the adopted consensus so that the information or blocks of the network are secured and the data are reliable. A global agreement reached by the nodes regulates consistency and the integrity of the blockchain's transactional information. The initial goal of endorsing nodes in a thriving common agreement is to make the information transparent (making sure that all nodes have identical data) and for network scalability efficiency. The following constraints are considered when reaching consensus: valid data structure, input, and output information must be exact (not empty as well), the value of the encryption output cannot be zero or negative one, the signed transaction must be legitimated. A more detailed view regarding consensus and its algorithms is described in Section 2.2.

2.1.3 How Blockchain Stores Information

Usually, the data inside the blockchain constitutes not only the transactional details (like sender and destination information and asset interchanging) but contains complete informational exchange from transactional initiation to storage and enquiring details. The complete constituted details of transactions inside a block are seen in Figure 2.7.

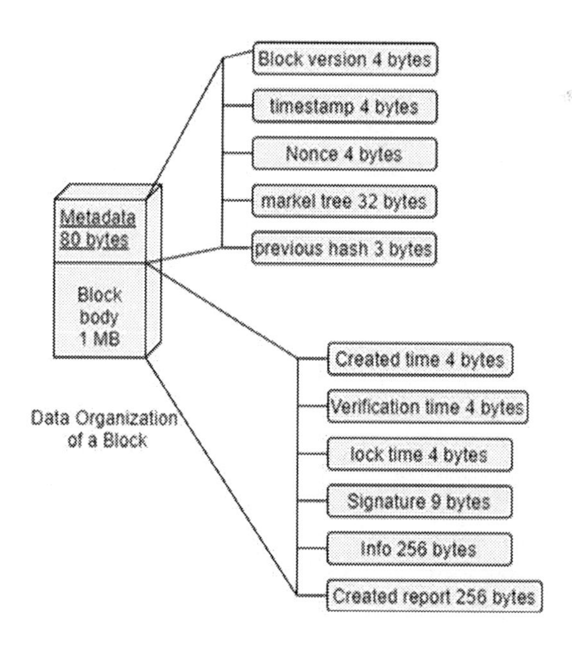

FIGURE 2.7 Data organization inside a block.

Typically, each block in a blockchain contains two containments, one for storing metadata and the other for actual data. The purpose of metadata block is to hold the address of former block along with the nonce adjusted value, created time stamp, and the data structure details (the popular framework developed by the authors in (Nakamoto et al., 2008) used Merkle trees as the data organizing structure, which is followed by most researchers). The actual body of the block holds the information value details as shown in Figure 2.7. On every successful block's authentication by the members of the network, it is sequenced (connected with a former block) by following the last block. The resulting blockchain is then stored in a local copy of the user as well as in the main global ledger (Belotti et al., 2019).

There are several steps involved in how blockchain stores information, which are described as follows:

- Generally, a participant in the network initiates a transaction by signing it off to the network from the blockchain framework application (in bitcoin through a wallet). A transaction typically contains cryptocurrency details from the sender to someone in the network. On successful signing off of a transaction, it is announced to the users or nodes (miners) of a network. Initially, every initiated transaction by a node is placed in a group of unconfirmed transactions waiting to get approved by miners. A node in the network is granted the unconfirmed transactional local groups so that they can mine or process it rather than a centralized giant group.
- The nodes of the network or miners are allowed to process the transaction by grabbing it to a block. A block can be constructed by many numbers of transactions. Each miner or node in a network can produce a block with any transactions. In special cases, the same transactions can be incorporated into a block or mined by more than two nodes or miners. Two miners in the network can incorporate different transaction and, in such cases, the network is less prone to the fork effect (having an identical previous address node associated with a newly confirmed block) because of the consensus it has followed.
- The transaction, once sealed to a block, is further incremented to the blockchain by solving a complex computational problem. The block needs a signature in order to get attached to the blockchain, and that signature is created with a computational workload performed by miners. Each block has its unique signature and different nodes will pick different transactions and form a block with its signature that is distinct from others. With computations to solve a complex mathematical problem that satisfies a condition (having a value with a specific number of zeros), this process is termed mining.
- Solving a complex mathematical problem is nothing but finding a unique signature through a hash value for the particular input. The unique signature or hash solution should start with some number of zeros. In general, finding a unique hash to the input is somewhat more complex than verifying its solution. Hashing will convert an arbitrary random input string with characters, digits, and special characters to a unique output depending on

the type of technique employed. Popular techniques are MD5 (128 output length), SHA1 (160 output length), SHA2 (256 output length), and SHA3 (256 output length).

- The process of mining is a continuous effort of matching consecutive zeros in the hash output by altering a changeable value called the nonce. Many researchers in the field referred to the process as complex, because an adjustable nonce will lead to an approximate output with a certain number of zeros that is time and energy (computational power) consuming. The nonce is adjusted until it reaches the prescribed number of zeros in the hash output. Adjustment of the nonce not only gives a random number of zeros but also provides a random number of unique outputs (signature).
- The miners of the network will continue altering the nonce value until the generated random output is matched with the prescribed signature defaults. An example in Figure 2.8 depicts the block organization in the blockchain.
- Figure 2.8 is an example of having 6 consecutive zeros and it is achieved by altering the nonce value; a random nonce will lead to an exact output. It requires a lot of computations by the miner for continuously guessing the nonce. Once an applicable signature is found by the miner, the block is broadcast along with its signature or hash output to the remaining nodes of a network. The nodes of the network capture the block with its associated hash value to verify the correctness of the prescribed consecutive number of zeros. Note that finding a signature for a particular input that satisfies the requirements is difficult whereas verifying the hash output value with its signature is relatively easy. Once authentication of a block is verified by the remaining nodes, the block is connected to the blockchain by reaching consensus. Once the block is attached to the blockchain, a copy of it is stored in the local space of all the nodes. It is successful when all the transactions in the blockchain history match with the new block.
- On addition of the successful block to blockchain, every addition of succeeding blocks on top of the new block is considered a confirmation. For instance, a block with number 203 is included with some transactions and

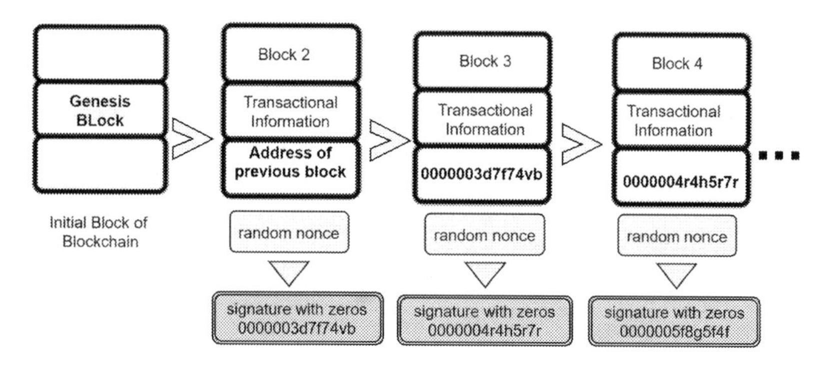

FIGURE 2.8 Block organization in the blockchain.

if its length is 208 then the transactional block has (208–203 = 5) has 5 confirmations. Confirmations merely indicate that the consequent transaction is agreed by nodes of a network on every new block entry because on every new addition of a block to the blockchain all the nodes should reach consensus. The more the confirmations of a particular block in the blockchain the more complex it is for an intruder to alter it.

2.1.4 Types of Blockchain Mechanisms

Most of the blockchain frameworks adopt protocols discussed in Section 2.2 that every node will reach a consensus. These protocols rely on the nodes with majority agreement. Generally, the majority is indicated as at least 51 percent of the nodes to validate and agree on it. Why do nodes follow only a specific protocol, why can't they follow another protocol which is less expensive than the specified protocol if an additional award is offered? This argument is raised to organize the system with authentic (good) nodes that are loyal. Blockchains are categorized into public (permissionless), private (permissioned), and consortium. Each variant is organized remotely (distributed nodes), has a common ledger, and will initiate, process, and validate transactions and create blocks. The categorized details are as follows:

- **Public**: Referred to as permissionless, it is open for every node who has certain resources, like proper connectivity, and computational sources to join the network. Any user can be signed into this type of systems and once it becomes a participant node, it is eligible to track the transactions and able to validate them. The basic motive of this type of blockchain is to exchange cryptocurrency. Blockchain frameworks like bitcoin (Nakamoto et al., 2008), Ethereum (Wood et al., 2014), and Litecoin are examples of open blockchains. The underlying consensus mechanism followed by most open blockchain systems is PoW that works in finding a hash output to meet a number of zeros. PoW is extensively followed by bitcoin and Litecoin blockchain systems. An alternative to PoW is proof of stake, in which the nodes will use their currency to become miners rather than showing computational expenditure. Blockchain systems like Ethereum (Wood et al., 2014) uses proof of stake, the details of which are explained in Section 2.2.3. The theoretical advantages of public blockchain are as follows: the details are transparent to every node, and the more stacked blocks in the blockchain the more complex it is to alter (immutability), making it secure. The drawback of this system is that they are relatively very slow; for bitcoin the number of transactions performed per second is 7, whereas in Ethereum, 15 transactions are performed per second. These numbers are desperately low compared with private networks like VISA with 24k transactions per second. These systems are prone to utilizing and consuming more system capabilities and electricity by the miners in finding the exact nonce.
- **Private**: In contrast to the public blockchain, in this type of closed blockchain the initial governance will decide the group of trusted members. The trusted members are allowed to agree with the consensus and have read and write capabilities to a transaction. Not all members of the network are trusted

members, as the governing body has to give that privilege. Members who are not trusted members can only read transactions. It is similar to open blockchain but with a restricted capability in the network. This type of system is well suited for digital voting systems, supply chains, authority assertion, digital identification, and so on. An example of closed blockchain is Hyperledger Fabric that has control over transactions in the network. No one other than the trusted nodes can access the transactions. The advantage of this system is that it is comparatively fast when compared with the public type of blockchain. With the minimal group of trusted members in the network, the consensus is reached rapidly, which results in faster validation of blocks. The number of transactions that can be reached per second is in thousands. These systems are more scalable, with limited nodes in the network they can be easily deployed and can manage addition or deletion of nodes more effectively. Besides these advantages, this system has a drawback in the need for a centralized authority system to allow its function. Initially, the centralized authority will have the administration role and has complete monitoring capability from adding nodes to deleting nodes. Security-wise it is less efficient; if any node in the network has access to the central management control then the system can be extensively altered and modified unofficially.

- **Consortium**: This is a special type of system that constitutes both public and private blockchains. A consortium blockchain is more similar to a private or permissioned or closed blockchain with a minor change. There are multiple governing bodies or centralized authorities in a consortium, and nodes that satisfied the consortium rules are added to the system. The generated transactions are only seen by the created nodes, making it secure and maintaining privacy. This function makes consortium different from public and private. These systems are feasible for banking sectors and government functions. A piece of broad information about the categorization of blockchain is captured in Table 2.1.

TABLE 2.1
Categorized Details of Blockchain

Property	Public or open blockchain	Private or closed blockchain	Consortium blockchain
Consensus decision	By all nodes	Governing or central authority node	Authorized nodes
Transactional immutability	Secured	Low secured	Low secured
Central authority	No	Yes	Partial
Transactional Speed	Low	Very High	Moderate
Network organization	Central and distributed	Decentral and distributed	Semi-central and distributed
Computational usage	Huge	Economical	Moderate
System efficiency	Low	Highly efficient	High

2.2 CONSENSUS AS THE MEANS OF AN AGREEMENT

As mentioned earlier, the miners should legitimate the validity of transactions. Block of transactions are added based on the agreement of nodes using consensus. Consensus is generally a protocol that achieves reliability of the data (all the nodes contain an identical copy of the data) and fault tolerance (solves the Byzantine problem). Initially, to make a reliable transfer of transactions, the authors in (Gray, 1978) tried to solve it by proposing 2 phase commit protocol, which enables the transactions to be processed automatically. However, failing a node in 2 phase commit protocol makes the consensus mechanism worthless, and the property of fault tolerance is not achieved. The primary property of fault tolerance is that the functioning of the nodes will not be halted even if communication failure or Byzantine problem (Lamport, 1983) occurs.

The Byzantine problem is very critical and common in communication systems in which sharing of transactions is involved among multiple nodes. It is referred to as a reliable data transfer problem, for instance, multiple generals are waiting remotely to attack the enemy city. They will only succeed if all the generals coordinate and fix a time to attack. Here, one major general takes the initiative to decide a time and sends the same to the other generals. To fix a time, the majority of the generals should accept it by acknowledging it to the major general. Even if an intruder manages to alter the time at one end, the system is not prone to get tapped because the majority of the generals voted for a specific time. It is nearly impossible for the intruder to alter every general's time. Hence this is called Byzantine generals' problem and the same theoretical scheme is implemented in blockchains consensus protocols.

To compress the effect of Byzantine failure, the authors in (Schneider, 1990) developed a "state machine replication (SMR)" technique that covers the effect of fault tolerance failure in the network. However, SMR is applicable in a centralized system with a common shared memory that is used for processing transactions read and written by the nodes. Because in decentralized systems like blockchain shared memory facility is not possible, every node has its local memory that is used for maintaining ledgers. Similar types of protocols are seen in the article (Pease, Shostak and Lamport, 1980), yet not much attention is drawn in this system. Since, achieving consensus in a distributed network depends on certain failures like crashing of the node due to network issue, information tampering by an intruder, or link damage in a network (Panda et al., 2019), (Bodkhe et al., 2020). Consensus protocols in blockchain not only deal with node failure and Byzantine problems but also balance properties like the consistency of the data, the integrity of the data, the agreement followed, and its validity. A detailed view of consensus protocols is presented in this section.

2.2.1 Practical Byzantine Fault-Tolerance (PBFT)

To mitigate the problem of the Byzantine general's problem, the authors in (Castro, Liskov et al., 1999) came up with a consensus mechanism called the PBFT protocol. The working criteria of PBFT resembles the tolerable Byzantine problem protocol proposed by the authors in (Pease, Shostak and Lamport, 1980). The motivation behind this is to make the transaction system in the distributed environment consistent and liable to withstand with the malicious attacks by the intruders. This protocol is suited for asynchronous systems deployed remotely that will enhance the security of transactions, providing fault tolerance. At most [f-1\3] nodes out of 'f' fault nodes are secured and could perform with better efficiency. The transaction is protected with cryptography; a transaction typically consists of a generated public key, the transaction code, and a unique hashing function generated output. The protocol is a kind of SMR (Schneider, 1990) that was designed by considering faulty nodes in communication, disturbance in the network that cause the secondary (replicated) servers to function inadequately. According to the authors, a system with 3n+1 secondary (replicated) nodes contains 'n' number of fault nodes that can have access over read, write operations. Popular blockchain systems that adopt PBFT are Hyperledger, Ripple, and Stellar. The rough working criteria of PBFT are as follow:

- Initially a client will request a primary server operation to perform a service.
- The primary server will broadcast the service details to replicated nodes.
- The replicated nodes process the request and acknowledge it to the client.
- The transaction is called successful if the initial client receives at least 'f+1' identical copies of acknowledgments.

The major components of the PBFT protocol are the client, primary server, and secondary or replicated server nodes that coordinately work to maintain consistency and fault tolerance. Hence consistency is marked as a major criterion in PBFT; the working scheme of PBFT protocol in maintaining consistency is illustrated in Figure 2.9.

Maintaining consistency is a vital property in PBFT. The transactions that are block sealed timely should reach the consensus by the nodes, conserving consistency property will make sure the consensus is working properly. Replicated node 3 from Figure 2.9 suffers from Byzantine failure; however, the PBFT protocol maintains consistency by eliminating the considerations from the faulty node without the transaction process being delayed or tampered. The depictions of Figure 2.9 constitute

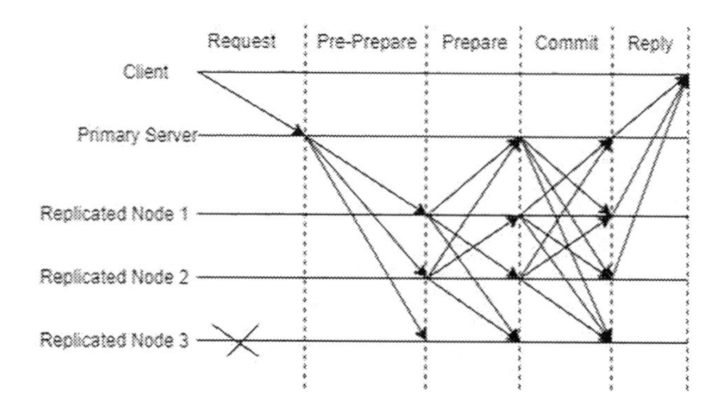

FIGURE 2.9 The process of maintaining consistency in PBFT.

three main operations, they are "pre-prepare", "prepare", and "commit". The details are as follows:

- "**pre-prepare**": broadcasting a "pre-prepare" alert from the primary server to the replicated nodes. Note that the replicated nodes enter into the "pre-prepare" state if they accept it.
- "**prepare**": the nodes will enter into the "prepare" state if they broadcast a "prepared" message. After broadcasting, the replicated nodes enter into the "commit" state.
- "**commit**": the nodes in the "commit" state wait for'2f+1' "commit" acknowledgements from the replicated nodes. After reaching the'2f+1' "commit" acknowledgements, the transactional block is added to the blockchain.

The system not only offers consistency but also provides stability (fault tolerance) by incorporating a "view-change" mechanism. The illustration of the "view-change" mechanism is depicted in Figure 2.10.

If a primary node fails to serve its purpose, then the replication changes its stage to primary and broadcasts a "view-change" certificate (indicates that the primary is not active) to the remaining replicated nodes. The node is called primary if it receives '2f+1' "view-change" certificates and acknowledgments from the remaining replicated nodes. The elected primary node then changes its state to "recent view" and checkpoints to continue the consistency mechanism in blockchain.

2.2.1.1 Merits and Demerits

Although the efficiency of the consensus mechanism in blockchain is extensively escalated by the use of PBFT when encountering the problem of Byzantine failure, there are still advantages and disadvantages in adopting it. The performance of the blockchain is accelerated by leveraging the consistency and fault tolerance with PBFT, resulting in reliability. However, it is limited to asynchronous nodes distributed remotely with a common shared memory; it is not suitable for the nodes in a network that act synchronously. Blockchains that opt for this consensus protocol have a vast number of transactions that impact speed differences. Normally for

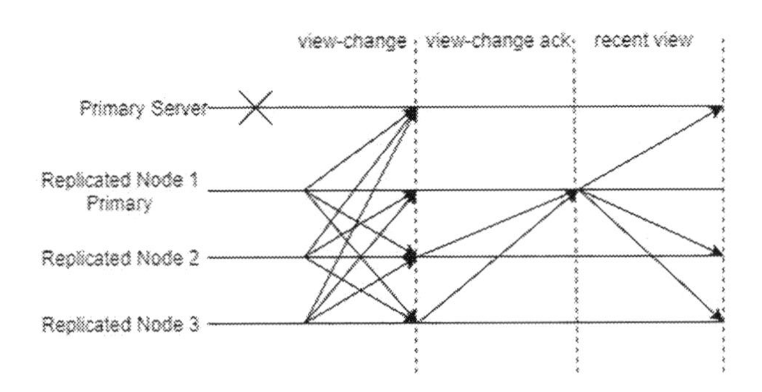

FIGURE 2.10 "View-change" mechanism in PBFT.

'n' number of nodes the trend will scale $O(n^2)$. Hence the transaction speed differs drastically when the number of nodes maximizes (Dinh et al., 2017). A number of schemes were proposed by the researchers in the field to enhance PBFT, the authors in (Luu et al., 2016) collaborated PoW (details given in Section 2.2.2) with PBFT to enhance the consensus process substantially.

2.2.2 Proof of Work (PoW)

The conceptual applicability of PoW came into existence when the authors in (Dwork and Naor, 1992) proposed a scheme to control junk emails in acquiring emails that are valid. The authors designed the consensus in a way that intruders needed to apply computational power in order to add a junk file in the transmission medium, which was not feasible with their expenditure. The ideology of (Dwork and Naor, 1992) was evolved by the authors in (Nakamoto et al., 2008) with a functionality to preserve the consistency and fault tolerance by the blockchains distributed consensus protocol. The working scheme proposed by the authors in (Nakamoto et al., 2008) has become a state of the art in blockchain research, hence termed the bitcoin blockchain, which has raised an enormous significance in the field.

Other than bitcoin, PoW was adopted also by Ethereum (Wood et al., 2014), Litecoin (Bradbury, 2013), and Monerocoin. The miners that take part in the mining process will participate in the consensus process and provide work proof before adding a new block to the blockchain. The miners are awarded an incentive for mining; mining refers in solving a mathematical problem via hashing algorithm. Typically, the miner that solves the problem is able to add the block to the blockchain. It requires the computational abilities of the system to find an appropriate hash value that matches the number of zeros appended before the encrypted data, the details of which were described in Section 2.1.3. The requirement of matching a certain consecutive number of zero's is altered by the modification of the nonce value, which shows an impact in the resulted encrypted string. The extensive computational capabilities of resources that is used for hashing is generally very expensive. The miners have to show the PoW to the network so that it is verified by the rest of nodes; this is the idea behind consensus. It is impractical for intruders to alter the value of a block in blockchain, because committing a new transactional block to a blockchain requires solving a mathematical computation via hashing in which the result should be validated by the rest of the network by reaching to a consensus, and its work flow is shown in Figure 2.11.

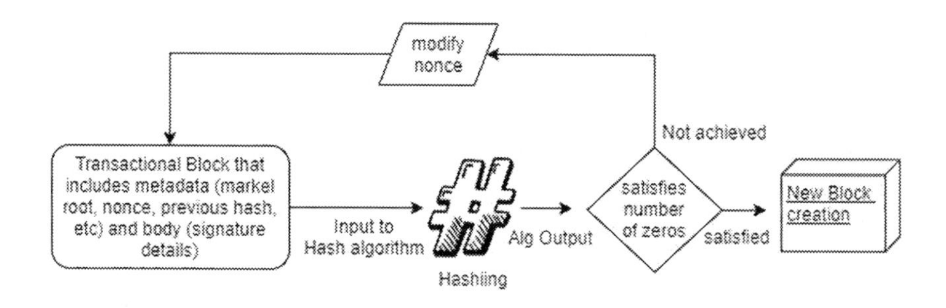

FIGURE 2.11 Proof of work's work flow.

2.2.2.1 Merits and Demerits

Proof of work provides system consistency as well as security, as it is impossible for an intruder to alter any blocks value. This system requires extensive use of computational resources on every adjustment of the nonce. The blockchain that follows PoW consensus is relatively very slow; the number of transactions performed per second is 7 for bitcoin and 15 for Ethereum, which are very low compared with a system like VISA with 24k transactions per second. These systems use and consume more system capabilities and electricity by the miners in finding the exact nonce. The time for confirmation of a transaction consumes too much *when a fork condition occurs.*

2.2.3 Proof of Stake (PoS)

PoS is computationally inexpensive compared with PoW and was introduced by King and Nadal (2012), with the modified consensus protocol used in "PpCoin". A network that works based on PoS will let the nodes to be designated as validators by placing or showing expenditure in terms of currency or in general tokens in which the nodes are selected randomly. The more the currency is, the more it is elected as validator and has the privilege to add a block to blockchain. With reference to computational power, validators are elected in PoS by showing expenditure in terms of the tokens it has. It has to solve the mathematical solution using Hashing, but the adjustment of the nonce is eliminated, and the energy consumption is minimal compared with PoW in terms of the computational power utilized to reach consensus.

The work flow of PoS is defined in Figure 2.12. Like bitcoin, "PpCoin" is a cryptocurrency that is used in blockchain based PoS consensus. In PoS consensus, with respect to the amount of coins or tokens shown, the time period of the tokens shown to become a validator are also counted in validating a block (King and Nadal, 2012).

For instance, the age of a node holding 20 coins in 11 days is determined to be 220. Once the elected validator has pushed a new block to the blockchain, its age will be set to zero.

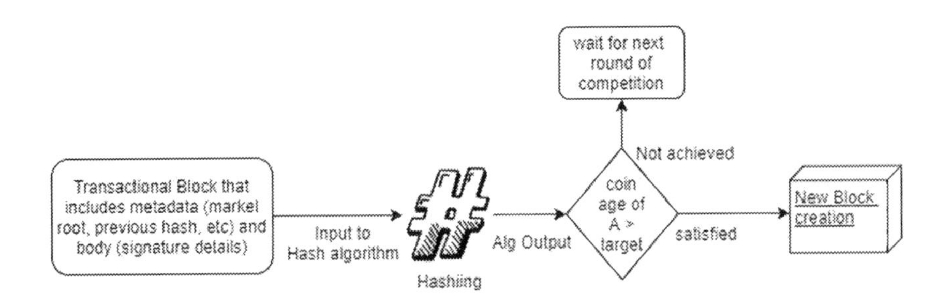

FIGURE 2.12 Work flow of PoS.

2.2.3.1 Merits and Demerits

The computational expenditure of PoS compared with that of PoW is very minimal. Although it is necessary for decrypting a puzzle using cryptography, the performance of PoS is efficient. The transactional speed of PoS is massive and the rate of transactions in a second is generally in the thousands. Besides the advantage, PoS is more prone to being affected by forking.

2.2.4 Delegated Proof of Stake (DPoS)

DPoS is an enhanced variant of PoS that deliberately works faster with minimum computational load. DPoS was proposed by the authors in (Larimer, 2014), and lets the nodes having stake value elect block producers by following consensus. The working scheme is similar to that of PoS, but with the digital democratic process, DPoS is eventually faster than PoS. A popular blockchain system following the DPoS consensus protocol is "EOS" in which the transaction speed is doubled. Any node in the network with their stake or tokens or currency can be elected as a validator or producer to add the block to blockchain. Depending on the stake value and the number of votes secured, any node can randomly become a validator. The decision regarding the validator selection depends on a couple of factors based on the blockchain system. One, the value of the stake they hold and the period it has been published, generally the higher the value the more likely they will be elected.

On every validation of a block by a node, the validators are shuffled randomly to make sure that balance is maintained across all the validators. Initially a two-third vote share is enough to become a validator, and every validator has to produce a minimum of one block every day. The block time is set to 3 seconds, adding an advantage to the system by making forking impossible. If any node is unable to participate in their validation time, the consensus is designed in such a way that they are removed from the validation process. The details are clearly detailed in Figure 2.13.

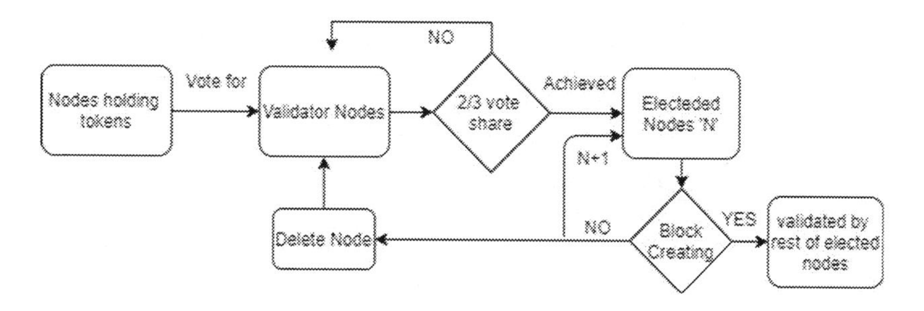

FIGURE 2.13 Work flow of DPoS

2.2.4.1 Merits and Demerits

DPoS is very efficient in transaction processing compared with that of other consensus mechanisms. The transactional speed per second is relatively faster than those of PoW and PoS because the computational cost is much lower than for the other consensus protocols as heavy computations are not involved. Forking is nearly impossible because nodes in the network will participate in voting and actively coordinate with each other in the validation process rather than competing for a block to be added into the blockchain.

2.2.5 Proof of Capacity (PoC)

The concept of PoC was determined by the authors in (Dziembowski et al., 2015) to allocate space for storing precomputed solutions in hard drives. In PoC, the miners will participate in mining by utilizing the storage capacity (Hard drive). The more storage capability the node has, the more likely it is to be elected as a validator similar to PoW (computational capabilities) and PoS (stake value). The process is similar in solving mathematical solution but the computations are carried out in local hard disk space. This process of storing solutions in the validator hard disk is called Plotting. This type of blockchain is intended to make the transactional cost cheaper with small miners.

2.2.5.1 Merits and Demerits

The advantage of opting for PoC is that the disks are cheaper to buy and store. The energy consumed is relatively less. The reward is awarded by comparing the storage capabilities. This protocol may suffer from attacks as hard drives are involved in storing precomputed hashing algorithms.

2.2.6 Proof of Elapsed Time (PoET)

This protocol designed under the observation of Intel group in 2016 is suitable for private blockchain systems like the Hyperledger framework. This protocol is concerned with the random period of time that decides whether or not a node is rewarded by generating a block. The process for a node to become a verifier to add a block to the distributed ledger involves the node being certified by Intel's Software Guard Extension (SGX) applications. The attestation provided by the SGX engine will allow the node to join in the network with a random period. The significance of the SGX technology is that it will effectively protect the data as well as the code without been accessed by anyone (safe from intruders). It also makes sure that verifiable code is running in a trusted environment.

The PoET consensus mechanism has two main phases, node verification through SGX application and joining the network with an elapsed time to produce a block. The details of the PoET consensus protocol mechanism are shown in Figure 2.14.

Initially, if a node is interested in joining a network, it has to be attested with a pair of keys by the SGX application. A node will verify itself by invoking trusted codes from SGX; SGX grants an attestation with public/private keys. Joining in the network is either confirmed or cancelled by the existing nodes in the network

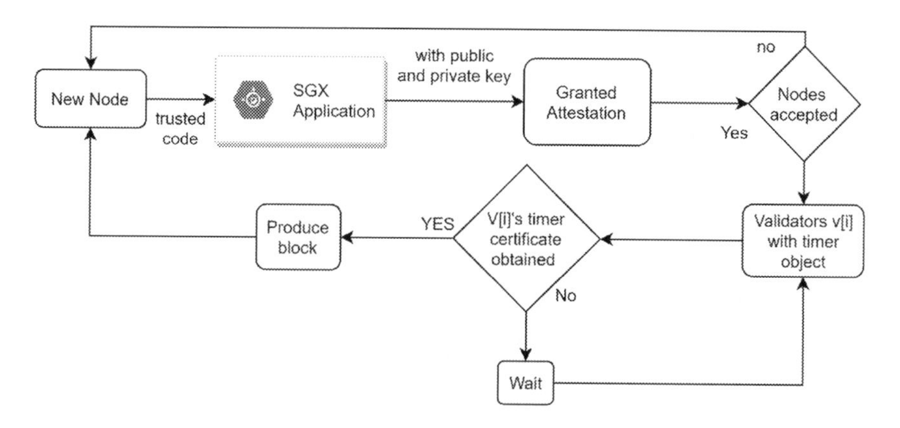

FIGURE 2.14 Working flow of PoET.

by verifying the nodes attestation certificate with the assigned keys. If the existing nodes approve its participation in the block validation process, the accepted node is allocated with a random timer object that determines when to produce a block. The accepted nodes are now eligible to participate in the consensus process. Each node waits for the times to finish, the certificate provided by the timer object will let the block be added indicating the expiration of its time. On authentication of the produced timer, certified blocks of transactions are successfully added to the blockchain.

2.2.6.1 Merits and Demerits

On integration of the SGX application, the application will ensure data security with trusted code. Use of PoET can make the system efficient with better scalability compared with prescribed consensus mechanisms. The major concern is that the systems security is only depending on SGX for the base protocol. PoET is an advantage to the "random leader selection process" problem that utilizes computational, stake, and storage capabilities to reach consensus. It is well noted for easy deployment and Hyperledger Sawtooth adopts this system with better efficiency and experimentations. With the advancements in technology, the SGX application in hardware is vulnerable to trust issues.

2.2.7 Delegated Byzantine Fault-Tolerance (DBFT)

DBFT is a consensus protocol that is identical to DPoS (process of electing a miner or validator via voting scheme). DBFT was popularized by its adoption as a consensus mechanism by the NEO cryptocurrency smart contract system from China. The election process is similar to the PoS mechanism, and any node can participate in voting regardless of the number of tokens they hold. Every node in the network holds some tokens; they are called token holders or citizens. To make the system run properly and consistently, a governing body called delegates is elected by the token holders. A delegate (randomly selected) from the elected delegates acts as speaker or leader and adds the block of transactions to the blockchain.

Any citizen can be promoted to delegate if specific requirements are met; these are good internet connection, an official identity, specific resources, and a GAS value of 1000. The responsibility of the delegate is to listen to the citizen or normal nodes. The value of GAS is determined by the award they receive in storing transactions consistently and ensuring proper functionality of the network. The more the computational resources, the more they are awarded GAS in the NEO system. Whenever a delegate posts a node with transactions, it is broadcast to the delegate members and every delegate in the network stores the details of the posted transaction consistently. By doing so, every detail of the transactional record is tracked and kept up to date in the network.

To build a recent block, a speaker is selected randomly from the group of delegates and given the privilege to add a new block. The speaker broadcasts the validated block before adding it to the blockchain. The delegates, after receiving the validated block, will cross-check the details with an existing copy from their local ledger. Consensus is reached by approval from two-thirds of the delegates. This is to make sure that every delegate maintains an identical copy so that it is nearly impractical for a malicious entry attempt.

2.2.7.1 Merits and Demerits

The significance of using DBFT is, even if the speaker and delegate are not loyal, meaning that if a speaker attempts to tamper with the transaction and broadcasts, consensus will be reached regardless of faulty delegates. However, every delegate holds an identical information and although tampered data appears, consensus can't be reached as two-thirds of the members reject it. The transactional speed of DBFT is moderately fast, 4 to 5 blocks can be added to a blockchain in a minute (15–20 seconds per block), and transactional speed per second varies in the thousands. As a 61% share is needed for a block to be added, forking is unfeasible.

2.3 APPLICATIONS EMPLOYING BLOCKCHAIN CONSENSUS

Blockchain has promoted a revolution in business applications that transformed from centralization to a decentralized model delivering equal business operations. In addition, the applications employing blockchain technology as a building framework can deliver business transactions with transparency, security, greater functional flexibility, better efficiency, and faster speed (Casino, Dasaklis and Patsakis, 2019), (Laroiya, Saxena and Komalavalli, 2020). The advent of blockchain technology is outstanding among industrialists, public and private organizations, and the financial sector that drives significant opportunities in the area. Researchers and scientists are attracted by this cutting-edge technological innovation, and have extensively improved the system day by-day-by addressing typical limitations. As per the adaptation statistics and according to the authors in (Laroiya, Saxena and Komalavalli, 2020), around 44 percent of global organizations are regulating their businesses using blockchain. Popular applications employing blockchain technology are listed following.

2.3.1 Smart Transport Based with Use Case

Transportation refers to the movement of goods and people from one region to another region. The field of transportation was greatly advanced by the implementation of blockchain. A detailed survey report made by the authors in (Carter and Koh, 2018) concluded that transportation systems are migrating to a higher dimension with great potential functionalities by building the system with blockchain. Smart Transportation offers services like smart charging facility, maintaining authentic details like insurance, number of kilometers driven, vehicular identity, smart notification services, autonomous vehicular services, motor security, and smart information exchange. Adoption of blockchain facilitates the sharing of real-time data regarding geographical locations, traffic situations, and any emergency circumstances.

The communication for data transfer is carried out by implementing ad-hoc networks ("VANET"). It is noted that around 50 to 60 percent of goods and 60 to 65 percent of humans are utilizing transportation services (Yang, 2019). This led to congestion problem resulting in emission of toxins by the vehicles into environment. With the use of blockchain, congestion can be reduced, yet some malicious vehicles may transfer false information blocks and disturb the functioning. This can be controlled by consensus in blockchain. So, the consensus plays a major role in blockchain technology. However, sharing of information in the transportation system has trust issues in which any user may feel unsecure to connect to an anonymous vehicle (Yang, 2019).

To mitigate the issues like privacy and security for smart transportation, the authors in (Hu et al., 2019) researched internet security issues of vehicles and incorporated blockchain technology with the internet of vehicles to enhance the communication equipment and gain better consensus. Many researches have proposed different vehicle internet protocols and the details were portrayed by the authors in (Bodkhe et al., 2020). Different consensus mechanisms have been used in intelligent transportation by researchers. To make the communication equipment efficient between connected nodes, the authors in (Lei et al., 2017) proposed a secured management of keys between connected peers. The sharing of keys among connected peers is based on the PoW consensus mechanism. Researchers in (Yuan and Wang, 2016) invented a layered blockchain architecture called "B2ITS" by incorporating the "ITS" system in securing leveraging. They justified that use of PoW, PoS is suitable to deliver good performance while considering security issues of the system with the fault-tolerance scheme. Different consensus mechanisms adopted by the intelligent transportation systems were surveyed and presented by the researcher in (Bodkhe et al., 2020).

Innovative companies like Arcade city and La'zooz are extensively working for Smart Transportation applications. Car company lease services appear to be easy but in practical it is a complex task because the supporting information systems are easily fragmented. Blockchain for smart transportation rental systems can make the details transparent where the nodes in the network can access, transfer, and monitor consistent reliable information (single agreed view of data) irrespective of the vehicle operated. IBM's solution for blockchain is working to improve rental services in smart transportation. It works as follows: the regulator creates and broadcasts new registration in blockchain. The regulator transfers the ownership of the car to the manufacturer through a secure smart contract by invoking a transaction that is verified by the nodes to reach consensus. The IBM blockchain allows the node validation process to be tailored to meet the needs of the business network. The manufacturer adds the vehicle identification number, make, and model number to the public ledger that has control for all the nodes. The details are transparent to everyone and dealerships can see available stock to transfer to the leasing company. At all states the security, integrity, and validity of the information is maintained.

Odometer frauds are most common in car sales, a car that has a 1 lakh reading is updated to 20k for sale purposes, resulting in a car being bought that is actually not worth buying. Replacing the regular odometer with smart odometers connected remotely with underlying blockchain technology can stop odometer frauds. This was developed by the researchers in Bosch's Internet of Things (IoT) lab (Chanson et al., 2017) and is currently in the testing stage with a hundred cars in Germany and Switzerland.

2.3.2 Supply Chain Management Based with Use Case

The commerce applications are transforming day by day from a centralized organization to an expanded distributed organization. As the process of product manufacturing takes places at every corner of the globe, transparency is very much needed for better product supply and supply chain management (Laroiya, Saxena and Komalavalli, 2020). Supply chain management (ScM) inculcates overall flow of the industry from product manufacture to supply of product from the management side, and from product order placing to product delivery from the customer side. Blockchain technology is particularly suitable for ScM systems, which involve real-time data transfer at every phase of business with tracking facility of product goods especially for the companies with several supply chains.

Business operations are majorly transformed with the use of blockchain in ScM that extends peer-to-peer to distributed decentralization through a global ledger. ScM includes logistics companies with truck loading, product dispatching and shipping functions, and all other transport phases. The need for transparency and traceability is much needed in these systems, and the smooth running of the process is only achieved by using distributed ledger technologies like blockchain. Current ScM lacks trust and customers are not sure whether the exact ordered package is delivered or not. In a streamline of the process, any package would be easily altered and take days to resolve, which is why the present systems have lag.

Using blockchain in ScM systems will enhance the existing system by eliminating its limitations. The systems performance is increased with less paperwork, faster identification of product tampering, and optimized courier prices and helps in better product delivery. For a giant multichain organization, it is ultimately a major task to keep track of transactional details of a product. Use of blockchain will ease this task by tracking the record of the package with implanted radiofrequency identification tags, resulting in transparency with improved quality of service and accuracy. As per the report noted by the authors in (Zhang, 2019), use of blockchain by the ScM vendors (for guaranteed delivery while maintaining time) are noted to be greater than 60 percent.

Various consensus mechanisms like PoW, PoS, PBFT, and some others are mostly used in blockchain-based ScM systems. Walmart (Kamath, 2018) recently collaborated with IBM and started working on blockchain-based ScM for better tracking of goods and delivery. They used blockchain to track food products from harvest to delivery of the product. It was observed that nearly 42 million lost their lives because of foodborne diseases. Blockchain could attach a digital certificate by attesting details like harvest details, origin of the food, where it's been, and delivery location. If the food is contaminated, tracing back to the roof of when it was contaminated is easy as is notify the user who bought the same food. Hence a record of all transactions sealed into a block is used for better traceability of products in ScM using blockchain.

2.3.3 IoT Based with Use Case

The world is organized with connected devices and internet everywhere, an emerging technology like IoT will provide the capabilities of sensing and communicating with any object in the world based on the application parameters. The present age of technology paradigm is the information age, it serves as an individual component technology that are instrumented to connect with each other with compatibility in cross technological platforms. These technological components are spread across the globe and are connected in a distributed manner with an adaptive nature and self-organized around the end users' needs, delivering seamless services. One of the best examples is the smart cities where different connecting components (sensor that makes objects communicate with protocols) are no longer centralized but are interconnected and organized around the end users' needs through the information network.

IoT is a transformative technology that lets the user interact with anything for a better understanding of the nature. It is emerging day by day from electronic chips intended to sense and transfer information, such as monitoring and diagnosing health through wearable devices. The use of IoT will transform a house to a smart house, a city to a smart city, a vehicle to a smart vehicle, a power grid to a smart grid, and many more. Over the course of the coming decades, the use of communication devices and smart devices to fulfil user needs will grow to a great extent. The quantity of information that the internet needs to handle will jump enormously as a huge capacity of connected devices have to communicate with each other to coordinate business processes for many applications such as transportation, supply chains, construction, infrastructure building, environmental monitoring, health devices, and many more.

It was estimated that the number of connected devices across the globe by 2020 will reach twenty to fifty billion devices (Rivera and Meulen, 2016) with constant streaming of data with secure micro exchange among the connected peers. This results in the infrastructure of the internet growing far beyond the current capacity of the internet in terms of dealing with the massive amount of secure transactional data between the edges of the network (part of next evolutionary internet age). A world with full-fledged IoT management infrastructure systems is complex for the economy. Because there are three major issues that should be considered while building an IoT project. They are firstly the bottleneck problem raised by the current centralized cloud IoT based systems, secondly the consequences surrounding the security of the system, and finally enabling automated dynamic allocation of resources between machines.

The present centralized IoT works as follows: the information from the sensing device is sent to a third-party cloud, the information in the cloud is processed using data analytics, and the result is sent back to the coordinated IoT devices. In a world of complex network connections, centralized IoT systems are incapable of processing huge amounts of data while coordinating nodes at the edges of the network. In decentralized peer-to-peer IoT systems, the coordination takes place by minimizing the bottleneck effect and security vulnerabilities even if the network scales. Security is a major demand and a huge issue in IoT. A promising IoT security will facilitate working with cryptocurrencies, providing assurance to the data. A decentralized approach like blockchain implemented in IoT could provide a solution to the preceding issues by processing hundreds to thousands of transactions per second that significantly reduces the costs associated with the infrastructure in centralized system. It will simply prevent the need for shutting down the whole system when a node failure occurs.

2.3.4 Digital Voting with Use Case

Digital democratic voting has the advantage of eliminating the flaws of the paper-based voting system, which could be easily manipulated and would take hours of time to finalize the result. E-voting-based systems have been actively researched by researchers trying to minimize the election cost by enhancing its security and privacy, thereby limiting fraud with vote traceability and verifiability (Hjálmarsson et al., 2018). The first E-voting system was introduced by the government of Estonia during the 2005 and 2007 elections (Barnes, Brake and Perry, 2016), following which many countries attempting to improve the system. The major flaw of E-voting is that it suffers from security issues. Because of the security issues, countries are adopting traditional paper-based system, because the E-voting systems are more prone to being affected or hacked by hackers. The risks associated with digital voting (security issues) are mitigated with the adaptation of blockchain technology.

Blockchain based E-voting systems are transparent as the voters can see the voting count for themselves and it would make them tamperproof. The Ontario-based researchers (Clark and Essex, 2012) have developed a consensus protocol called "CommitCoin" that has enhanced the E-voting system based on blockchain by securing the person's vote. Agora, a Swiss based company, is working on a similar issue using blockchain. The possible challenges of this type of systems are as follows: voter identification by omitting privacy issues and allowing voters to participate in elections from their local systems, which are prone to being affected with malwares designed for tampering. The working criteria of the system is as follows: generally

Elliptical curve cryptography (ECC) (asymmetric cryptography) is used for digital vote creation; ECC uses both private and public keys in decryption and encryption of data.

During voter registration to the system, the voter creates two pairs of ECCs. One pair is used to reveal the voter's identity to the verifier and the second pair is used by the voter to register for casting their vote virtually. The voter signs off the vote with their private key and is verified whether the signature is valid or not; all votes are verified by the voter's public key, which is a legitimate action. This is how a blockchain helps in creating a secure and transparent voting scheme. Start-up companies are extensively working on blockchain-based e-voting systems and Adam Ernest, the founder of FollowMyVote, and his team are working to introduce blockchain for casting votes electronically. Other researchers have released a white paper that describes the enhancement of current e-voting system through the use of blockchain, called "BitCongress" that uses a similar consensus protocol to PoW by replacing the token with votes for recording tamper proof votes.

2.4 POPULAR BLOCKCHAIN FRAMEWORKS

Because of the wide range of applicability, blockchain technologies have gained tremendous popularity with industrialists, scientists, and researchers. At first glance, blockchain was initially developed for financial transactions in the industry. Presently, it has been widely used in many applications. The framework of blockchain allows it to develop distributed decentralized ledger-based applications (that can be public (permissioned-less) or private (permissioned)) intensively. This framework allows the developers to deploy and host blockchain-based applications. A number of frameworks were developed by scientists, and some of them are open source; this section describes the popular frameworks.

2.4.1 Working Criteria of Bitcoin Blockchain

Bitcoin, when divided into two parts, becomes a "bit" that is understandable by computers and a "coin" that shows currency in coins. These are virtual coins stored and transacted by computers from one person to another with tamper proofing. The tamperproof security of the transactions is gained by masking with cryptography. So, the transaction of value is encrypted at one end and decrypted at the other end, hence the terms cryptocurrency or digital currency. Bitcoin blockchain was invented by the authors in (Nakamoto et al., 2008) for secure digital currency transactions by the nodes deployed in a distributed manner. It is a public blockchain and uses PoW as its consensus protocol. The concept was referenced by a number of researchers and led to a revolution in the development of different blockchain applications.

The primary motive for developing bitcoin blockchain was to eliminate the need for the centralized authority in used in conventional systems till date. For instance, if a person wants to send money to someone, the process of debiting and crediting depends on third party authority (bank officials) with some minimum charges per transaction depending on the amount. Suppose if the servers are unavailable, then the transactions are not committed in spite of being charged and debited with money. The end user has to wait until the servers are back to operating and the money debited will be reflected in a specified period of time, generally it would take 3 to 4 business days. The entire control will revolve around the central authority that has to read, write and update the access permissions. Bitcoin is a solution in building a system with no central authority in which the security is added by developing the application with cryptography, in a distributed and decentralized manner.

Distributed in the sense that every node in the distributed network can act as client and server, each node in the network can initiate a transaction (sending bitcoins from sender to receiver) and can validate the transaction via block. This makes the information transparent. Cryptography is added to make a transactional value secure; note that every node in the network maintains every detail of information similarly—if an intruder tries to tamper with any transaction value, the update should be validated by every node in the network which is impossible. Generally, transactions are sealed into blocks and each block of transactions is linearly organized and stored in the global ledger or database in which every node in the network has access to it.

The transaction commitment (block of transactions added to blockchain) is done through the nodes in the network reaching consensus. Initially, when a wallet is created in the bitcoin applications, the creator is granted a pair of public and private keys that represent the authenticity of a node. The public key is needed to send bitcoins from one node to another. Bitcoins are accessed using the private key. A user will initiate a transaction by signing it off via the bitcoin wallet application, and the transaction is sealed to a block with a signature and the public key. The nodes of the network will validate the correctness of that block by having network miners solve a mathematical puzzle. If the details are authenticated, then the block is added to blockchain by reaching a consensus. But the current use of the consensus protocol has a major drawback of utilizing unlimited computational expenditure.

2.4.2 Working Criteria of Ethereum Blockchain

Ethereum is an open-sourced platform that lets the developers develop blockchain-based applications apart from providing cryptocurrency services. The Ethereum (Wood et al., 2014) platform was developed by Vitalik Buterin in 2013. The working criteria of Ethereum is exactly as in bitcoin from initiating a transaction and sealing it to a block with cryptographic security to storing a block of transactions in a distributed ledger in the network. But what makes Ethereum different from Bitcoin? The goal of Bitcoin is to eliminate the need for centralized currency transfer controlled by an authority. It is just an update registry with a set of simple commands controlled by the nodes of the network via the use of cryptography, distributed environment, decentralization and reaching consensus. Ethereum, the second largest cryptocurrency system after Bitcoin, has its own crypto digital currency called "Ether".

Ethereum also has do-it-yourself applications called "Dapps", that lets the user not only use cryptocurrency but also develop their own decentralized applications. Ethereum has its own programming language called "Solidity" by which the developer can build blockchain applications. Once the returned application is deployed into the system, the nodes will execute it as coded. The consensus adopted by Ethereum is an advanced version of the PoW used in Bitcoin that follows "Memory Hardness", which reduces the use of heavy computational resources as in PoW. "Solidity" in Ethereum is used to develop Smart Contracts, these contracts (deals with enforcement, management, performance, and payment) are organized as a set of If (condition) Then (action) rules through programs or Dapps that run as a virtual machine. Smart contracts are self-executing programs which act as an uneditible source file. Once the smart contract application is deployed in Ethereum, it cannot be edited or corrected even by the developer of application.

However, maintaining perfect privacy of the contracts is very challenging in every possible way of contracts executed. The consensus adopted is EthHash PoW, a modified version of PoW that has major burdens. The platform has plans to adopt PoS as the consensus protocol for better running of the smart contracts by covering the need for computational expenditure required in the current adopted consensus mechanisms. It still has potential drawbacks in which it can be modified with the protocol expressed in Section 2.4.2.1.

2.4.2.1 Casper Protocol

Casper by the authors in (Zamfir, 2015) is an enhanced version of the PoS-based protocol that Ethereum is planning to adopt, making the system efficient with scalability and security. The nodes in the Casper protocol are referenced as validators, if a validator is willing to participate in consensus, he has to deposit a portion of Ether (the cryptocurrency in Ethereum) as a stake. To add a block of transactions to a blockchain, the validators have to be elected. The election of validators depends on the amount of stake they showed in a walled application. Suppose if node 'a' deposited 32 ethers and node 'b' deposits 64 ethers, then node 'b' has the higher probability to be elected as a node validator from the participant validators selected by the validators committee. The validator committee is selected randomly and has the privilege to vote for a validator that adds a block to the blockchain. If a validator is recognized as malicious, then its privilege will be cancelled immediately. This makes the system secure from attacks.

2.4.3 Hyperledger Fabric

Hyperledger Fabric is one of the popular blockchain technologies that is used widely; it is a private version of blockchain in which the individual wanting to join the network has to be validated by the system, unlike in open blockchains wherein any stranger can join the blockchain. Hyperledger Fabric is generally an enterprise collaborative system in which different business models exchange transactions seamlessly and efficiently. Considering the custom requirement of different companies working via blockchain, Hyperledger Fabric provides a modular design through which businesses can use different required functionalities. A fabric in Hyperledger fabric records transactions (in a linear order) in the way blockchain usually does. For instance, a ledger in Bitcoin holds the transactions from sender to receiver. However, the transactions transfer in fabric is associated with assets rather than coins. Usually, an asset holds everything of value like food, and transports anything of value.

Hyperledger Fabrics allows the users to determine the value of assets by themselves. An asset is viewed as a group of state change key-value pairs that are recorded as transactions on the ledger. The modification of assets in Hyperledger Fabric is done using chain codes. They define an asset and instructions to modify an asset. Smart contracts will execute chain code in order to be deployed in Hyperledger Fabric. The businesses will share their business logic that change the ledger so that all the participants will sign off the changes to be made. The members of the network will access the ledger by invoking chain code, through deploying a new contracts business logic or through interacting with the existing deployed contract transactions. The participants of the network are authenticated to be added into the private network through the membership identity service (MIS) provided by Hyperledger.

MIS acts as an authentication service that manages all the participant identifications in the network. It also provides an additional layer to the permissioned network through an Access control list that is authorized by a specific network. To be elaborative, a network here may contain sub private networks that all will collaboratively work together. The network roles are assigned by Hyperledger depending on the node type so that specified nodes can access the chain code but can't deploy it. Nodes in Hyperledger are categorized into two types: peer nodes and ordering nodes. Peer nodes are assigned in executing and verifying a transaction, ordering nodes will order and propagate transactions by maintaining the exact history of operations in the network. This feature makes the fabrics efficient and scalable with simultaneous processing of multiple transactions.

A legitimate transaction is created by ordering nodes that are responsible for network following consensus. The ledger of fabric is constituted with two components, a blockchains log that stores the immutable transaction organized sequentially and the state of the database that maintains the blockchain's exact state. Maintaining the current state of the block allows the member of the network to monitor it there by increasing acceleration of the network. The purpose of the log is to track the state of an asset, as assets are generally exchanged among a number of nodes. Tracking details contain the time and place of the asset's creation in spite of every exchange that occurred. It enhances the visibility of verifying ownership of the asset. A distributed ledger typically contains all the transactions and every business node can have access over it, even the nonparticipants of a specific transaction can monitor the transactional details performed by other nodes. It is a deal breaker in which companies expect their transactions to be private from nonparticipants, but it can't be omitted in public blockchain because every node monitors transactional details.

Hyperledger solves this by creating private channels through implementing restricted message paths that provide privacy and confidentiality for a subset of the network. All data including members transactions are invisible to the network. This serves as a major advantage for the business to make transactions inaccessible and invisible to the other nodes of the network that are not involved in such operation.

2.5 CONCLUSION

Blockchain is an empowering technology with distributed database documents or public ledger of transactions shared amongst organizations or individuals to associate with trust and transparency. The decentralized distributed feature on top of cryptography makes blockchain grant higher safety than other systems. The significance of blockchain is to record transactions that are consensually shared and synchronized across multiple nodes of a network. Besides these facts, maintenance of data, fault tolerance of the system, scalability, and security are crucial concerns to be considered in upholding the reliability of the network. Consensus mechanisms play a vital role in maintaining these concerns in which the current chapter demonstrated different consensus protocols with their merits and demerits. Following consensus, the applications employed by blockchain, its industrial practices were depicted to allow the reader to get a complete view on distributed ledger technologies like blockchain. Major consensus protocols and their significance were portrayed so that understanding of concept allows researchers to draw insightful information. Finally, considering the advantages of blockchain compared with conventional centralized systems, its applicability in a wide variety of applications and the concerns of providing secure and scalable applications by adopting different consensus mechanisms, researchers have opportunities to fine-tune the system with novel architectural designs of blockchain and authentic consensus protocol designs that make the system a state-of-the-art model.

REFERENCES

Barnes, Andrew, Christopher Brake, and Thomas Perry. 2016. *Digital Voting with the Use of Blockchain Technology*. Team Plymouth Pioneers, Plymouth University.

Bayer, Dave, Stuart Haber, and W. Scott Stornetta. 1993. "Improving the efficiency and reliability of digital time-stamping." In *Sequences Ii*, 329–334. Springer.

Belotti, Marianna, Nikola Božić, Guy Pujolle, and Stefano Secci. 2019. "A vademecum on blockchain technologies: When, which, and how." *IEEE Communications Surveys & Tutorials* (IEEE) 21: 3796–3838.

Bodkhe, Umesh, Dhyey Mehta, Sudeep Tanwar, Pronaya Bhattacharya, Pradeep Kumar Singh, and Wei-Chiang Hong. 2020. "A survey on decentralized consensus mechanisms for cyber physical systems." *IEEE Access* (IEEE) 8: 54371–54401.

Bradbury, D. 2013. "Bitcoin's successors: From Litecoin to Freicoin and onwards." *The Guardian*. 25.

Carter, C., and L. Koh. 2018. "Blockchain disruption in transport: Are you decentralised yet?" *CATAPULT Transport Systems* 1–48.

Casino, Fran, Thomas K. Dasaklis, and Constantinos Patsakis. 2019. "A systematic literature review of blockchain-based applications: Current status, classification and open issues." *Telematics and Informatics* (Elsevier) 36: 55–81.

Castro, Miguel, Barbara Liskov, et al. 1999. "Practical Byzantine fault tolerance." *OSDI*. 173–186.

Chanson, Mathieu, Andreas Bogner, Felix Wortmann, and Elgar Fleisch. 2017. "Blockchain as a privacy enabler: An odometer fraud prevention system." *Proceedings of the 2017 ACM International Joint Conference on Pervasive and Ubiquitous Computing and Proceedings of the 2017 ACM International Symposium on Wearable Computers*. 13–16.

Chaudhry, Natalia, and Muhammad Murtaza Yousaf. 2018. "Consensus algorithms in blockchain: Comparative analysis, challenges and opportunities." *2018 12th International Conference on Open Source Systems and Technologies (ICOSST)*. 54–63.

Clark, Jeremy, and Aleksander Essex. 2012. "Commitcoin: Carbon dating commitments with bitcoin." *International Conference on Financial Cryptography and Data Security*. 390–398.

Davenport, R. A. 1978. "Distributed database technology—a survey." *Computer Networks (1976)* (Elsevier) 2: 155–167.

Dinh, Tien Tuan Anh, Ji Wang, Gang Chen, Rui Liu, Beng Chin Ooi, and Kian-Lee Tan. 2017. "Blockbench: A framework for analyzing private blockchains." *Proceedings of the 2017 ACM International Conference on Management of Data*. 1085–1100.

Dwork, Cynthia, and Moni Naor. 1992. "Pricing via processing or combatting junk mail." *Annual International Cryptology Conference*. 139–147.

Dziembowski, Stefan, Sebastian Faust, Vladimir Kolmogorov, and Krzysztof Pietrzak. 2015. "Proofs of space." *Annual Cryptology Conference*. 585–605.

Gray, James N. 1978. "Notes on data base operating systems." In *Operating Systems*, 393–481. Springer.

Haber, Stuart, and W. Scott Stornetta. 1990. "How to time-stamp a digital document." *Conference on the Theory and Application of Cryptography*. 437–455.

Hjálmarsson, Friðrik, Gunnlaugur K. Hreiðarsson, Mohammad Hamdaqa, and Gísli Hjálmtýsson. 2018. "Blockchain-based e-voting system." *2018 IEEE 11th International Conference on Cloud Computing (CLOUD)*. 983–986.

Hu, Wei, Yawei Hu, Wenhui Yao, and Huanhao Li. 2019. "A blockchain-based Byzantine consensus algorithm for information authentication of the Internet of vehicles." *IEEE Access* (IEEE) 7: 139703–139711.

Kamath, Reshma. 2018. "Food traceability on blockchain: Walmart's pork and mango pilots with IBM." *The Journal of the British Blockchain Association* (The British Blockchain Association) 1: 3712.

King, Sunny, and Scott Nadal. 2012. "Ppcoin: Peer-to-peer crypto-currency with proof-of-stake." Self-published paper, August 19: 1.

Kroll, Joshua A., Ian C. Davey, and Edward W. Felten. 2013. "The economics of Bitcoin mining, or Bitcoin in the presence of adversaries." *Proceedings of WEIS*. 11.

Lamport, Leslie. 1983. "The weak Byzantine generals problem." *Journal of the ACM (JACM)* (ACM New York) 30: 668–676.

Larimer, Daniel. 2014. "Delegated proof-of-stake (dpos)." *Bitshare whitepaper.*

Laroiya, Chetna, Deepika Saxena, and C. Komalavalli. 2020. "Applications of blockchain technology." In *Handbook of Research on Blockchain Technology*, 213–243. Elsevier.

Lei, Ao, Haitham Cruickshank, Yue Cao, Philip Asuquo, Chibueze P. Anyigor Ogah, and Zhili Sun. 2017. "Blockchain-based dynamic key management for heterogeneous intelligent transportation systems." *IEEE Internet of Things Journal* (IEEE) 4: 1832–1843.

Luu, Loi, Viswesh Narayanan, Chaodong Zheng, Kunal Baweja, Seth Gilbert, and Prateek Saxena. 2016. "A secure sharding protocol for open blockchains." *Proceedings of the 2016 ACM SIGSAC Conference on Computer and Communications Security*. 17–30.

Nakamoto, Satoshi, et al. 2008. "Bitcoin: A peer-to-peer electronic cash system." *Journal of Cryptology* 3: 1–9.

Narayanan, Arvind, Joseph Bonneau, Edward Felten, Andrew Miller, and Steven Goldfeder. 2016. *Bitcoin and Cryptocurrency Technologies: A Comprehensive Introduction*. Princeton University Press.

Panda, Soumyashree S., Bhabendu Kumar Mohanta, Utkalika Satapathy, Debasish Jena, Debasis Gountia, and Tapas Kumar Patra. 2019. "Study of blockchain based decentralized consensus algorithms." *TENCON 2019–2019 IEEE Region 10 Conference (TENCON)*. 908–913.

Pease, Marshall, Robert Shostak, and Leslie Lamport. 1980. "Reaching agreement in the presence of faults." *Journal of the ACM (JACM)* (ACM New York, NY, USA) 27: 228–234.

Rivera, J., and R. Meulen. 2016. *Forecast Alert: Internet of Things—Endpoints and Associated Services, Worldwide*. Gartner.

Schneider, Fred B. 1990. "Implementing fault-tolerant services using the state machine approach: A tutorial." *ACM Computing Surveys (CSUR)* (ACM New York) 22: 299–319.

Sturm, Christian, Jonas Szalanczi, Stefan Schönig, and Stefan Jablonski. 2018. "A lean architecture for blockchain based decentralized process execution." *International Conference on Business Process Management*. 361–373.

Tschorsch, Florian, and Björn Scheuermann. 2016. "Bitcoin and beyond: A technical survey on decentralized digital currencies." *IEEE Communications Surveys & Tutorials* (IEEE) 18: 2084–2123.

Wang, Wenbo, Dinh Thai Hoang, Peizhao Hu, Zehui Xiong, Dusit Niyato, Ping Wang, Yonggang Wen, and Dong In Kim. 2019. "A survey on consensus mechanisms and mining strategy management in blockchain networks." *IEEE Access* (IEEE) 7: 22328–22370.

Wood, Gavin, et al. 2014. "Ethereum: A secure decentralised generalised transaction ledger." *Ethereum Project Yellow Paper* 151: 1–32.

Yang, Fan. 2019. "Research and application of control algorithm based on intelligent vehicle." *Procedia Computer Science* (Elsevier) 154: 221–225.

Yang, Wenli, Erfan Aghasian, Saurabh Garg, David Herbert, Leandro Disiuta, and Byeong Kang. 2019. "A survey on blockchain-based internet service architecture: Requirements, challenges, trends, and future." *IEEE Access* (IEEE) 7: 75845–75872.

Yuan, Yong, and Fei-Yue Wang. 2016. "Towards blockchain-based intelligent transportation systems." *2016 IEEE 19th International Conference on Intelligent Transportation Systems (ITSC)*. 2663–2668.

Zamfir, Vlad. 2015. "Introducing casper the friendly ghost." *Ethereum Blog* URL: https://blog.ethereum.org/2015/08/01/introducing-casperfriendly-ghost

Zhang, Jian. 2019. "Deploying blockchain technology in the supply chain." In *Blockchain and Distributed Ledger Technology (DLT)*. IntechOpen.

3 Structure, Security Attacks, and Countermeasures in the Blockchain Network

Karthikeyan Saminathan, Hari Kishan Kondaveeti, and Sathya Karunanithi

3.1 INTRODUCTION

Blockchain is a system of validating and storing information in a way that makes it extremely difficult or impossible to hack, cheat, or change. Fundamentally, blockchain is like a spreadsheet or data structure of transactions that is copied and distributed to all the computer systems on the blockchain, called nodes. Every block has a link to the previous block, attesting to the fact that all blocks in the blockchain are chained together. A new transaction can only be added to the block when up to 51% of the participating nodes have validated and approved it through a consensus mechanism. Once the transaction is validated, and the block is approved, all other nodes participating in the blockchain update their ledger. Validation, approval, duplication, and distribution of transactions is what makes blockchain difficult, if not nearly impossible, to hack or cheat.

3.1.1 Structure of Blockchain

A blockchain is often a series of blocks that includes a complete collection of transaction information as well as logs of all transaction details. Every block of a blockchain network includes a block header, a payload, and a transaction counter (logs of all transactions). The block header is used to identify a particular block on an entire blockchain and is hashed repeatedly to create proof of work for mining rewards. Within the same network, the block without the parent (first node) is called the "Genesis block." The block header includes items such as (i) block version, (ii) Merkle tree root hash, (iii) time stamp, (iv) nonce, (v) nBits, and (vi) hashes of previous blocks.

(i) **Block version:** Block version is a 4 bytes number and parent block hash size of 32 bytes.
(ii) **Merkle tree root hash:** A 256-bit hash of all transactions that will be updated when the transaction is accepted in the block and usually has a data block size of 1024 bytes.
(iii) **Time stamp:** The block header size is 4 bytes and updated every few seconds.
(iv) **Nonce:** 4 bytes field will always initialize with 0 and for each hash calculation it will be increased.

(v) **nBits:** nBits is unique to each block used to validate a block containing 4 bytes of block header size.
(vi) **Hash pointer:** binds each block in the network to the previous block resulting in the creation of a chain.

3.1.2 Types of Blockchain

In general, there are three major types of blockchains that are defined by blockchain application as follows:

- Public blockchain
- Private blockchain
- Hybrid blockchain

3.1.2.1 Public Blockchain

Public blockchains are fully decentralized, an open-source network in which anyone can access, read and write without any permission, i.e., they allow anyone to join, participate, and act as users, miners, developers, or community members. All transactions information on public blockchain topology are completely transparent and made available for everyone to view the transaction details. The main advantage of public blockchain is the lack of control (no one will have full control over the network). All nodes connected to this blockchain network have almost the same authority and therefore this public blockchain is said to be fully distributed.

Examples: Bitcoin, Litecoin, and Ethereum are public blockchain examples.

3.1.2.2 Private Blockchain

Private blockchains are also considered *"restrictive or permissioned blockchains."* This type of blockchain network can be used within an organization where the write permission is kept centralized and monitored by an authority and the read permission may be either public or restricted. The main advantages of this type of blockchain are as follows:

1. Selected participants are allowed to join the network and the transaction details are visible to those participants only.
2. Private blockchain are more centralized.
3. The number of nodes within the network is controlled by an authority that decides who can participate in the transaction.
4. Highly secured.
5. Helps to store sensitive information of the organization.

Examples: MONAX, Hyperledger, Corda, and Multichain are the examples of private blockchain.

3.1.2.3 Hybrid Blockchain

The hybrid blockchain is a combination of public and private entities. With such a hybrid network, users can control who can access transaction details in the blockchain. Only a section of data or records selected from the blockchain will be allowed to go public by keeping it private on the private network.

3.2 SECURITY ANALYSIS OF EACH LAYER OF BLOCKCHAIN

The blockchain architecture includes the network layer, data layer, consensus layer, execution layer, and application layer.

3.2.1 Network Layer

Blockchain's communication is primarily based on a peer-to-peer (P2P) network. It is a decentralized communication model used to store, distribute, and encode data on a network This P2P network depends on neighboring nodes for the transmission of data, which must expose each other's IP. Maybe there are attackers on the network; it would be very easy for them to cause security threats to the other nodes. Network layer security is enhanced mainly by two factors: P2P network security and network authentication.

3.2.2 Data Layer

Virus signatures, politically sensitive content, etc. on the blockchain are considered to be malicious attacks on information. With blockchain's data unleashed feature, it's pretty difficult to delete information once it's written to the blockchain. When malicious information is detected in the blockchain, there are a number of issues with it.

3.2.3 Consensus Layer

The consensus approach enables blockchain to distinguish itself from other P2P technologies. Commonly used mechanisms of consensus are Proof of Work (POW), Proof of Supply (POS), and Delegated Proof of Stack (DPOS). Attacks such as long-range, accumulation, bribe, precomputing, and Sybil attacks are possible attacks mostly on the consensus layer.

3.2.4 Application Layer

Application layer security further involves security issues besides the centralized nodes such as transactions like cryptocurrency transactions and handling of large amounts of funds. These nodes can fail the decentralized blockchain network at any time, and the attack output is high and the cost is low, which is the target the attacker prefers.

3.3 SECURITY ATTACKS AND COUNTERMEASURES

Blockchain has a decentralized nature and uses a cryptographic algorithm to keep transaction information/records in the digital ledger secure from cheats and hacks.

Although this security structure makes hacking a blockchain close to impossible, there have been several cases in which transactions in the blockchain have been compromised. This development should not come as a surprise, as blockchains are targeted by cybercriminals because transactions cannot be reversed as much as they can in a traditional financial system.

Blockchain like every other technology has its strong points and its vulnerabilities. Before the sudden creation and boom of several new cryptocurrencies, it would have been safe to say that blockchain was unhackable. But with the number of cryptocurrencies now available, hackers now have more ground to perpetrate their attacks. The total worth of cryptocurrency stolen by hackers since the beginning of 2017 adds up to $2 billion—and that includes only figures released to the public.

At this point, asserting that the blockchain technology is "unhackable" is a misconception and a dangerous state of mind. It's therefore important for us to notice the inherent vulnerabilities and understand how they could affect the future of digital assets and blockchains. To do this, the various types of blockchain security attacks have been described in Figure 3.1 and their countermeasures will be discussed.

3.3.1 Blockchain Structure Attacks

Just like the name implies, these attacks are related to the design constructs of the blockchain. This means that blockchain structure attacks emerge as a result of the vulnerabilities of the blockchain structure or system, which can in turn compromise a blockchain-based application. In Figure 3.2, the attacks in relation to this are subsequently classified into two types.

3.3.1.1 Blockchain Fork

A blockchain fork is a system update that is collectively agreed upon. In a blockchain community composed of both developers and miners, they may struggle to agree with one another on the direction of the blockchain. This is evident when a group of nodes in the community is hell-bent on a particular system change and the other groups say No. This group may decide to go their own way. When this type of scenario occurs, it is considered a fork.

Because there has been a split in the system, the chain then duplicates and splits thereby allowing each group to create and implement their own design solutions.

In addition, but forks can also be caused by malicious purposes such as "Sybil nodes" and also by deliberate selfish mining. Another form of fork is the deliberate creation of what is called a child application from the parent application. A perfect example of this was the emanation of Bitcoin cash from Bitcoin. Bitcoin cash was the child application. The fork can be categorized into two types, hard forks and soft forks.

Hard fork*:* This is a situation whereby the fork is not compatible with the older or original versions of the software. The change is irreversible, meaning that all users would have to upgrade to the new version of the protocol system.

Soft fork*:* This type of fork is directly the reverse of the hard fork. In this type of fork, users of the old protocol system or addresses are not included in the update to

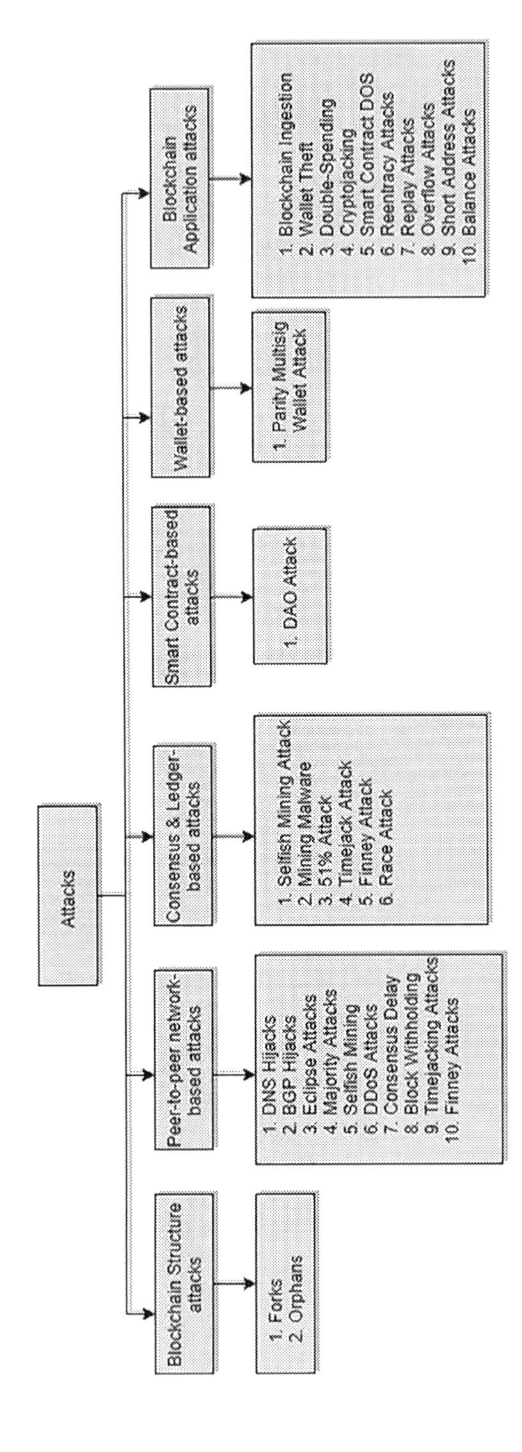

FIGURE 3.1 Types of security attacks in blockchain.

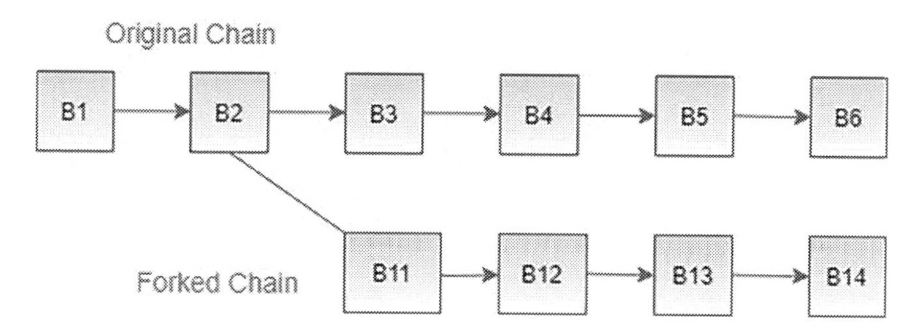

FIGURE 3.2 Blockchain fork.

the newest version. When an upgrade is backward compatible, it is considered a soft fork. A perfect example of this is the Bitcoin SegWit.

The bottom line is, blockchain fork attacks arise as a result of agreement issues and inconsistency among the nodes that can then be exploited by adversaries to cause confusion, distrust, and fraudulent activities within the network.

3.3.1.2 Orphan Block and Stale Block

These are two inconsistencies that can leave valid blocks out of the blockchain. A stale block is a cryptocurrency block that has been successfully mined but is not acceptable in the current best blockchain. This form of block occurs mostly in public blockchains because of something called a race condition. In the race condition, the network accepts one of the winning blocks and discards the others. As a result of this, the discarded or unaccepted blocks become stale blocks because they couldn't get attached to the main blockchain. Another form is the orphan block. An orphan block is a type of block on cryptocurrency space whose ancestry/parent block is unknown or nonexistent. These types of blocks emanate from the older version of cryptocurrency software. The term orphan block is used when describing a valid mine block that has been discarded.

These two forms of inconsistencies have always been a problem because, in a blockchain, only one block would be acceptable, this will then lead to more orphaned blocks in the system.

3.3.2 Peer-to-Peer Network-Based Attack

The whole idea behind the creation of blockchain is to create P2P money without banks. This means that people in different jurisdictions and countries who have no legal bindings with each other can now interact with each other without using any third-party agents, such as banks, or any clearing institutions. Peers communicate in a P2P network through the use of gossip protocols. The most common P2P attacks will be discussed next.

3.3.2.1 Eclipse Attacks

Peer A joins the network before any other peers, from there they connect and broadcast any new messages to Peer B, Peer C, and Peer D, which are preconfigured, and they, on the other hand, broadcast the new messages to the other nodes. Thus, Peer A's view on the distributed ledger system is then dependent on Peers B, C, and D. An attack is made possible because of Peer A's isolation from the rest of the network; as a result of this, Peer A's view can be manipulated by an attacker. What an Eclipse attack does is it prevents peer nodes from talking to other honest nodes, instead it makes sure the information the node gets would be from the attacker's nodes.

3.3.2.2 DNS Hijack

The Domain Network System (DNS) basically connects or links a website and its IP address through its database. The DNS is an online directory protocol that ensures the safety of online traffics. DNS hijacking is also known as DNS redirection. What it does is simple and yet suffocating, it hijacks traffic and intercepts DNS queries and then redirects them to a malicious site. The aim of this attack is for phishing or pharming. When it comes to phishing, attackers disguise fake sites as the original, and when unsuspecting internet users click on it, it gives attackers access to data, and even cryptocurrency could be wiped away.

To avoid this type of attack, one has to be very vigilant of the site one clicks.

DNS hijackers can redirect the domain server name through the use of the following:

- Cache poisoning
- Local DNS hijacking
- Router hijacking
- The man-in-the-middle attack
- Rogue DNS server

How to prevent DNS Hijacking?

- Be vigilant and avoid clicking on suspicious websites
- Avoid using public Wi-Fi to access sensitive sites
- Consider using VPN
- Install antivirus

3.3.2.3 Sybil Attack

The Sybil attack was named after a lady, Sybil Dorsett, who was treated for a rare condition called multiple personality disorder. The Sybil attack is a type of cybersecurity threat where one person creates multiple accounts or nodes with the purpose of taking over the blockchain network. A Sybil attack causes a lot of problems on the blockchain network and they include:

- When multiple nodes are created by a single person, it gives the attacker the chance to outvote the honest nodes on the network.
- A 51% attack can be launched by an attacker if they manage to control the majority of the network. When this happens, it can change how a transaction happens. That is, the attacker might delay transactions and also reverse a transaction. This can ultimately lead to double-spending.

3.3.2.4 BGP Hijack

Border Gateway Protocol (BGP) is a routing protocol that is concerned with connecting networks on the internet. BGP peering is used by networks to know about the existence of another network; this is made possible because the networks have to be manually configured before they can be aware of one another. The BGP is then the illegal or illegitimate takeover of IP addresses by corrupting the routing protocol that is concerned with connecting the networks. Although BGP attacks are minimal because through proper configuration, the network won't peer with just about anyone, BGP hijacking is still very possible. Among the cryptocurrencies, Bitcoins are the most frequently hijacked, and this can cause significant revenue loss and a wide range of exploits.

3.3.2.5 Selfish Mining

This is an "unjust" way of creating a block. This happens when a selfish miner instead mines the block continuously and then fails to publish it on the network. The miner did all this while still maintaining its track (stealth mode). A selfish miner usually receives more mining power than the honest miners because it can maximize block rewards by publishing a longer chain [11][12][13].

This concept has been a reoccurring problem in the Bitcoin system. It is believed that selfish mining affects close to 15% of the Bitcoin community. Because if a miner doesn't disclose their mined block to the entire blockchain community, they give themself a greater chance of mining the next block. There has always been a debate about this practice, but nothing tangible has been done to curb it.

3.3.2.6 Time Jacking Attacks

Time jacking attacks increase the chances of double spending, drain other nodes' resources, and also slow down the transaction process. One might be wondering how time jacking attacks happen; an attacker can speed up or slow down the nodes network time counter. This is done by connecting multiple nodes with different time stamps to the network. Although network time is always used when accepting new nodes into the system, the attackers would then set their new block a time stamp that is ahead of the real-time, which is usually 190 minutes. There are solutions to this attack, which are mostly just prevention methods.

- Always use the node time stamp instead of the network time stamp
- Your acceptance time range needs to be further tightened
- Be vigilant of your network health, and if you can, shut down once you notice anything suspicious
- You should prioritize using only trusted peers
- Before accepting a transaction, it would be better if you request more confirmations

3.3.2.7 Block Withholding

This is a form of P2P attack in which an attacker exploits a P2P network of cryptocurrencies and can be used to create conflict about the blockchain network. The attacker uses malicious nodes to withhold or forge important information that needs to be broadcast across the network, thereby making the mining pool lose all bitcoin reward that is in the block

There are usually two forms of this attack and they are the Finney attack and the block withholding (BWH) attack. In countering BWH, different experts such as Schrijve and Rosenfeld have postulated different schemes to curb BWH on the blockchain network. The former introduced incentive-compatible reward schemes that would discourage this attack from taking place. The later introduced a program called "Honeypot"; the goal of this technique is to lure the attacker into a trap.

3.3.2.8 Consensus Delay

Consensus delay is another type of P2P attack whereby attackers inject fake blocks to prevent peers from reaching consensus about the state of the blockchain. This form of attack was noticed by Karame et al. (2015).

3.3.2.9 DDoS Attacks

Distributed denial of service (DDoS) attacks are one of if not the most common blockchain attacks [1]. It mostly attacks Bitcoin and Ethereum despite the fact that the blockchain-based applications are P2P systems. DDoS attacks in a different way depending on the blockchain-based applications nature, peer behavior, and network structure. For example, a 51% attack can lead to DDoS attacks; this happens when a miner has significant hashing power and this can lead to a delay or invalidation of transactions, failure in the blockchain network, and can also lead to an intentional fork.

There are different forms of DDoS attacks and they include:

- **Mempool flooding**: Also known as memory flooding, it is a cache for unconfirmed transactions. When blockchain-based applications (DDoS) attacks this, it causes an increase in the mining fee of cryptocurrencies.
- **Stress testing**: This is another form of DDoS attack that arose as a result of the limited number of transactions that a blockchain can process in a given time frame.
- **DDoS attacks on private blockchains:** This specifically targets private blockchains whose network size is known to the other participating nodes.

A number of solutions have been procured by experts to combat DDoS; Silva (2009), in particular, proposed that there should be a cap on the minimum amount that a sender can have before he could increase his block size to perform more transactions. Another technique proposed is to increase the block size for an easy transaction.

3.3.3 Consensus and Ledger-Based Attacks

In the previous section, we looked at different blockchain attacks on a P2P system, their purposes, and the proposed plans to mitigate them. In this section, we will expose you to yet another section of blockchain attacks, such as 51% attacks, Finney attacks, Race attacks, etc.

These attacks are ledger/mining-based attacks and just like the former, they can delay or redirect transactions and also make the loss of resources possible.

3.3.3.1 51% Attacks

This type of attack is made possible when a miner on the blockchain controls 51% or more of the mining pool in the blockchain network. The success of this attack is dependent on the size of the network; if a miner or group of miners control well over 51% in a relatively small blockchain network, an attack is very possible. But if the miner controls that same figure in a relatively large network, the chance of an attack would be very low.

If a miner controls the majority of the mining pool and transactions on a blockchain network, they could:

- Prevent transactions from taking place. That is, they can render them invalid.
- Reverse transactions and thus allow double-spending.
- Fork the entire blockchain community and cause a split in the network.
- Deliberately delay other miners from finding blocks.

A 51% attack is also known as a majority attack.

3.3.3.2 Mining Malware

This is a type of attack done by cybercriminals. Malware is a program used by hackers to siphon cryptocurrencies from unsuspecting computer users. In a figure released by China some time ago, they claimed that well over a million computers have been infected by malware, and close to 26 million tokens of different types of cryptocurrencies have been mined without the knowledge of the victims.

Malware comes into the system or computer in different ways,

- Via an email that is disguised as something else
- From installing applications from nonsecure websites, etc.

There are various ways by which this can be prevented,

- Perhaps the best way to prevent this is to be extra vigilant of the sites we access and properly monitor our network traffic.
- By enabling an email filtering option to properly vet the emails that are coming in.
- By installing antivirus software. Try to install quality antivirus software and do periodic updates of these antiviruses.

3.3.3.3 Finney Attack

The Finney attack is a type of attack whereby a miner or group of miners delays block propagation just so they can double-spend their transaction.

How is this done? When a miner generates a transaction, they then compute a block, but instead of sending the newly generated block to the recipients, they would instead generate a duplicate of their past transactions and then send it to the recipient. After the recipient delivers the products, the attacker would then publish their original block that contains their actual transactions.

This type of attack is a variant of a double-spending attack.

3.3.3.4 Race Attack

This type of attack is a bit similar to Finney attacks, although there is a slight difference between this attack and Finney attacks. The only difference between the two is, unlike the former, the attackers don't really have to pre-mine the block with the transaction that they intend to double spend. What the attacker does is submit to an unsuspecting victim an unconfirmed transaction that they have broadcast to the blockchain community.

The attacker tricks the victim into believing that their transaction is the first. And unknown to the victim, the attacker has and will not submit the victim's transaction to the blockchain network.

A perfect illustration of this is, if Mr. A offers to pay Mr. B a bitcoin in exchange for a particular good, it is still possible for Mr. A to again offer the same bitcoin to Mr. C in exchange for goods at the same time. And because the system will only accept one out of the two transactions, that is what is called a race attack.

3.4 BLOCKCHAIN APPLICATION ATTACKS

Blockchain applications are also faced with constant attacks. And because there are a lot of blockchain applications out there, each of them would be subjected to different types of attacks. In this section, we are going to discuss different application attacks such as blockchain ingestion, double spending, wallet theft, cryptojacking, smart contract denial of service (DOS), replay attacks, etc.

3.4.1 CRYPTOJACKING

This can be seen as the illegal use of people's computers to mine cryptocurrency. This act of illegal mining is usually done by hackers to unsuspecting computer users. This form of attack can be initiated manually or through the internet. Manually in the sense that a person who has access to your computer might manually load crypto mining codes onto your computer. This can also be achieved by sending malicious links to unsuspecting internet users either by email or by internet ads. Whichever way the crypto mining code is inserted into the computer, what it does is work in the background while the unsuspecting user uses their computer as normal. Examples of cryptojacking include,

- Spear-fishing PowerGhost
- Graboid
- Malicious Docker
- BadShell
- MinerGate

How to prevent cryptojacking:

- Install an ad-blocker on your PC
- Be vigilant before clicking on a link
- Install an anti-cryptomining extension

3.4.2 Double Spending

Simply put, double spending is a major flaw in digital currency networks. As the name says, this is a process whereby certain money is being spent twice. I called it a major flaw because with physical cash/money, spending the same money twice on two different things is not possible. But in the blockchain network, it is possible to do that. When you buy something with digital cash, your transaction is being broadcast to all the nodes in the system. And this is where the problem comes in—before the nodes receive it and confirm it, it usually takes a long time in most cases.

Because blockchain is a P2P system and the bitcoins, Ethereum, etc. are virtual monies, there is nothing like a central bank or authority to listen to and settle disputes that might arise.

Blockchain has actually done well to mitigate this issue,

- When a transaction has been made, blockchain will notify all the nodes.
- While proposing a solution to double spending, Satoshi Nakamoto proposed that a time stamp should be used.
- Another way Bitcoin in particular has helped to mitigate this issue is by maintaining the records of all the transactions that everyone makes.

3.4.3 Reentrancy Attack

This attack is one of the most devastating attacks out there; it is mostly used to attack developing smart contracts. A reentrancy attack is made possible when there is an external call to an untrusted contract. These untrusted contracts are what attackers seize to initiate an attack. Reentrancy attacks have two types, single function reentrancy attacks and cross function reentrancy.

- **Single function reentrancy attacks**: These are easy to prevent because the function it is trying to call is usually the same as the vulnerable one.
- **Cross function reentrancy attacks**: Unlike the single function reentry attack, this type of attack is hard to detect and defend against.

Reentrancy attacks can be prevented in different ways;

- Mark the untrusted functions
- Mutex
- Check-effects-interactions.

3.4.4 Replay Attacks

This is a type of attack that can still be called a playback attack. This type of attack occurs or is made possible when two transactions are made by different blockchains. When there is a fork of cryptocurrencies in two different currents, the user keeps both as a form of ledgers. So, when relay attacks happen, the attacker takes the transaction of one of the ledgers and replays it in the other. For example, in Ethereum, a transaction data of one is valid throughout the blockchain. Ethereum has done wonderfully well to mitigate relay attacks by the incorporation of chainID in transactions.

3.4.5 Wallet Theft

Because blockchain is a digital wallet, that leaves unsuspecting accounts vulnerable to attacks. Credentials such as the key that is associated with the P2P system are all stored in the digital wallet. If an attacker gets access to your wallets, they could mine all your cryptocurrencies and data away.

Key exposure: When an attacker has the private key that belongs to an unsuspecting user, the attacker could log in on their behalf and generate new transactions.

Software Clients Vulnerabilities: The likes of Bitcoin and Ethereum use an open-source network that allows their users to connect to the internet. Users are vulnerable to attacks when they do not update their software to the latest version [8][15].

How to prevent wallet theft?

- Backups
- Wallet Insurance

3.4.6 Overflow Attacks

In a smart contract, overflow happens when a number is increased above its maximum value [7]. This is a programming vulnerability that is written in Solidity in a smart contract. Let us take, for example, that Solidity can only handle 256-bit numbers; if the bit gets increased by one, it would be called an overflow. What danger is associated with this? If an overflow happens, it will be impossible to get the required number of tokens.

3.5 SMART CONTRACT-BASED ATTACKS

In the previous sections, we looked at the types of attacks that all face, the vulnerabilities in P2P networks, application networks, etc. The advent of smart contracts, which belong to the generation of Blockchain 2.0, has brought about a new attack type. Smart contracts are automated contracts.

3.5.1 DAO Attack

Decentralized autonomous organization (DAO) [14] is not only an organization, it is in fact a smart contract. What is a smart contract? A smart contract involves two things, digital assets and two people; no middleman is involved. The DOA attack is arguably the largest heist ever committed in a blockchain. Here is how DOA works:

- A group of people write the smart contract codes that govern the whole organization.
- People purchase tokens on DOA by add funds. This called crowdsale.
- The DOA begins operation when funding ends.
- Proposals on how the money should be spent would be made to the DOA while members vote to approve these proposals.

The first decentralized cryptocurrency on DOA is bitcoin, because it runs on the Ethereum network. How did the DOA attack happen? After raising well over $100 million in the first month of operation, there were claims by token holders that there were vulnerabilities in the DOA system. They were in the process of fixing this when an unknown hacker had already crept in and drained well over 3.5 million ethers into a "Child DOA." The Child DOA had the same structure as the DOA.

This big attack was made possible by the poorly run state of the DOA because all the ether was in the same address. And also, the codes had flaws.

The attacker only stopped when ta fork was proposed.

3.6 WALLET-BASED ATTACKS

Crypto wallets don't necessarily store cryptocurrencies. What it stores mostly are tools, keys used in interacting with a blockchain. Wallets get attacked also and if care is not taken, one could lose both public and private keys. A perfect example of this type of attack is the Multisig wallet attack.

3.6.1 Parity Multisig Wallet Attacks

Multisig is a short word for Multisignature. Multisignature wallets are also smart contracts that were created to manage digital assets through the consent of many wallet owners. Multisig is like keys required by multiple users to authorize a transaction in the blockchain.

In a Multisig wallet attack, a Multisig exploit-hacker (MEH) looks for a vulnerability in an unsuspecting user wallet contract, then an attack is initiated whereby the attacker takes over the victim's wallet in just a single transaction and then drains their funds.

3.7 COUNTERMEASURES

3.7.1 Countermeasures against Blockchain Structure Attacks

3.7.1.1 Forks

Fixing soft forks to a blockchain network structure is a relatively simple procedure. Both peers in the blockchain will reach a consensus on the actual state and then resume operations. Solving hard forks is more difficult because the chains last up to a time of conflict with transaction activities.

3.7.1.2 Stale or Orphan Blocks

Because of the change in decentralized networks, the number of orphan blocks present in the blockchain network recently decreased. The solution to prevent stale or orphan blocks is to dynamically adjust the difficulty of the network. In the Bitcoin network, network difficulty is fully adjusted every two weeks from the 2016 block. During this time, if the network hash rate increases/more miners are added then the expected time to find new blocks decreases. Stale/orphan blocks are therefore highly likely to be produced. Additionally, adjusting the dynamic difficulty can significantly help to reduce the number of old and orphan blocks.

3.7.2 Countermeasures against Peer-to-Peer Network-Based Attack

3.7.2.1 Eclipse Attack

Ethan Heilman (2014) delivered eight countermeasures are as follows:

1. Deterministic random eviction,
2. Random selection,
3. Test before evict,
4. Feeler connections,
5. Anchor connections,
6. More buckets,
7. More outgoing connections,
8. Ban unsolicited addressed messages,
9. Diversify incoming connections,
10. Anomaly detection.

These countermeasures are inspired by Botnet architectures that help to make eclipse attacks much more difficult. This will show how an attacker can perceive any targeted victim with sufficient IP addresses and time, regardless of the status and new tables the victim has attempted.

3.7.2.2 DNS and BGP Attacks

Numerous researches have been conducted to equip blockchain technology with a DNS attack defense to prevent DNS attacks [3]. Maria Apostolaki et al. (2017) proposed countermeasures (short-term and long-term solutions) against routing attacks.

The short-term solutions to prevent the routing attacks in the network are listed following:

1. Increase the diversity between nodes.
2. Select the best bitcoin peers while taking routing into transaction.
3. Monitor round-trip time (RTT).
4. Monitor more statistics.

5. The churn of acceptance, i.e., the churn has indeed a major impact on block propagation in Bitcoin.
6. Use access point in different autonomous systems.
7. Consider peers hosted in the same AS and/24 prefixes.

The following are the recommendations considered as long-term solutions to detect and prevent attacks on the Bitcoin network.

- Bitcoin Communication Encryption and/or adopt MAC.
- Use independent control and data channels.
- Use the user datagram protocol (UDP) heartbeats.

The counter measures listed prior take advantage of (i) selecting a bitcoin peer to increase the diversity of internet paths and limit routing attacks and (ii) monitoring the behavior of peers in the network to check for unexpected disconnects and delays in block discovery.

3.7.2.3 Sybil Attack

Consider your node is heavily attacked by a Sibyl node, because it has only one connection to the honest node connected to that same actual bitcoin network. Here, the single honest peer sends true blockchain data to your entire node, it is clear that any Sibyl attacker is trying to deceive your data and your node will ignore them.

3.7.2.4 Selfish Mining Attack

Researchers have suggested several solutions to solve the problem of selfish mining or block-withholding attacks. To minimize the probability of this attack, Siamak Solat and Maria Potop-Butucaru (2017) proposed a scheme called the "*zero-block scheme*" that provides lifetimes for blocks, preventing self-mining attacks by selfish miners. In the zero-block algorithm, a maximum acceptable time (MAT) interval is calculated for each block in the Bitcoin network according to its expected delay and information propagation time. Within this MAT interval, the honest miner either acquires/removes a new block or it generates a dumpy block (zero block). In the occurrence of a self-mining attack, the selfish miner captures the block and then later sends it to the public chain.

Thus, it helps the selfish miners with private blocks to have more value than the maximum acceptable time. Furthermore, the expected lifetime of the new block expires and the block is then rejected by the network.

Ethan Heilman (2014) introduced a defense scheme solution called "*Freshness Preferred*" (FP) to minimize the probability of selfish mining attack. It uses the following rules

- The FP miner prefers to choose blocks with a more recent valid time stamp.
- If block A and block B are from the same branch of equal length, accept the block with the more recent time stamp, otherwise reject it.
- If block A and block B have equal time stamps, then the FP miners accept the block that reaches it first.
- Otherwise, rejects the blocks from the same branch with varied length.

3.7.2.5 Timejacking Attack

The following measures are to prevent timejacking attacks:

- Use the system time of the node instead of the network time for creating blocks and also determine block time stamps
- The appropriate time of the network node should be limited to 30 minutes
- Use trusted peers to secure nodes
- Monitor network health and shut down if there is suspicious activity
- Further confirmations are required before accepting the transaction
- Use the median blockchain time specifically when verifying blocks

3.7.2.6 Distributed Denial of Service (DDoS) Attacks

The countermeasures proposed for DDoS attacks on blockchains as follows

- Increasing throughput
- Increasing the block size
- Limiting the transaction size

Muhammad Saad et al. (2019) suggested countering DDoS attacks on the memory pool with two architectures, fee-based and age-based. This architecture design can effectively improve the size of the pool and counteract the impact of DDoS attacks.

1. *Fee-Based Design*

When the mempool size is greater than the threshold size, all incoming transactions shall be accepted by the mempool with a minimum relay and a minimum mining fee. This scheme recognizes an attacker technique that removes spam transactions and reduces the size of the mempool.

2. *Age-Based Design*

Consider N as the lists of parent transactions of current transaction, initialize the average age as zero and the average age of each transactions is calculated as follow

$$Average\ age = \frac{\sum_{i=1}^{N} Parent(i)}{N}$$

Apply "*minimum age limit*" as a filter on the memory pool. If the transaction has greater than the average age then it gets updated in the memory pool, otherwise the transaction is rejected.

3.7.3 Countermeasures against Consensus and Ledger-Based Attack

3.7.3.1 51% Attack

Users are considered to be safe if they don't receive coins from unknown parties. Usually, the target attackers are exchanging the coins between the parties. In the

51% attack, the attacker can do anything with the network and prevention involves (i) waiting for enough confirmations, which increases the resource cost to perform the attack; (ii) users should be aware of and careful about large transactions; and (iii) prior analysis of the attackers is required to be prepared.

3.7.3.2 Finney Attack

The Finney attack is an attack on a single-confirmation vendor. To overcome this attack, the vendor should wait for several confirmations until the product is delivered or the service quality is given to the client. Thus, it really does not preclude a double-spending attack, and therefore reduces the risk and makes it more difficult for the attacker to spend more than one coin at a time. Waiting for multiple confirmations acts as a countermeasure to minimize the risks from the Finney attack.

3.7.3.3 Race Attack

In race attacks, the attacker sends two distinct transactions, resulting in double-spending threats. To overcome this, Ghassan O. Karame et al. (2015) use different strategies such as (i) listening period, (ii) adding observers, and (iii) communicating double-spending notifications among peers to resist double-spending on the Bitcoin network.

(i) **Listening period:** During the "listening period," each vendor compares each received transaction and tracks all transactions received from the attacker (A) during that time. When the listener does not see a double-spending effort during the listening period, the vendor can offer the product or provide the service.
(ii) **Inserting observers:** Vendor V inserts one or two nodes to act as an "observer" that relays all transactions received directly to Vendor V. It helps Vendor V to detect a double-spending notification within seconds, either by Vendor V or by its observers.
(iii) **Communicating double-spending alerts among peers:** Here, peers within this network send alerts once they receive commonly shared inputs and from two or more transactions, and this will be considered an effective countermeasure for double-spending attempts on fast Bitcoin payment.

Ghassan O. Karame et al. (2015) discussed "*Forwarding double spending attempts*" to detect double-spending attacks on Bitcoin efficiently. This method uses peers in the Bitcoin networks to forward the transaction notification about the double spending of the same coins. In particular, whenever new transactions take place in the Bitcoin peers, it scans whether the transaction seems to use coins not spent on any other transactions in the network and its memory pool. Although coins aren't yet spent on a transaction, the peers attach the transaction in the memory pool and forward it to the network. In fact, peers discover that there might be another transaction in the memory pool where the same coins are spent on different recipients, and then colleagues forward this double-spending transaction to their neighbors. This approach not only blocks this attack, but also stops it in advance, alerting the vendor about the double-spending transaction. It takes advantage of the fact that the vendor can detect the attack before sending or delivering the service.

3.8 BLOCKCHAIN THREATS

This section provides a detailed study of individual security risks in blockchain. These include double-spending, wallet, network, mining, and finally smart contract security threats. The threats and security are listed in Table 3.1.

3.8.1 Double-Spending Security Threats (DSST)

DSST is when the same cryptocurrency is used multiple times for a single transaction. That is a given currency is spent multiple times for a single transaction. For example, Jack sending money to Jill (the merchant) to get an item, and then Jill sends the products to Jack, because if Jack makes a long tail with a cloud reverse transaction, the nodes will always receive a long tail verified transaction. In the same input environment, Jill leaves her money and her products. There are various ways to generate the dual cost attack vectors or dual cost attacks mentioned in Table 3.1.

3.8.2 Mining/Pool Threats

Mining pools are created for a team of miners to work together, mobilize their resources, contribute to the production of a block, and share the block reward according to the additional processing power. In the case of bitcoin, when a pond is settled in a block, the 12.5 PTC (bitcoin) generated by the settlement reward of that block is split and distributed among the mines. The pool was distributed among the participants. However, the current difficulty of bitcoin is that it is not possible for a miner to create a profitable mine for soloists without adequate computer resources. The pool operator or manager controls the mines.

The following chart (Figure 3.3) gives the Bitcoin network hash rate as of March 10, 2018, in most mining pools in the world.

TABLE 3.1
Taxonomy of Blockchain Security Threats.

Security threats	Attack vectors	Cause
Double-spending threats	Race attack, Finney attack, vector76, history attack	Transaction verification mechanism
	51% attack	Consensus mechanism
Pool Threats	Selfish mining, block with holding, bribery, fork after withhold, pool hopping attack	Consensus mechanism
Wallet threats	Vulnerable generation, flawed key generation, bugs & malware	Poor randomness
Network threats	DDOS, Sybil, eclipse, refund, balance attacks	Flaw in blockchain protocols and consensus mechanism
Smart contracts threats	Vulnerabilities in contract codes, blockchain, byte codes	Program design flaws

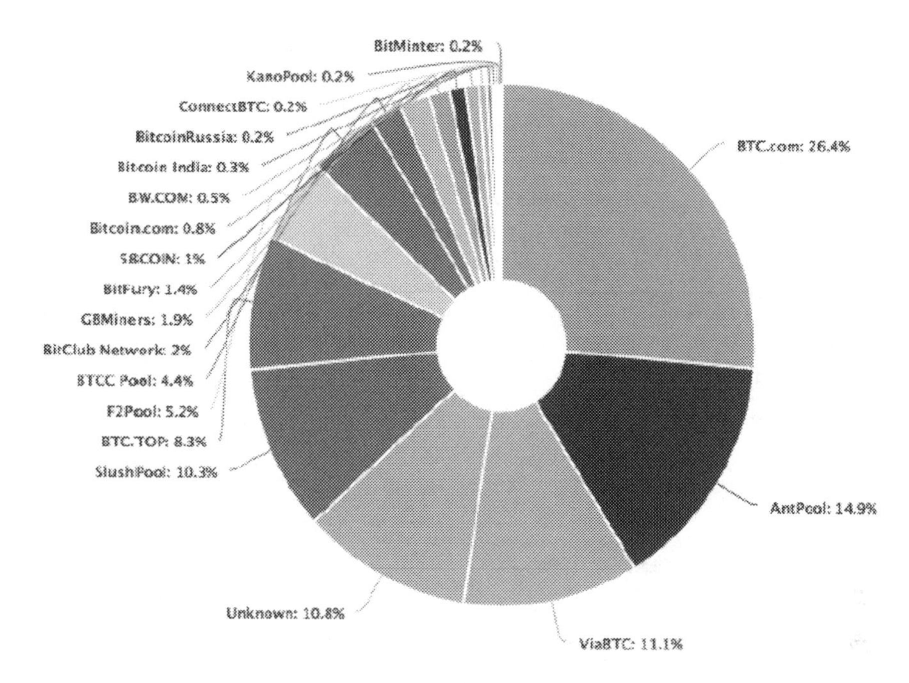

FIGURE 3.3 Bitcoin Hashrate distribution during 2018 across the world.

3.8.3 Wallet Security Threats

A variety of major management plans are used in the wallet blockchain ecosystem. Generally, wallets can be hot (connected to the internet/online) or cold (disconnected from the internet/offline). Hot & cold (online and offline) wallets can be divided into four types as follows:

- **Software Wallet**: This wallet stores the private keys locally (Bitcoin Core, XT, Armor)
- **Hardware Wallet**: This wallet stores the private keys offline (Laser Nano, Treasure, PI)
- **Paper Wallet**: This wallet prints part of your public and private key (QR Codes)
- **Online Wallet**: This wallet stores the private keys in the cloud

3.8.4 Network Security Threats

Threats can occur because of the nature of blockchain P2P networks and the services that come with blockchain networks, including all nodes that run and execute blockchain protocols. In the case of Bitcoin, there are two types of nodes: those that accept incoming DCP connections and blockchain (miners) and those that create modules on the other nodes (users) and only generate transactions and store them in the Bitcoin network. Also, various attacks can occur on the blockchain network layer for various reasons.

3.8.5 Smart Contract Security Threats (Ethereum Case Study)

A smart contract is a program or script that executes automatically when certain conditions are met. Ethereum blockchain is designed as a smart contract platform and is a well-known and used framework for smart contracts. Ethereum virtual machine (EVM) is a runtime environment where smart contracts run on Ethereum. Ethereum explored the security vulnerabilities of smart contracts and delivered them as Solidity and EVM Bytecode. Table 3.2 provides the 12 types of security vulnerabilities and their causes at various levels.

3.9 VULNERABILITIES IN BLOCKCHAIN

Blockchain technology has many cyber security vulnerabilities. Some of these vulnerabilities are specific to specific blockchain processes, whereas others are general. Blockchain cyber security vulnerabilities are divided into five categories.

3.9.1 Customer Vulnerabilities

Because blockchain addresses are not linked to one person, there are concerns about how people interact with blockchain and do not need to disclose participant identities

TABLE 3.2
Taxonomy of Vulnerabilities in Ethereum Smart Contracts.

Number	Vulnerability	Cause	Level
1	Call to the unknown	Called function does not exit	Solidity
2	Gasless send	Callee's fallback function is invoked	
3	Exception disorders	Irregularity in exception handling	
4	Type casts	No expectation is thrown, if type-mismatched	
5	Reentrancy	A non-recursive function reenters before termination	
6	Keeping secrets	Contracts private fields, secrecy is not guaranteed	
7	Immutable bugs	Defective contracts cannot be patched or recovered	EVM
8	Ether lost in transfer	Ether sent to orphan address is lost	
9	Stack size limit	The number of values exceeds 1024 in the stack	
10	Unpredictable state	The actual state of the contract is changed before invoking	Blockchain
11	Generating randomness	Malicious miner biases the outcome of random number	
12	Time constraints	Time stamp of the block is changed by malicious miner	

in the transaction process for all transactions conducted. Vulnerabilities may include digital signature vulnerabilities; hash function vulnerabilities; mining malware; address vulnerability; and software flaws.

3.9.2 Consensus Methods Vulnerabilities

These relate to modules that require an efficient and secure consensus algorithm to install, distribute and publicize the digital ledger across multiple nodes. 51% of victims may be affected; attacks include alternative history attack, Finney attack, and many more.

3.9.3 Mine Vulnerabilities

Mining pools use "stocks" to monitor the activities of each miner. Attackers can use different tactics to get more shares because they will get the bulk of the reward. There are issues like the BWH attack [2] and the bribery attack.

3.9.4 Network Vulnerabilities

There are several tricks that attackers can use to affect the normal functioning of a blockchain network, such as the transaction mellitability attack, which modifies the enemy transaction identifier (TXID) without canceling the transaction. Thus, the enemy can continue to resend. Mt. Gox was one of the largest exchanges in the history of Bitcoin, declaring bankruptcy after losing more than $450 million worth of coins. The attackers carried out a transaction compliance attack to steal coins from the exchange, which forced the exchange to freeze the user's account and stop the withdrawal.

Smart Contract Vulnerabilities: These include EVM bytecode vulnerabilities and solid vulnerabilities.

3.10 CONCLUSION

Blockchain, despite all the attacks, is still here. Digital money, digital wealth, etc. are the future of the economy. Most of the attacks I mentioned previously can be prevented by raising awareness, being careful when you do not understand certain things and, most importantly, seeking professional help. Blockchain builders should prioritize safety over other things.

REFERENCES

Apostolaki, M., Zohar, A., & Vanbever, L. (2017). Hijacking Bitcoin: Routing Attacks on Cryptocurrencies, In *2017 IEEE Symposium on Security and Privacy (SP)* (pp. 375–392). San Jose, CA, USA. doi: 10.1109/SP.2017.29.

Chicarino, V., et al. (2020). On the detection of selfish mining and stalker attacks in blockchain networks. *Annals of Telecommunications* 1–10.

Heilman, E. (2014). One weird trick to stop selfish miners: Fresh bitcoins, a solution for the honest miner. In *International Conference on Financial Cryptography and Data Security*. Springer, Berlin, Heidelberg.

Karame, G. O., Androulaki, E., Roeschlin, M., Gervais, A., & Čapkun, S. (2015). Misbehavior in bitcoin: A study of double-spending and accountability. *ACM Transactions on Information and System Security (TISSEC)*, *18*(1), 1–32.

Saad, M. et al. (2019). Countering Selfish Mining in Blockchains. *2019 International Conference on Computing, Networking and Communications (ICNC)* 360–364.

Silva, P. (2009). *DNSsec: The Antidote to DNS Cache Poisoning and Other DNS Attacks*. A F5 Networks, Inc. Technical Brief, Seattle, WA, USA.

Solat, S., & Potop-Butucaru, M. (2017). Brief announcement: Zeroblock: Timestamp-free prevention of block-with holding attack in bitcoin. In *International Symposium on Stabilization, Safety, and Security of Distributed Systems*. Springer, Cham.

ADDITIONAL READING

Bag, S., Ruj, S., & Sakurai, K. (2016). Bitcoin block withholding attack: Analysis and mitigation. *IEEE Transactions on Information Forensics and Security*, *12*(8), 1967–1978.

Bissias, G., Ozisik, A. P., Levine, B. N., & Liberatore, M. (2014). *Sybil-Resistant Mixing for Bitcoin* (pp. 149–158). ACM Press, New York, NY, USA

Eyal, I., & Sirer, E. G. (2014). Majority is not enough: Bitcoin mining is vulnerable. In *International Conference on Financial Cryptography and Data Security*. Springer, Berlin, Heidelberg.

Grincalaitis, M. (2017, September). *The Ultimate Guide to Audit a Smart Contract*. USA.

Kędziora, M., et al. (2019). Analysis of blockchain selfish mining attacks. In *International Conference on Information Systems Architecture and Technology*. Springer, Cham.

Peng, T., Leckie, C., & Ramamohanarao, K. (2007). Survey of network-based defense mechanisms countering the DoS and DDoS problems. *ACM Computing Surveys (CSUR)*, *39*(1), 3-es.

Saad, M., Thai, M. T., & Mohaisen, A. (2018, May). POSTER: Deterring DDoS attacks on blockchain-based cryptocurrencies through mempool optimization. In *Proceedings of the 2018 on Asia Conference on Computer and Communications Security* (pp. 809–811). Asia.

Sapirshtein, A., Sompolinsky, Y., & Zohar, A. (2016). Optimal selfish mining strategies in bitcoin. In *International Conference on Financial Cryptography and Data Security*. Springer, Berlin, Heidelberg.

Siegel, D. Understanding the DAO attack. www.coindesk.com/understanding-dao-hack-journalists/.

Sybil attacks: www.apriorit.com/dev-blog/578-blockchain-attack-vectors; Sybil attacks: https://coincentral.com/sybil-attack-blockchain/

Velner, Y., Teutsch, J., & Luu, L. (2017, April). Smart contracts make bitcoin mining pools vulnerable. In *Financial Cryptography and Data Security* (pp. 298–316). Sliema, Malta, USA.

4 Electronic Voting—Cloud Storage—Smart and Collaborative Transportation—Blockchain and International Trading—Blockchain Business Models

Yogesh Sharma, Balusamy Balamurugan and Nidhi Sengar

CONTENTS

4.1 INTRODUCTION

The blockchain technology came into existence in 2008 when Satoshi Nakamoto came up with the concept of cryptocurrency known as Bitcoin (Nakamoto, 2008). The blockchain technology for many years focused on the cryptocurrency. There are two generations of the blockchain. In the first generation, the blockchain was introduced along with the concept of the Bitcoin. The primary application of this generation of the blockchain was the electronic cash or better known as cryptocurrency using the Bitcoin. The Bitcoin network made use of the blockchain technology in order to record the transactions that transfer the Bitcoin cryptocurrency.

In 2013, Vitalik Buterin proposed the second-generation blockchain network known as Ethereum. Ethereum is an open-source and programmable blockchain platform (Buterin, 2014). In the white paper, Buterin proposed that instead of separate blockchain networks for different types of cryptocurrencies, a single programmable blockchain network can be used to develop different types of applications. The applications developed on this second generation take the form of a smart contract.

After the development of cryptocurrencies, many industries and organizations have found the benefits of the blockchain technology. It was the time when the blockchain technology moved ahead and was incorporated into many different applications in many sectors like healthcare, banking and finance, supply chain, insurance, manufacturing and other important business sectors. The biggest benefit that blockchain provides to industry is the absence of third-party involvement (Ganne, 2018). Because the blockchain is a decentralized and distributed ledger technology, all nodes have access to the ledger, which makes the nodes updated all the time. Once the ledgers are updated with the information, none of the nodes can make any changes or modifications to the ledger; this makes blockchain technology an immutable technology. The user of the blockchain technology can now use the technology from anywhere, anytime as the only thing required is a working internet connection.

Blockchain technology works to decentralize the transactions of assets to the internet network through a chain of networks known as a ledger (Pilkington, 2016). This ledger is shared among all the nodes of the network, which is why it is called a decentralized ledger, and works in a peer-to-peer mode. The two types of blockchain, permissionless and permissioned, also play an important role in providing the security and privacy to the data stored in the blockchain. The business models built using this technology would be more secure and preserve the privacy of both the company and the customers. The customers using the business models would also require that their critical information—like PAN number, Aadhaar Number (in India), phone number or other sensitive information—if required by the company for their business model, should not get leaked or misused. Thus, the blockchain technology would be a perfect technology for the creation of such types of business models.

4.2 TYPES OF BLOCKCHAINS

There are different types of blockchains that are employed depending upon to the usage of the blockchain.

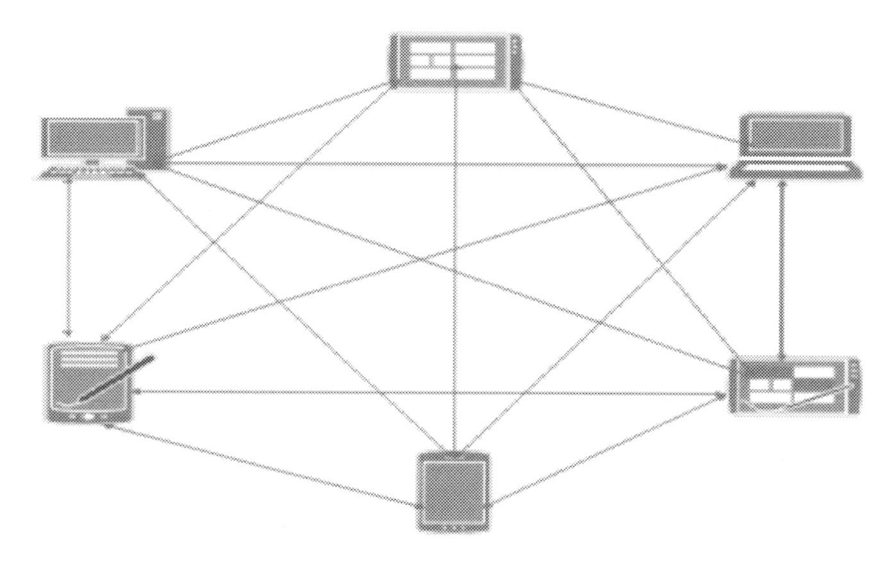

FIGURE 4.1 A public blockchain.

4.2.1 Public Blockchain

A public blockchain is also called as permissionless blockchain as there is no need to obtain permission to get into the blockchain network. A public blockchain is open to all, meaning anybody wanting to read and write onto the network can join the network. A public blockchain is a decentralized type of network, which means that there is no control by a single authority on the network and that the provided data should not change once it is written on the blockchain as shown in Figure 4.1.

In a permissionless blockchain, no one single user is allowed to change the rules of the blockchain on their own. Any information regarding the transaction is authenticated by the users of the network based on a mutual consensus among the users of the blockchain. All the users of the blockchain work in a trust-based manner, that is, each member of the public blockchain trusts each other in order to verify a transaction before the block of transactions is added into the chain of blocks.

The problem with the public blockchain comes when a group of >50% of users of the network verifies a wrong transaction, which is popularly known as a 51% attack (Zhang, Xue and Liu, 2019); however, according to different studies, this type of attack was found to be unrealistic and require a lot of expenditure.

The public blockchains are mostly useful in generation of cryptocurrencies, such as Bitcoin and Ethereum, which are available to anyone anywhere with a computer and access to good internet speed.

4.2.2 Private Blockchain

Unlike public blockchains, in which the chain is open to all and anyone can join and read and write into the blockchain, Figure 4.2 shows a private blockchain,

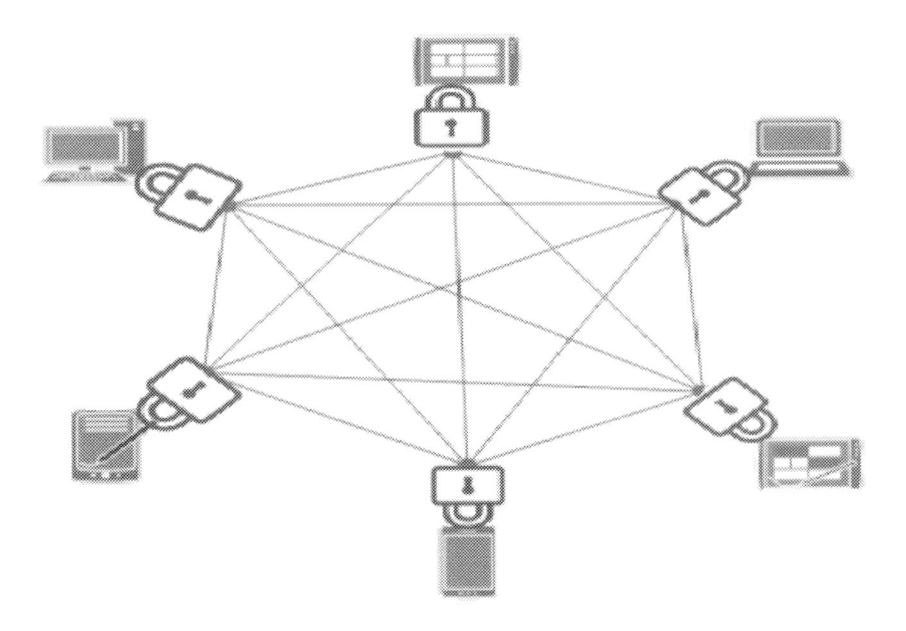

FIGURE 4.2 A private blockchain.

another type of blockchain that is not open to all—a user needs permission to join this blockchain. This type of blockchain is known as a private or permissioned blockchain.

In a private blockchain, users are not allowed to join the network freely, neither are they allowed to see the recorded transaction nor allowed to issue any transaction on their own. These types of blockchains are preferred mainly by companies, industries or government where only authorized persons are allowed to enter into the blockchain and join the blockchain. Only these approved persons can verify a transaction, issue a transaction or even execute a transaction using smart contracts.

In contrast to the public blockchain, in a private blockchain an organization running the blockchain can easily change the rules of the blockchain, can reverse a transaction and can even modify the balance. If the speed of the two blockchains are compared, then the private blockchains are faster than the public blockchains as in the private blockchain there are not so many barriers with high trust levels and also for the verification of a transaction (Dinh et al., 2017).

Private blockchains are used in various industry sectors such as the banking sector, healthcare sector, government sector, or various business sectors including pharmaceuticals, agriculture, retail chain management or any other supply chain management, which we will be discussing in further topics.

4.2.3 Consortium Blockchain

A consortium blockchain can be consider a semi-private blockchain. In a private blockchain, the members of the blockchain are connected in a permissioned environment. The private blockchain is owned by a single company or an industry that is more specifically described as a centralized system but with strong cryptographic methods attached.

A consortium blockchain has the same benefits that a private blockchain provides, but in a consortium blockchain, the ownership is not under the control of a single company or person but rather operates under the leadership of a group (Li et al., 2018).

4.3 SMART CONTRACT

Blockchain technology is very useful when it comes to the transactions to be made between the parties involved in the blockchain network. For this purpose, one of the key elements of blockchain is smart contracts, a concept given by Szabo in 1997 (Szabo, 1997). A smart contract is a computer program that executes automatically and is enforced when a transaction is completed between the two parties; it acts as an agreement between the parties, whether to agree or not to agree depends on the parties but the use of the smart contract will definitely eliminate any third party. A smart contract is a self-executable and self-verifying code written in Solidity or Python and integrated into the blockchain network. Because the smart contract is on blockchain, it will be immutable, which means the code once written on the blockchain cannot be altered or changed. The smart contracts provide a high level of security in a blockchain network.

Figure 4.3 shows the structure of a smart contract, which includes a value, address, state and functions. When an input is given to a smart contract, the corresponding function of the smart contract executes with a predefined output.

A smart contract can be used in a bidding process (Chen, Chen, and Lin, 2018) in which the smart contract structure will use the address as the address of the auctioneer and the current winner, the state will include the current auction time and the value field could be the current value for bidding.

Let's see how a smart contract works. Let us take up an example where in the network three parties are connected in a blockchain network used for paying rent to the owner. So, all the parties have agreed upon a rent agreement made in the form of a smart contract. Now this smart contract will be triggered by the software or a computer according to some conditions in terms of days, time+amount of money that need to be triggered maybe after a month or a year. A unique address may be used to trigger the condition in the smart contract and, once triggered, the amount will be paid directly to the owner of the house as a rent. Similarly, the smart contract is very useful in transferring important documents from one party to another without the involvement of any third party.

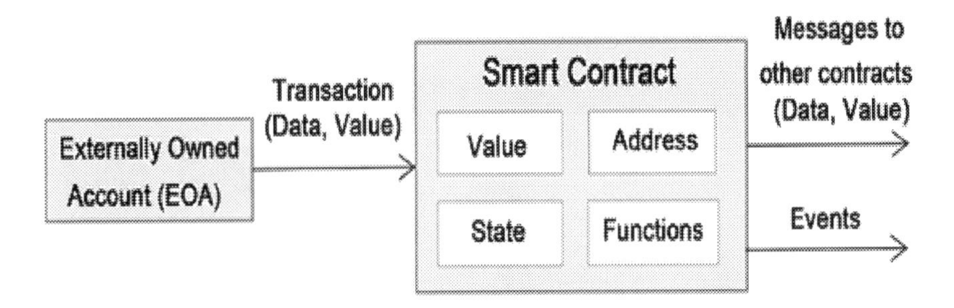

FIGURE 4.3 Basic structure of a smart contract (Bahga and Madisetti, 2016).

4.4 BLOCKCHAIN: ADDING VALUE TO A BUSINESS PROCESS

Blockchain technology has seen a remarkable growth with the technology being used for various different purposes other than just cryptocurrency. The technology is being used in various sectors such as education, health care, or in improving the quality of a business process. For many different companies, the matter of concern is with the design, execution, monitoring and improvement of business processes and for this, many different companies are using multiple systems that support the execution of the process. A business company works on a mutual trust with the customers and other business partners. Blockchain thus provides an environment where the interorganizational process can operate and execute in a trustworthy manner as now you don't have to trust any single person but you are trusting the data on the blockchain. The blockchain network can provide end-to-end traceability of the product and real-time auditing can be done using the time stamps and digital signatures (Litke, Anagnostopoulos, and Varvarigou, 2019). Transactions, which are a very important part of a business process, can use the smart contracts of the blockchain technology where the transactions execute in a transparent manner to other participants of the blockchain network. Smart contracts are small pieces of code or small programs on a blockchain. The smart contracts execute automatically whenever certain conditions are fulfilled. Thus, smart contracts can be used as an agreement between the users of the blockchain network for verification of the transaction without the interference of any third party and the rules, once created, cannot be changed or modified which allows the business parties to work in a transparent way.

Blockchain can benefit a business by tracing back the items or any product that is being traded (Mendling et al., 2018). We know that it become very difficult if the item we are dealing with gets lost or becomes untraceable. Blockchain thus comes into the picture as the complete information on the item with the exact date and time the item was sent or received are recorded on the blockchain, and there is no possibility that the data gets lost and cannot be recovered. However, every business needs to take precautions as to when to invest in blockchain technology, as it is not guaranteed that every business may benefit form a blockchain network in both design and process improvement and any unorganized or unorderly structure that is implemented on blockchain may lead to strategic failures (Bertrand, Vlasov and Bani, 2020). Thus, companies must first determine which part of the business or which application needs to be introduced into blockchain. Hence, we can say that from a business perspective, the right approach at the right time may gain the maximum benefit in a business process.

4.5 ELECTRONIC VOTING USING BLOCKCHAIN

In any democratic country, casting a vote is one of the most important events that allows the citizens of the country to use their power by casting votes and electing their representatives. By casting their vote, the citizen can protect their rights. Conducting fair elections is the basic prerequisite for any country. The election commission of the country works hard to have a fair election all over the country as every vote counts in any type of election. It has also been seen that many adult citizens do not vote for various reasons. Sometimes the citizen enjoys a holiday and stays at home, or they feel that the polling center is very far away or they may be far away from their voting city. Some might not go because they feel and believe that their vote doesn't count because of unfair election results. Standing in long queues could also be one of the reasons that a citizen does not go to cast their vote.

In the current system in some countries, the traditional paper ballot system is being used for voting. The concept of the paper ballot system is very simple; a person marks his or her vote on a piece of paper and puts it in the ballot box. After the election process ends, the votes are counted and whoever gets the most votes is the winner. The fraud that can happen in this type of voting is someone can mix incorrect voting papers with the correct ones, which may change the result of the election (every vote counts). Thus, there is a need for a system that will replace the traditional voting system that could limit fraud, which would in turn make the complete election process as well as the counting process of votes more transparent and give the deserving candidate to the country.

Further, the system developed must make voting convenient and user friendly for the voters. The system should bear minimum cost for conducting elections as the amount used by the election is huge. A system should be developed that allows a person to vote while travelling. Today we are living in the digital era and prefer to live online. Everyone prefers to buy anything from online stores be it ordering food, booking cabs, shopping for daily groceries and even finding our life partners online. So, there could be a system by which a person could cast his voting and selecting representatives whose profiles are visible with a few taps or clicks on a screen. A system should be developed in which voters can vote online while doing their regular work at home. Although many online voting solutions have been proposed in the past, none of them have become a reality. There may be various challenges that such online e-voting or remote voting systems may face (Dahlberg, 2018). Let us first have a look at these challenges—there is a need for tight security actions for remote electronic voting in the voting process (Rubin, 2002). Because in this case, the risk of large-scale manipulation is simply too high. There is a possibility of hacking of the server or the machine, which will directly affect the voting process and will lead to the wrong election result (Bannet et al., 2004). There could be chances of forced or influenced voting, possibly in large numbers (Haynes, 2014). There is no way till date to verify if your vote has been counted in this online voting system. Blockchain technology is the one technology that would definitely build the type of system that is required for the stringent process of the election (Shah, Kanchwala and Mi, 2016). Blockchain technology can help in the implementation of a system that can serve as an electronic voting system, which would be immutable, transparent and secure and preserve the privacy of the system. The stored information regarding the voting process cannot be hacked into in order to modify the results. A blockchain-based voting process could be more effective than other remote e-voting systems to conduct fair elections (Curran, 2018).

The initial step in this blockchain-based voting system is the authentication process that will authenticate the identity of the voter (Hjalmarsson et al., 2018), (Bulut et al., 2019) shown in Figure 4.4. This step is to make sure that the correct person is casting the vote and that there is no fake user because every vote counts in the voting process and will affect the election process. To avoid this risk, the user needs to download a simple application on their phone or a laptop or any such device. After that, the user needs to upload his documents, which will identify the user and authenticate the process of voting after the verification from the organization or government body conducting the election.

The documents uploaded by the voter before the voting process are matched with the list of voters in the database of the organization in order to determine if the person is registered and eligible for the voting process or not. After this, all the information

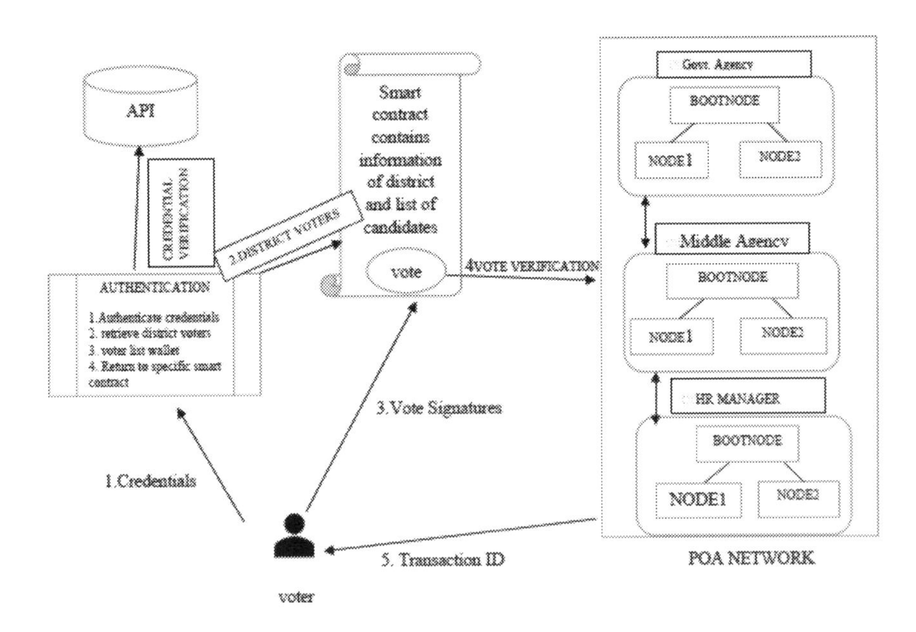

FIGURE 4.4 Voting authentication and vote casting process.

Source: Hjalmarsson et al. (2018)

of the voter is added onto blockchain in a secure manner. Once the identity of the voter is verified, a smart contract in terms of a ballot box can then be issued to the voter. The voter can vote and submit the vote in this smart contract-based ballot box.

This type of voting system based on blockchain technology will ensure that a user votes only once and not multiple times. In a blockchain-based voting system, when a voter votes, the polling stations will look into the blockchain of the voters and make sure that the voter has not already cast his vote (Zou et al., 2017). If the user's vote is valid, the polling station accepts his vote and in the case in which the vote is found to be invalid, the vote gets rejected by the polling station. Thus, blockchain will help in avoiding multiple votes being cast by an individual voter.

Once the voting process is over, all the votes become transactions, which will be stored in the blockchain after the encryption process. As we know, blockchain is immutable so that once the vote is cast and becomes a transaction, it will not be possible to change or modify the transaction.

However, the voter will have an option to obtain a printed receipt that they have cast their vote. Using the blockchain technology, the voter would be able to check whether their vote is cast and counted in the counting process. Because the blockchain technology is a distributed technology, this feature of blockchain will help the voter audit the ballot boxes, which will confirm the accuracy of the election results without interfering with the privacy of other voters. Further, if we go by the traditional voting process, then there may be the chance of some human error in the counting process, which will impact the election process. But with blockchain, the election results can be declared immediately after the voting is over without any chance of human error. All the voters on the blockchain will receive notification of the election result on their devices, because the ledger is updated with the result. With the blockchain technology, the user will have the flexibility to login and vote from anywhere (Yu et al., 2018). The only requirement would be a phone and an active internet connection. This would encourage more and more people to vote and become a part of the election process from any part of the world, where the opinion of every person matters. With this system, the valuable votes won't get wasted and the right candidate will win. The system made using blockchain technology would definitely be a cheaper method as compared with the current way of conducting elections. In many countries and states, blockchain technology is being adopted slowly and steadily for secure voting. For example, West Virginia and Sierra Leone tested mobile voting through blockchain in 2018. Apart from state elections, this type of blockchain-based voting system could also be very useful in the voting processes in private companies, organizations and college elections. This type of secure voting system could also be used in the process of voting in reality shows like talent hunt shows where the votes of the viewers matter the most.

Andrew, Brake, and Perry (2016) in their report discussed the three types of authentication process in regard to the voting process and the method of casting votes. These distinct values discussed are the voter's identification number (e.g., Indian Citizens have their Aadhaar Number or Voter ID number), the password supplied on registration and their ballot card, which contains a QR code, because there could be two methods of voting, through a web portal or physically going to the polling station. There could be different ways a user will put in the authentication details. However, in order to vote, the person must provide all three pieces of information. Each polling station will have its own URL, using which the voter can login to the website where they can cast their vote. This URL will also be present on the ballot card.

(Bulut et al., 2019) has proposed a voting system in a levelled structure. This system has been designed into levels. If the country is designed into a single blockchain, then the synchronization of the system will degrade the performance of the system because of the huge number of ballot boxes present all over the country and the distance between the polling centers. The distance factor will cause a kind of latency in the system if a single blockchain is built for the whole country. To decrease the latency in the system, it is divided into chain levels from the lowest level to the highest level.

At the lowest level, the chain will consist of nodes that can be represented by either machine or voting centers where voters will cast their votes. At the second lowest level, clusters of chains will store the data that are coming from the level below. Communication between the levels is guaranteed using communication protocols. This type of communication should be done at regular intervals so that there will be time delay in synchronization of the levels.

The levels in this system can be increased based upon the population or voters in order to decrease the collisions between the transactions.

4.6 CLOUD STORAGE AND BLOCKCHAIN

Blockchain is a system in which information is stored in blocks about who's participating in a certain transaction. A good example of this can be seen in Bitcoin exchanges. In Bitcoin exchanges, what happens is that you would make a transaction and then for every single party that was involved in this transaction, a copy of the information would be stored in the blockchain (Underwood, 2016). If one party decides to delete it or get rid of it, there's still other copies of the information available. That's the basic of blockchain. Cloud service basically is a service made available to users on demand via the internet from a cloud computing service provider not by a company owned on-premises.

The benefit of combining cloud storage with blockchain is that we can store more data online in the simplest way possible, and it is less vulnerable and less prone to tampering over the cloud.

Cloud servers are still hacked almost on a daily basis and information gets leaked by big companies; client data gets leaked from the cloud almost on a daily basis (SelfKey, 2020). But this will not happen with blockchain because every time there's new information, all participants will have a unique digital key in their computers and without the access of all of them, none of the information can be released, which is why blockchain is superior to the cloud in terms of security (Park and Park, 2017). Blockchain keeps track of transactions, so blockchain is preferred for keeping track of things like online purchases and trade.

The cloud storage market is increasing day by day and is dominated by companies like Amazon, IBM, Google and Microsoft (Sikeridis et al., 2017). The environment in the cloud storage is still centralized as these service providers have complete control of the data that are stored in the cloud. Service provider companies charge a significant amount for providing the services. They can use the data that are stored in their cloud to provide directed, tailored advertisements to the data owners, which is intrusive. The main disadvantage is that they are based on a server model that is prone to failure because of its centralized feature.

The solution of all these problems is a decentralized cloud with blockchain that has already started giving answers to processes of business and companies (Barenji et al., 2018). Cloud storage that is a decentralized cloud is like people renting out spare rooms in their house and other people staying there. Decentralization is enabling people to rent out spare space on their hard drives that can then be used for cloud storage and in fact can offer 150 petabytes of cloud storage but with no disk drives of their own (Mearian, 2015). All the disk drives that are part of the network are run by individuals or data centers who have spare space on their hard drives. This decentralization can be applied to computing and networking also. Communications will be secure and reliable and compliant if they are built by individuals or built right initially by companies. Decentralized and cloud computing of course is really popular, but cloud computing is, if anything, far more centralized. The decentralized cloud is a fundamentally different technical and economic model for delivering infrastructure where things are not controlled by any single individual or company and it is built off of open source and the principles that underlie the internet. When building a decentralized system, what we are trying to do is build a system that is very different in its approach from a centralized system. If Airbnb could be the largest hotel chain without owning a single hotel room and if Uber can be the largest taxi company without owing one single taxi, cloud storage with blockchain can build the largest and ultimately most secure, resilient, best performing and economical cloud storage service without owning or operating a single data center (Golub, 2019).

Storj and Siacoin are the most famous companies that are providing a decentralized cloud solution that provides much better security than server model cloud providing companies (Storj Labs, 2018), (Brett Kotas, 2017). Storage is a cloud shared by the community. Cloud storage with blockchain is potentially the largest, cheapest and most secure cloud available. The client gets what they have shared. Clients could even be paid by renting their extra space. Distributed cloud is more secure because each file is shredded, encrypted and spread across the network until it is ready to be used again (Sen and Labs, 2010). Files are safe because the keys are in the client's pocket, not a company's. Clients have access to their stuff because the network is shared and there is no worry about slow download speeds coming from one place. Clients renting out empty hard drives have a cloud with security, no downtime and speed at a fraction of the cost.

The main aim of the decentralized cloud storage is to provide functionality, efficiency and security to their clients (Müller, Ludwig, and Franczyk, 2017). The meaning of functionality is to provide data updates and queries, efficiency is to have the server respond to queries in the minimum possible time and security is that searchable data should be encrypted.

The decentralized cloud providers should meet following conditions:

1. No single authority should control all services.
2. Clients should get the same data back from the cloud providers that they have stored. Clients should be able to delete the data from their computers.
3. The clients should be able to see or access their data. The service providers should not be allowed to access or see the data of the clients.
4. It is very important that clients should be assured that their data is safe with inbuilt redundancy and cannot be destroyed without their permission.
5. The solution should be very much cost effective.
6. An incentive-based system for customers and storage providers should be built for better business perspective.

The persons having more computer storage can rent their resources in exchange for some charge is the idea in a decentralized cloud with blockchain. This type of cloud service provides efficient security and safety of the data.

Distributed cloud storage with blockchain working can be explained in a simple manner:

1. The most basic function is to share the files in a P2P network, i.e., peer-to-peer network, and keep check on storage, retrieval, encryption and smart contract administration using a blockchain technology.
2. The client controls their data with their own encryption keys. They can encrypt their file.
3. The file is then split into many fragments and duplicates of each fragment are generated to maintain redundancy.
4. The fragments are then stored in multiple computers in the peer-to-peer network.
5. Each fragment is encrypted already, so they cannot be accessed by the computers in the peer-to-peer network. Even hackers can only see an encrypted fragment only.

It is not possible for the hackers to reconstruct the whole file from one encrypted fragment.

6. When clients want to get back their file, they will use a hash table that is based on blockchain with their encryption keys. The whole file can be rebuilt from the fragments.
7. The client will use the encryption key to decrypt the file.
8. All terms and conditions are governed by the smart contracts in the blockchain.

In cloud storage based on blockchain, no service provider or any other node in the network can access the content. In this, every data is encrypted and only part of the file is stored at one node, not the whole file. So, the data are secured better.

Storj is the most famous blockchain company providing a decentralized cloud solution (Marco e. G. Maltese, 2015). Their basic model is to be highly distributed, ridiculously resilient, easy to use, faster than AWS or equivalent and be more durable at a fraction of the price. There are three main actors in the system. There are customers who have applications; these can be s3 compatible applications or native applications. There are the people who operate the nodes called storage node operators or snows and there are 150,000 or so around the world. Then there's a big network of satellite metadata services that know about the hashes and the locations but don't know anything else about the data. To understand the working, imagine that you have an application that needs storage. Before you upload anything to the network, it gets encrypted locally using client keys that only the client will have. The clients don't have to trust anybody, including storage providers. The data is split up into pieces and those pieces are distributed across the network. To split up the pieces, Reed–Solomon coding is used, which is not a new technology. Basically, it allows division of the encrypted file into 80 pieces. Then each one of those 80 pieces is put onto a different drive in the network. Each one of those drives is run by a different person on a different power supply in a different geography on a different network with different software. They use peer-to-peer technology so that the nodes are making sure that other nodes are up and also do frequent and random cryptographic checks to make sure that people are storing what they claim to have stored. This makes the system more robust and durable. It mainly uses three tools. The first is StorjShare. This tool is used so that nodes that are willing to set up their own machines or storage can share. The graphical interface is provided that takes the parameters like the storage amount they will share, storage location and payment address. The second tool is Bridge, which will negotiate the smart contacts with clients who want a service (Alex and Levi, 2018). All client interactions are done by Bridge. The third tool is Storj API, which provides the main service of the decentralized cloud service.

SIA coin is also one of the technologies in this field and is one of the top cryptocurrencies (Harsh, 2020). In SIA coin, decentralized means that your files are encrypted then distributed in redundancy across an array of devices all around the world. You're the only person who can access your own files, not even somebody sitting on one of these devices like the personal computer renting out their hard drive space has access to an entire file at any given time so it's actually very secure. These decentralization solutions provide services at cheap rates compared with other providers. Like for $2 a month you'd be able to store one terabyte of data on the cloud as opposed to $23 a month at Amazon.

SINOVAT is an entirely decentralized innovation community based on blockchain technology. It aims to provide the power and authority of data transparently to humans. It offers innovative and low-cost big data solutions by developing its infrastructure with its own blockchain and distinctive architecture. Decentralized blockchain-based cloud data storage within infinity nodes. Direct messaging that protects personal privacy that is not registered anywhere. Blockchain-based email proof can be stored in the chain, and the data can be stored in the nodes (Patil and Puranik, 2019). The possibility of keeping the proof of official and essential documents of all kinds in the blockchain and the data distributed in the nodes for the desired time. Blockchain, by its nature, is not suitable for information storage, and there is also the problem of scaling. The IDS (Incorruptible Data Storage) technology developed by SINOVATE enables data to be stored in nodes and evidence in the chain (Sinovate n.d.). So how do these nodes provide stability? For a node to be stable, it must be online for a long time. This is a costly process. The user must be able to earn income. In traditional master nodes, node owners can deactivate their nodes as soon as they earn income or whenever they want. For this reason, it is not possible to store data in traditional master nodes in any way, considering the stability. In SINOVATE technology, infinity nodes become active with the burning of coins. These nodes have a 12-month limited life. The node holder has to be active for 12 months in order to get back the coins he has burned and also to obtain passive income. This 12-month continuous period ensures that the data in the nodes are transferred to the active nodes before the nodes expire. Thus, the data can be transported quickly and safely, completely decentralized, at a fraction of the cost of centralized cloud solutions.

Iagon is also a decentralized cloud service that's optimized by artificial intelligence (Dhaliwal, 2018). It uses a network of computers and laptops, including home PCs, a decentralized cloud of storage for data which has been sharded. The owners of these computers act as miners who are then paid in tokens for the use of their storage. So, in a sense, you can be a miner if you with your laptop can provide storage-based processing power to Iagon. In return you get tokens. The company earns a 10% commission from the profits its miners make and uses artificial intelligence to learn about its miners. They are learning behavior in terms of performance, availability, cost-benefit and trust ability. The information is used to give a miner a cloud score and based on that cloud score miners can price services higher or lower depending on the client's needs. Its decentralized service solves three major problems with cloud storage, including security. They have no central point of attack and completely secure and that's the main advantage. The other is privacy as client can only access data or files through a private key, and only they have access to that. And because it doesn't need one giant computer for storage and processing, it's much cheaper. They able to decrease the cost by up to 80%. Iagon guarantees that because the data is sharded, it only accepts the client that can access it making it hack proof. In order to get the full file, you need all the shards combined together and there's no way of telling from the hacker point of view where the rest of the shards are. They use the blockchain as a security layer to store the hashes. The decentralized cloud model is also used for companies or individuals looking for additional processing power. Virtual reality (VR) needs a lot of processing power, the gaming industry needs a lot of processing power so they provide that processing power from these miners.

Decentralized cloud service for storing and backing up data gives us many reasons to use this as compared with the present day cloud service (Mrvosevic, 2019). The first reason is that your files are always there and you are able to be accessed at any time. The second reason is being able to quickly and easily access all of your files and data from anywhere on any device. The third reason is that we are able to access thousands of your photos, songs and videos from anywhere in the world from any device. The cloud storage on Facebook or Google Drive is in a centralized location and technically these companies have the keys to your data. They can access it at any time. If a government agency requests your data, they may be obligated to give that data over to the government agency. Decentralized cloud services no longer store data in a central location but instead it is encrypted, split into pieces and distributed amongst different machines all around the world. This means only you will have access to your digital possessions. The next reason is that the decentralized cloud is more secure than any traditional cloud. Today's cloud is vulnerable to a variety of attacks, which can lead to encryption walls being bypassed and makes your personal information accessible to hackers (National Security Agency, 2020). With distributed cloud, your stored personal information is also a part of the distributed network meaning your personal information is no longer just sitting in one central location but is split up and distributed across thousands of locations. If hackers know where your personal data is, then they can do all kinds of things to hack into that one location and once they have access, then all of your files are vulnerable. A decentralized cloud service is faster and cheaper. The fact that they're able to use thousands of machines for computing power means that your files are going to be delivered to you much faster than with traditional cloud services and it's much cheaper because they're able to utilize all the machines on the network.

We believe that open source, blockchain and decentralization will fundamentally change cloud computing in the same way that open source and internet changed computing over the past two decades.

4.7 SMART AND COLLABORATIVE TRANSPORTATION WITH BLOCKCHAIN

The world is changing fast as cities are growing and the urban population is also increasing. The need for transport of goods and people is increasing but congestion, pollution, road accidents and climate change are also increasing with it. Today, we need to increase mobility but in a smarter way. Like electric vehicles charged simultaneously with the loading of goods. We need dedicated bus lanes where drivers enter to lead the connected convoy on to public roads. Autonomous connected vehicles create flexible, safe and modular vehicle convoys that are adapting to transportation needs. It can adjust schedules, update the flexible convoy and switch goods on the way–it will be a mobile consolidation center. The smart transportation is empowered through a connecting vehicle, integrated cloud platform, people, logistics partners and infrastructure (Wang and Li, 2016). In the smart city, sensors of self-driven vehicles can sense the movements in the surroundings and react quickly and automatically to any unusual events like accidents, making sure passengers are safe and preventing accidents. Bus platooning in city transportation improves efficiency and capacity. With emission-free and silent vehicles, public transport becomes efficient and brings people closer. This will create new prospects for smart city planning. Charging from the electric grid in a fast and efficient manner in just a minute. For cities and infrastructure, building tomorrow in a sustainable way can be achieved by autonomous vehicles, electric and with zero emission, low noise vehicles and integration of technologies like blockchain, internet of things and cloud computing technologies.

One of the ways to achieve a smart city is with the smart transportation that can enhance the lives of people and increases sustainability. This involves:

1. Information system: This system gathers data related to traffic and use of different modes of transport and vehicles. These systems help make efficient and accessible public transport and also optimize the usage of private cars.
2. Smart city technologies: These technologies include new and also improved existing types of transport, e.g., mobile apps, connected cars and more.

The elements of smart transport are:

1. Connected vehicles
2. MaaS (mobility-as-a-service)
3. ATMS (Advanced Traffic Management system)

4.7.1 Connected Vehicles

Modern vehicles are now furnished with IoT (Internet of things) that integrates it with other systems that manage the traffic. This is called V2I or vehicle to infrastructure (Ubiergo and Jin, 2016). These vehicles share surrounding conditions like roads, rerouting of traffic etc. They can also communicate with other cars and can share information about speed, location and the direction.

4.7.2 MaaS

MaaS combines functionalities like payments, planning and booking. It provides a single interface for different types of transport. Passengers can choose from many services available that can help them reach their destination with convenience, speed and less cost. It has the potential to improve the public transport system in terms of quality and accessibility.

4.7.3 ATMS

This is a management system for traffic in smart cities and it takes information from street lights, smart roads, toll booths and traffic lights. This system controls traffic lights and can adjust charges on tolls and give traffic information to other control centers. It helps in providing information to drivers about real-time traffic conditions, can help in optimizing the flow of traffic and reduce congestion.

Some of the features of smart transportation are contactless fare payment, shared vehicle system, smart parking, autonomous vehicles and on-demand service.

The main purpose of smart cities is to provide efficient services to their citizens and decrease administrative costs (Lima et al., 2020). In smart cities, all the vehicles are connected to each other by use of IoT and the internet so issues of security and privacy can be resolved by blockchain. By this real-time position, traffic conditions and accident information can easily be shared between vehicles in a secure manner. Cooperative intelligent transport system and smart transportation increase road traffic sustainability with the use of smart, autonomous and connected vehicles (United Nations, 2015). These vehicles can completely change traffic networks by improving security and safety, reducing energy consumption and reducing pollution. Driverless cars, electric vehicles will completely change the transportation sector and new business models are emerging where mobility as a service is a goal (Pangbourne et al., 2018). The trust in vehicular networks can be achieved by blockchain technology; this can support deployment and development of autonomous systems that are mobile where all transactions are secure and immutable in distributed ledgers. Traceability in transportation effectiveness in blockchain reduces food waste or food scandals. Smart contracts can enhance the market for electric cars with a reduction in emissions that can benefit the environment. The data sharing among drivers with incentives mechanisms can reduce the traffic in cities.

Smart transportation involves a blockchain decentralized network that is not owned by any company but on which every company can participate, collaborate and innovate. DAV is Decentralized Autonomous Vehicle Network that works with autonomous vehicles.

The DAV open protocol defines how vehicles and users communicate with each other, discover each other and the token through which they transact with each other (Copel and Ater 2017). The DAV open-source platform is designed to be easily integrated into any autonomous vehicle platform; it's written in any language the developers are used to working with and it enables them to connect their vehicles to the blockchain without knowing anything about the blockchain. This is one example of an application developed using the DAV open-source platform. The DAV Works on Ethereum like blockchain along with Kademlia Decentralized Hash Table that is used for discovery of node. One example of an application developed using the DAV open-source platform is drone delivery. User requests bids for drones nearby by submitting the parameters for the mission in a drone app, then drones submit their bids, submit their pricing in DAV tokens. The user accepts the bid, smart contracts are signed and the mission is on the way, and when it's finished the tokens are transferred. DAV alliance is a list of commercial companies that have integrated the DAV protocol and the DAV platform into their vehicles.

Some of the biggest challenges faced by shippers in today's market are about capacity. Short order lead times aren't changing, small order sizes aren't changing and so there's only so many consolidation opportunities possible within a given company's outbound and inbound flows. The next logical evolution is about collaborative logistics. Now in order to leverage collaboration, you have to know your network backwards and forwards. You also have to have visibility across both your outbound and inbound flows. Collaborative logistics don't necessarily have to be with other shippers they can often times be within your own company or across divisions. Understanding where you have consistent freight, understanding where you have order sizes that would be conducive to outbound and inbound complementary flows be it within your network or across a network of shippers that's terribly essential to your transportation strategy. Developing a transportation strategy to leverage a collaborative network takes a conscious effort; it's really about understanding and seeing what's happening first and foremost within your own network. If you don't have visibility, if you don't understand, if you can't see what's happening between your outbound and inbound operations or across different divisions even within your own company then collaborative logistics are a bit of a stretch, but once you get there, there's lot of shipping that's still happening. There's a lot of 80% full trucks that could be 100% full via collaborative opportunities both within your network that is with other divisions, with your suppliers even with your carrier partners but also across networks. Now in order to do it with other shippers, you truly have to see what's happening in your own network to understand where you have consistent flows, understand where you have freight that's conducive to cross loading and transportation consolidation opportunities. Being able to see in real time not only the planning and execution happening within your own supply chain but having visibility to what's happening within other networks. But being able to see is not enough, having advanced algorithmic and optimization capabilities that will help identify collaborative logistics opportunities is the only way. The collaborative logistics comes to fruition with optimization and transparency.

Blockchain has a tremendous potential in improving transportation, which makes companies more profitable and efficient (Yuan and Wang, 2016). Major issues that are faced by transportation industries are disputes for payments, high administration cost because of paper transactions and industry struggle of matching the demands of shippers with carriers (demand and supply). Blockchain enables coordination among documents by using distributed shared ledger leads to less paperwork. A smart contract enables fast processing by making approvals and clearance fast. Secure, updated and authenticated data are provided to the companies to make the right decisions by trusting blockchain. Authentication and tracking of order are done immediately with the help of the blockchain that is making the system scalable. BiTA is a United States company that is based on blockchain for collaborative transportation (Oriold, 2011). Blockchain solves the problem of transportation by resolving payment disputes, reducing administration costs and tracking goods that are temperature controlled like vaccines. Modifying any data in the blockchain is almost impossible, making it more trustworthy.

Freight tracking is much more efficient with the blockchain because it provides data authentication and validation (Irannezhad, 2019). Blockchain with IoT and artificial intelligence are used for monitoring the capacity. Sensors in vehicles can detect the amount of space occupied in a shipment and determines the cost and transport all the data to the blockchain. Individual vehicle performance can be tracked with the help of blockchain, aiding in the buying and selling of vehicles. As shown in fig 4.5, Blockchain and IoT are used for V2V (vehicle-to-vehicle communication) so that vehicles can communicate with each other, which will improve safety and fuel efficiency (Dey et al., 2016). Technology totally removes the middleman and increases trust among parties by authenticating data and also preventing duplications. Smart contract removes the need for all administrative work, cuts costs and leaves no room for errors. With the smart contracts, payments are automatically paid to the shipper when the parcel reaches a destination.

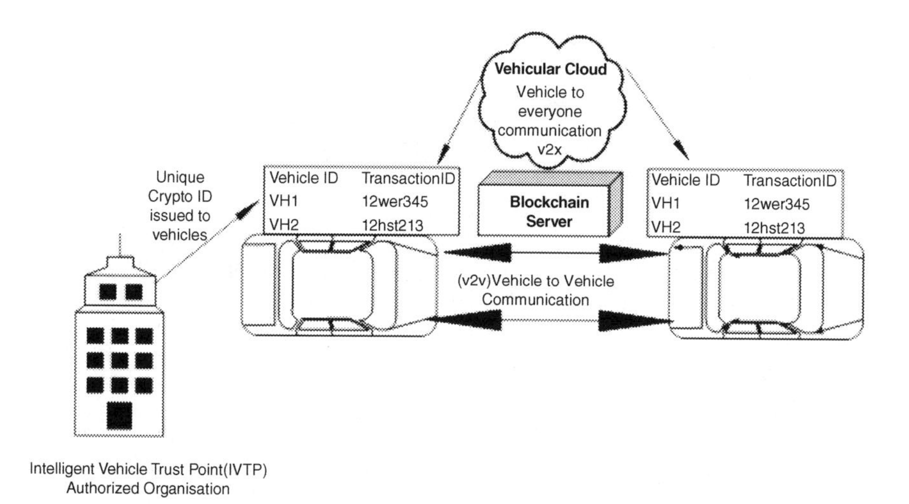

FIGURE 4.5 Smart transportation.

Source: Singh and Kim (2018)

The Toyota company is using distributed ledger and blockchain for a fast-track growth in autonomous driving technology (ETAuto, 2017). They are working in fields like data & transaction sharing and insurance. Blockchain enables people and companies to share driving data securely in the market place. Vehicle sensors totally aware of their surroundings and also connected to the cloud are producing lots of data. Blockchain creates an environment for sharing by preserving the privacy and ownership of the data. Blockchain-based tools have the potential to permit the owner of the vehicle to use their asset by selling their cargo space or rides. These tools can store all the information about the drivers, passengers and owners. The use of a smart contract can validate the payments between parties without any middle-man. To reduce the hassle and fraud in the insurance, blockchain can be used for storage and maintain the transparency between owners, drivers and the insurance companies.

Porsche also implemented blockchain to improve the autonomous driving technology. They have tested that with an application we can lock and unlock a vehicle, so it can be used for authorization. According to them, services based on blockchain are fast and secure. Access and authorization are distributed and can be done quickly, securely and remotely. All communication between vehicles and other participants are very much secure. They have made a network using blockchain for electric vehicle charging and payments that is secure. This technology in combination with IoT make it possible that self-driving cars can update the road conditions on the network, provide traffic, accident and weather alerts and traffic congestion alert in the V2V network.

IBM is also using distributed ledger and blockchain, which enables secure payments so that vehicles can pay their battery charging, toll charges and parking charges. This will allow a vehicle to respond to its surroundings without human intervention.

In smart cities and smart transportation, blockchain enhances the trust between the participants, exchange of money and data security. It reduces the malicious attacks on vehicles by increasing security, improving management of keys and also can provide legal advice in the case of accidents. It can also provide incentives to the drivers who share information about accidents and traffic. Supply of energy for recharging electric vehicles can be managed well by blockchain. The advent of smart transportation catalyzes an efficient, unified and connected new mobile ecosystem. This system revolves around consumers, system and service focused business models. The conjunction of IoT, blockchain and cloud are the main pillars of the decentralized mobile platform.

4.8 INDUSTRIAL TRADING BUSINESS MODEL USING BLOCKCHAIN

Blockchain essentially is a digital database that is decentralized; there is no central entity. The blockchain technology relies on a peer-to-peer network and it's distributed so the records on the database are shared with all participants in the network. Blockchain is a fast-changing environment and it is a very key element for the regulators to know what's going on and to understand the technology. Ganne (2018) in his book builds a bridge between the blockchain community and the trade community and try to explain in simple terms, to trade officials, what this technology is about, what it can do, and what it cannot do. Blockchain could enhance the transparency and the traceability of supply chains without a policy environment, which allows the technology to thrive. We may be losing the opportunity to make international trade more efficient and more inclusive. In fact, according to Gartner, in a recent study has been estimated that the gains from blockchains could deliver three trillion dollars of value worldwide by 2030 (ConsenSys, 2019). Although the technology opens interesting opportunities, it also raises legal, regulatory and policy issues that deserve our attention. Blockchain opens many different opportunities when it comes to international trade because the transactions are linked and are time stamped thus the technology makes traceability easier. The technology also enables facilitation of a number of processes, border procedures for example, because there are actors that have access to the same information in real time. These actors do not talk to each other, they work in silos, they do not share data and do not necessarily trust each other. So, it's very powerful to have a technology that allows these different actors to work together in a way and trust each other. Except that the trust comes from the technology and it's very important for regulators to keep that in mind and to see what their role should be and what they can do to allow the technology to really reach its full potential and to really make a difference.

Blockchain technology will revolutionize international trading. SAP is demonstrating how different parties can save time and money through digitized accelerated processes. In international trade, it's not just the seller and buyer involved.

There are banks, insurers, carriers, freight forwarder, agents, brokers and also there will be authorities involved. Today, parties exchange data and documents in an isolated peer-to-peer manner. There is no consistent view of the process or document versions. Individual interfaces between partner systems are costly. Manual handling of paper-based documents is inefficient and delivering paper documents via express courier services adds another significant cost factor. Blockchain for ocean shipping resolves these issues and delivers joint benefits for all parties in the business network. Blockchain for ocean shipping can be operated by SAP Cloud Platform Blockchain Service (SAP, 2018). The various parties can use a cloud application to access the blockchain or connect any system via web services. The trade is documented and the involved parties are invited and enrolled, and relevant documents like the letter of credit are shared and signed electronically. The shipment process is managed jointly. The shipper submits documents relevant for export customs. Once the customs clearance status is received, the ocean carrier posts the bill of lading. Using a mobile app, electronic documents can be digitally signed. The ownership of a bill of lading is securely transferred. Banks, insurers and buyers always have a clear picture of what's going on. At the destination country, it's about imports customs clearance and getting the container released from the port of discharge. SAP's highly secure container release process includes Mobile QR code scans, two factor authentication and a check against the blockchain-based entitlements. Blockchain for ocean shipping aims to establish a new level of security, trust and transparency as it transforms ocean shipping processes, enabling digital document sharing, signing and approval. In short, this new solution brings cost savings, time efficiency as well as the elimination of fraud and stolen freight. SAP is working in close cooperation with customers in a co-innovation model for this new solution.

4.9 CONCLUSION

It will not be an understatement if we say that blockchain technology is the technology of the decade. The technology has grown greatly since 2008–2009. The features of blockchain, like security and privacy, has attracted many organizations to the technology. The technology has been used in many different industries and organization for various kind of transactions and also in the creation of many business models. Many countries are working on online voting based on blockchain technology, which will be easier for the voters and more secure and preserve the privacy of the electronic voting, which will boost the economy of any country. Because the data is very secure and there could be large amount of data, blockchain-based cloud storage could be very useful. The smart and collaborative transportation, which includes smart vehicle and smart transportation, could also be possible using the blockchain technology. International trading, which includes many different entities in the network, can now be connected in a secure blockchain network that will not only enable the users of the network to track the item for trading but also enable the payment and transaction among the users of the node through the smart contract of the blockchain.

REFERENCES

Alex, Lipton, and Stuart Levi. 2018. "An Introduction to Smart Contracts and Their Potential and Inherent Limitations | Insights | Skadden, Arps, Slate, Meagher & Flom LLP." *Harvard Law School Forum on Corporate Governance and Financial Regulation*. The Harvard Law School Forum on Corporate Governance. https://corpgov.law.harvard.edu/2018/05/26/an-introduction-to-smart-contracts-and-their-potential-and-inherent-limitations/

Andrew, Barnes, Christopher Brake, and Thomas Perry. 2016. "Digital Voting with the Use of Blockchain Technology." www.economist.com/sites/default/files/plymouth.pdf.

Bahga, Arshdeep, and Vijay K. Madisetti. 2016. "Blockchain Platform for Industrial Internet of Things." *Journal of Software Engineering and Applications* 09 (10): 533–546. https://doi.org/10.4236/jsea.2016.910036.

Bannet, Jonathan, David W. Price, Algis Rudys, Justin Singer, and Dan S. Wallach. 2004. "Hack-a-Vote: Security Issues with Electronic Voting Systems." *IEEE Security and Privacy*. https://doi.org/10.1109/MSECP.2004.1264851.

Barenji, Ali Vatankhah, Hanyang Guo, Zonggui Tian, Zhi Li, W. M. Wang, and George Q. Huang. 2018. "Blockchain-Based Cloud Manufacturing: Decentralization." *Advances in Transdisciplinary Engineering* 7: 1003–1011. https://doi.org/10.3233/978-1-61499-898-3-1003.

Bertrand, Copigneaux, Nikita Vlasov, and Emarildo Bani. 2020. "Blockchain for Supply Chains and International Trade Report on Key Features, Impacts and Policy Options." https://doi.org/10.2861/957600.

Brett Kotas. 2017. "The Decentralized Cloud and the Future of Data are Here—Influencive." October 22, 2017. www.influencive.com/the-decentralized-cloud-is-here/

Bulut, Rumeysa, Alperen Kantarci, Safa Keskin, and Serif Bahtiyar. 2019. "Blockchain-Based Electronic Voting System for Elections in Turkey." In *UBMK 2019 — Proceedings, 4th International Conference on Computer Science and Engineering*, 183–188. Institute of Electrical and Electronics Engineers Inc. https://doi.org/10.1109/UBMK.2019.8907102.

Buterin Vitalik. 2014. "Ethereum White Paper: A Next Generation Smart Contract & Decentralized Application Platform." *Etherum*, January: 1–36. https://github.com/ethereum/wiki/wiki/White-Paper

Chen, Yi Hui, Shih Hsin Chen, and Iuon Chang Lin. 2018. "Blockchain Based Smart Contract for Bidding System." In *Proceedings of 4th IEEE International Conference on Applied System Innovation 2018, ICASI 2018*, 208–211. Institute of Electrical and Electronics Engineers Inc. https://doi.org/10.1109/ICASI.2018.8394569.

Copel, Noam, and Tal Ater. 2017. *DAV White Paper*. DAV Foundation, Zug, Switzerland.

Curran, Kevin. 2018. "E-Voting on the Blockchain." *The Journal of the British Blockchain Association* 1 (2): 1–6. https://doi.org/10.31585/jbba-1-2-(3)2018.

Dahlberg, Chris. 2018. "Challenges in Designing an Electronic Voting System." *Cs.Hmc.Edu*. www.cs.hmc.edu/~mike/public_html/courses/security/s06/projects/chrisd.pdf.

Dey, Kakan Chandra, Anjan Rayamajhi, Mashrur Chowdhury, Parth Bhavsar, and James Martin. 2016. "Vehicle-to-Vehicle (V2V) and Vehicle-to-Infrastructure (V2I) Communication in a Heterogeneous Wireless Network—Performance Evaluation." *Transportation Research Part C: Emerging Technologies* 68 (July): 168–184. https://doi.org/10.1016/j.trc.2016.03.008.

Dinh, Tien Tuan Anh, Ji Wang, Gang Chen, Rui Liu, Beng Chin Ooi, and Kian Lee Tan. 2017. "BLOCKBENCH: A Framework for Analyzing Private Blockchains." In *Proceedings of the ACM SIGMOD International Conference on Management of Data*, Part F1277: 1085–1100. New York: Association for Computing Machinery. https://doi.org/10.1145/3035918.3064033

Dhaliwal, Navjit, Elad Harison, and Claudio Lima. 2018. "IAGON." www.iagon.com/pdf/Iagon Whitepaper v4.0.pdf

ETAuto. 2017. "Toyota Research Institute: Toyota Research Institute Explores Blockchain Technology to Develop New Mobility Ecosystem, Auto News, ET Auto." https://auto-stage.economictimes.indiatimes.com/news/auto-technology/toyota-research-institute-explores-blockchain-technology-to-develop-new-mobility-ecosystem/58802271

Ganne, Emmanuelle. 2018. *Can Blockchain Revolutionize International Trade? Can Blockchain Revolutionize International Trade?* https://doi.org/10.30875/7c7e7202-en

Golub, Ben. 2019. *Decentralized Cloud? How the Intersection of Blockchain, Decentralization and Open Source is Impacting Cloud Storage*. Storj Labs Inc, Atlanta, GA.

Harsh, Agrawal. 2020. "Siacoin (SC) Cryptocurrency: An Ultimate Beginner's Guide." June 2, 2020. https://coinsutra.com/siacoin-cryptocurrency-sc/.

Haynes, Peter. 2014. "Online Voting: Rewards and Risks." www.itif.org/files/2011-e-id-report.pdf

Hjalmarsson, Friorik P., Gunnlaugur K. Hreioarsson, Mohammad Hamdaqa, and Gisli Hjalmtysson. 2018. "Blockchain-Based E-Voting System." In *IEEE International Conference on Cloud Computing, CLOUD*, 2018 July: 983–986. IEEE Computer Society. https://doi.org/10.1109/CLOUD.2018.00151

Irannezhad, Elnaz. 2019. "Is Blockchain a Solution for Logistics and Freight Transportation Problems?-Review under Responsibility of World Conference on Transport Research Society Blockchain for Supply Chain and Logistcis View Project ScienceDirect is Blockchain a Solution for Log." www.sciencedirect.comwww.elsevier.com/locate/procedia2352-1465

Li, Zhetao, Jiawen Kang, Rong Yu, Dongdong Ye, Qingyong Deng, and Yan Zhang. 2018. "Consortium Blockchain for Secure Energy Trading in Industrial Internet of Things." *IEEE Transactions on Industrial Informatics* 14 (8): 3690–3700. https://doi.org/10.1109/TII.2017.2786307

Lima, Evandro Gonzalez, Christine Kowal Chinelli, Andre Luis Azevedo Guedes, Elaine Garrido Vazquez, Ahmed W. A. Hammad, Assed Naked Haddad, and Carlos Alberto Pereira Soares. 2020. "Smart and Sustainable Cities: The Main Guidelines of City Statute for Increasing the Intelligence of Brazilian Cities." *Sustainability (Switzerland)* 12 (3). https://doi.org/10.3390/su12031025.

Litke, Antonios, Dimosthenis Anagnostopoulos, and Theodora Varvarigou. 2019. "Blockchains for Supply Chain Management: Architectural Elements and Challenges Towards a Global Scale Deployment." *Logistics* 3 (1): 5. https://doi.org/10.3390/logistics3010005

Mearian, Lucas. 2015. "New Service Wants to Rent Out Your Hard Drive's Extra Space | Computerworld." January 13, 2015. www.computerworld.com/article/2867040/new-service-wants-to-rent-out-your-hard-drives-extra-space.html

Marco e. G. Maltese. 2015. "Blockchain-Based Decentralized Cloud Storage: Storj and Competitors." November 29, 2015. https://cointelegraph.com/news/blockchain-decentralized-cloud-storage-storj-and-competitors

Mendling, Jan, Ingo Weber, Wil Van Der Aalst, Jan Vom Brocke, Cristina Cabanillas, Florian Daniel, and Søren Debois, et al. 2018. "Blockchains for Business Process Management—Challenges and Opportunities." *ACM Transactions on Management Information Systems* 9 (1). https://doi.org/10.1145/3183367

Mrvosevic, M. 2019. "Blockchain Based Decentralised Cloud Computing." https://medium.com/@eternacapital/blockchain-based-decentralised-cloud-computing-277f307611e1.

Müller, André, André Ludwig, and Bogdan Franczyk. 2017. "Data Security in Decentralized Cloud Systems—System Comparison, Requirements Analysis and Organizational Levels." *Journal of Cloud Computing* 6 (1). https://doi.org/10.1186/s13677-017-0082-3.

Nakamoto, Satoshi. 2008. "Bitcoin: A Peer-to-Peer Electronic Cash System." www.bitcoin.org.

National Security Agency. 2020. "Mitigating Cloud Vulnerabilities." https://media.defense.gov/2020/Jan/22/2002237484/-1/-1/0/CSI-MITIGATING-CLOUD-VULNERABILITIES_20200121.PDF

Oriold, Frank. 2011. "Interoperability, Technical." In *SpringerReference*. https://doi.org/10.1007/springerreference_62308.

Pangbourne, Kate, Dominic Stead, Miloš Mladenović, and Dimitris Milakis. 2018. "The Case of Mobility as a Service: A Critical Reflection on Challenges for Urban Transport and Mobility Governance." In *Governance of the Smart Mobility Transition*, 33–48. Emerald Publishing Limited. https://doi.org/10.1108/978-1-78754-317-120181003.

Park, Jin Ho, and Jong Hyuk Park. 2017. "Blockchain Security in Cloud Computing: Use Cases, Challenges, and Solutions." *Symmetry* 9 (8): 1–13. https://doi.org/10.3390/sym9080164.

Patil, Sukesha Subhash, and Y. L. Puranik. 2019. "Blockchain Technology." *International Journal of Trend in Scientific Research and Development* 3 (4): 573–574. https://doi.org/10.31142/ijtsrd23774.

Pilkington, Marc. 2016. "Blockchain Technology: Principles and Applications." In *Research Handbooks on Digital Transformations*, 225–253. Edward Elgar Publishing Ltd. https://doi.org/10.4337/9781784717766.00019.

Rubin, Aviel D. 2002. "Security Considerations for Remote Electronic Voting." *Communications of the ACM* 45 (12): 39–44. https://doi.org/10.1145/585597.585599.

SAP. 2018. *Blockchain in Ocean Shipping*. SAP, Barcelona.

SelfKey. 2020. "All Data Breaches in 2019 & 2020—An Alarming Timeline." *SelfKey*. https://selfkey.org/data-breaches-in-2019/.

Sen, Jaydip, and Innovation Labs. 2010. *Security and Privacy Issues in Cloud C Loud Computing*. Kolkata: Innovation Labs, Tata Consultancy Services Ltd. ABSTRACT, no. iv.

Shah, Sagar, Qaish Kanchwala, and Huaiqian Mi. 2016. "Block Chain Voting System." www.economist.com/sites/default/files/northeastern.pdf.

Sikeridis, Dimitrios, Ioannis Papapanagiotou, Bhaskar Prasad Rimal, and Michael Devetsikiotis. 2017. "A Comparative Taxonomy and Survey of Public Cloud Infrastructure Vendors." *ArXiv Preprint ArXiv:1710.01476*. http://arxiv.org/abs/1710.01476.

Singh, Madhusudan, and Shiho Kim. 2018. "Branch Based Blockchain Technology in Intelligent Vehicle." *Computer Networks* 145 (November): 219–231. https://doi.org/10.1016/j.comnet.2018.08.016

Sinovate. n.d. "Innovate With SINOVATE: Creating A Decentralized Cloud Platform for Efficient Data Storage | by SINOVATE | Medium." Accessed August 4, 2020. https://medium.com/@sinovatechain/innovate-with-sinovate-creating-a-decentralized-cloud-platform-for-efficient-data-storage-cf5955ecdeb5

Storj Labs. 2018. "Storj: A Decentralized Cloud Storage Network Framework." 1–90. https://github.com/storj/whitepaper

Szabo, Nick. 1997. *The Idea of Smart Contracts*. Satoshi Nakamoto Institute. no. c: 2–3.

Ubiergo, Gerard Aguilar, and Wen Long Jin. 2016. "Mobility and Environment Improvement of Signalized Networks through Vehicle-to-Infrastructure (V2I) Communications." *Transportation Research Part C: Emerging Technologies* 68 (July): 70–82. https://doi.org/10.1016/j.trc.2016.03.010

Underwood, Sarah. 2016. "Blockchain Beyond Bitcoin." *Communications of the ACM* 59 (11): 15–17. https://doi.org/10.1145/2994581

United Nations, ESCAP. 2015. "Intelligent Transportation Systems for Sustainable Development in Asia and the Pacific." 1–38. www.unescap.org/sites/default/files/ITS.pdf

Wang, Xiaoxia, and Zhanqiang Li. 2016. "Traffic and Transportation Smart with Cloud Computing on Big Data." *International Journal of Computer Science and Applications* 13 (1): 1–16.

Yu, Bin, Joseph K. Liu, Amin Sakzad, Surya Nepal, Ron Steinfeld, Paul Rimba, and Man Ho Au. 2018. "Platform-Independent Secure Blockchain-Based Voting System." In *Lecture Notes in Computer Science (Including Subseries Lecture Notes in Artificial Intelligence and Lecture Notes in Bioinformatics)*, 11060 LNCS: 369–386. Springer Verlag. https://doi.org/10.1007/978-3-319-99136-8_20

Yuan, Yong, and Fei Yue Wang. 2016. "Towards Blockchain-Based Intelligent Transportation Systems." In *IEEE Conference on Intelligent Transportation Systems, Proceedings, ITSC*, 2663–2668. Institute of Electrical and Electronics Engineers Inc. https://doi.org/10.1109/ITSC.2016.7795984

Zhang, Rui, Rui Xue, and Ling Liu. 2019. "Security and Privacy on Blockchain." *ACM Computing Surveys* 52 (3): 35. https://doi.org/10.1145/3316481

Zou, Xukai, Huian Li, Feng Li, Wei Peng, and Yan Sui. 2017. "Transparent, Auditable, and Stepwise Verifiable Online E-Voting Enabling an Open and Fair Election." *Cryptography* 1 (2): 13. https://doi.org/10.3390/cryptography1020013

5 Blockchain-Based Smart Supply Chain Management

Annapurani Kumarappan, Annie Uthra and S. H. Shah Newaz

CONTENTS

5.1 INTRODUCTION

Internet of Things (IoT) is the network of physical objects (called things) which are connected to each other through the internet. The devices connected to the IoT platform integrates analyses and shares the data to other devices in order to pave the way to the development of a 'smart' environment. IoT enables seamless communications across the devices used in day-to-day life like cars, household appliances, fleet management, etc. Research findings impart that the number of IoT devices reached 26.66 billion in 2019. It is worth noting that every second 127 new IoT devices are connected to the web.

5.1.1 Internet of Things

The watchword Internet of Things was devised in 1998 by Kevin Ashton (Procter & Gamble) and developed by the Auto-ID Center of MIT in 2003. The hardware-oriented network called the Internet was invented by Vinton Cerf in 1973. Tim Berner Lee invented the software called the web in 1989. Web 1.0, internet of documents is a one-way communication that is used to publish and push the content to the user's system. Web 2.0, internet of multimedia is a two-way communication that supports dynamic content creation and social interaction. Facebook and blogs are a few examples of Web 2.0. The boon of the modern era, the IoT was introduced in Web 3.0. Internet and Web are the dual faces of the IoT, and the web world introduces communications like machine-to-machine communication, man-to-machine communication or the IoT. Internet of things can be explained as, the things called machines, sensors and actuators communicate between each other either through wires or a wireless network depending on the application. It is predicted that the things that have been connected to the internet will be 30 times more than the humans connected to the Internet during the year 2020. The things are called smart devices and the network is called the smart network. Smart grid, smart city, smart building and utilities, smart transportation and environment, retail, health care and life sciences, consumer and home, safety and security, industrial and manufacturing, energy and resources are some of the developing fields where IoT plays a significant role.

The network devices in the internet are connected using the TCP/IP protocol. The first version of IP was IPv4, the internet protocol that was deployed in 1983 for the ARPANET that is the extensively used IP version until today. These protocols help in identifying the devices on a network using an addressing system. IPv4 is a 32-bit addressing protocol that allows storage of 2^{32} addresses that is around 4 billion addresses. Until now, IPv4 was the primary internet protocol that handles 94% of internet traffic. IPv6 is the new version of the internet protocol, which was initiated by the Internet Engineer Task Force (IETF) in early 1994 in order to provide more Internet addresses. With 128-bit address space, IPv6 can handle 340 undecillion (10^{36}) unique address spaces, which caters to the need for the connectivity of a large number of devices in the IoT. IPv6, also called IPng, may be called the next generation internet protocol. China is adopting IPv9 on the Internet. This IPv9 internet protocol automatically generates an address where the digits 0 to 9 are used to represent 4-bit binary numbers with a maximum of 32 bits (IPv9). It may generate a size of 128 64-bit interface identifier which outfit better solutions to generate a separate IP address for each Thing in the world.

The operations and functionalities of IoT can be broadly classified into four pillars of IoT in terms of control, network, objects and devices and communications. The four pillars of IoT depicted in Figure 5.1 are: supervisory control and data acquisition

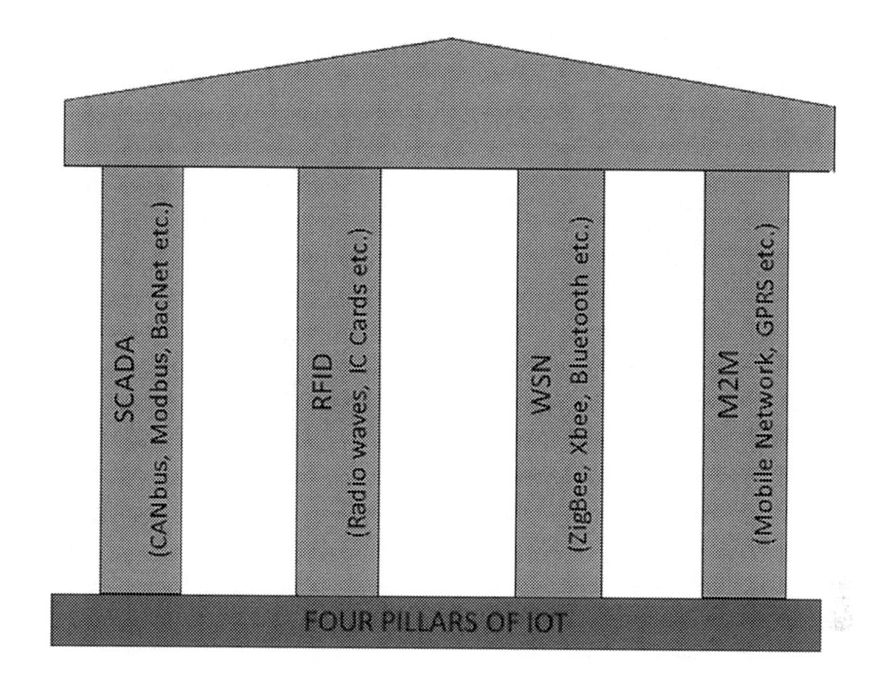

FIGURE 5.1 Four pillars of IoT.

(SCADA), the Internet to Control; RFID, the Internet of Objects; wireless sensor network (WSN), the Internet of Transducers; and machine-to-machine (M2M) communication, the Internet of Devices. Many protocols are adapted for things to communicate like long-time evaluation (LTE), GSM, CDMA, machine-type communication (MTC), Wi-Fi, ZigBee and so on based on the four pillars of IoT.

Devices connect and manage (DCM) is the three-layer architecture of IoT as shown in Figure 5.2. The three layers are the Device layer, Network layer and Application layer (DNA). The device layer comprises things like sensors and actuators, embedded middleware and local/Ad-hoc sensor networks. The devices are things that can talk and pass information like the humidity, temperature, vibration, noise level, chemical component value, pressure, etc. The network layer consists of pervasive networks, edge middleware and machine-type communication. The application layer consists of data management, server-side middleware and vertical applications.

Middleware in the IoT is a hidden translation layer that enables communication and data management in distributed applications. It provides authentication, authorization, soft switching, certification and security. The application programming interface links middleware to the operating system and application. Games middleware (Autodesk), peer-to-peer middleware (JXTA), messaging middleware (MOM, MQ), radiofrequency identification middleware (smart card), security middleware (site minder), real-time CORBA middleware (real-time CORBA) are some examples of middleware. M2M middleware is 3GPP (Third Generation Partnership Project). SCADA uses middleware

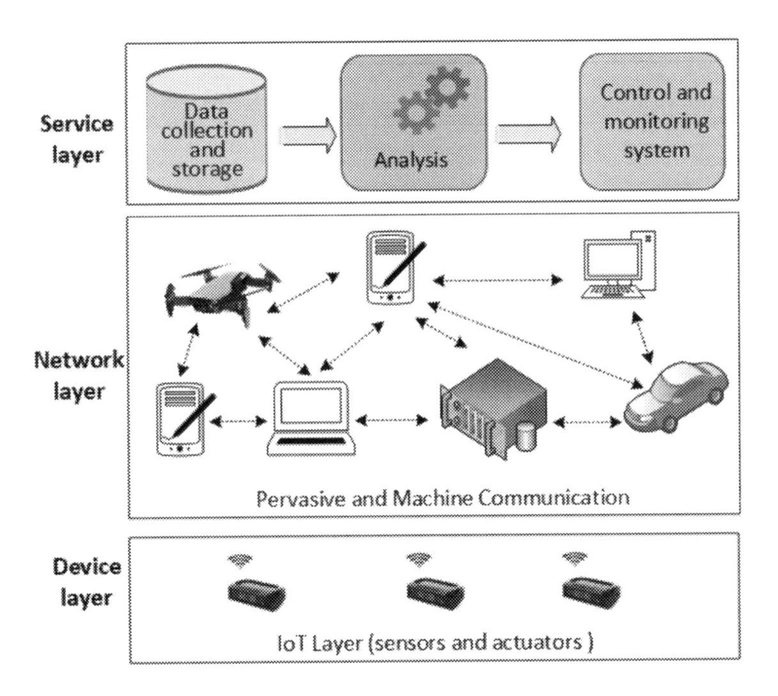

FIGURE 5.2 DCM three layer architecture of IoT.

like ModBus TCP/IP, Ethernet, Profibus, Controlnet, etc. Edge middleware is used for RFID. MagnetOS and IMPALA are some of the middleware used for WSN. IoT plays a major role in the revolution of cyber-physical system (CPS), a system mainly used for communication, computation and control. This data can be controlled or monitored by anyone, anytime, anywhere, anyhow and for anything.

5.1.2 Blockchain

Blockchain is also called distributed ledger technology (DLT) and is defined as an open, distributed ledger that records all transactions between two parties efficiently, verifiably and permanently. Blockchain is a peer-to-peer (P2P) network that cooperatively follows a procedure for communication among the nodes and a consensus algorithm to validate new blocks and the state of the blockchain. Blockchain is structured such that the data stored once cannot be modified. The data can be altered by modifying the blocks of the blockchain after ensuring consensus of more >50% of the nodes in the blockchain network. Blockchain can perform exceptionally well to reach a consensus even when some nodes respond with false information or no information at all. Blockchain comprises three components: blocks, miners and nodes.

5.1.2.1 Blocks

Every chain consists of multiple blocks and each block has three fundamental components. The first component represents the data. The second component is the nonce, a 32-bit randomly generated whole number. The nonce is used to generate a block header hash. The hash is a 256-bit number wedded to the nonce that starts with a large number of zeroes and a cryptographic hash created during the creation of the first block of a chain. The data in the block is signed and permanently tied to the nonce and hash unless it is mined.

5.1.2.1.1 Structure of a Block

An individual block in the blockchain is considered as a page in a ledger. A block contains numerous components that are broadly categorized into two parts, the block header and block body.

5.1.2.1.1.1 Block Header

A block header serves as an identity to recognize a particular block on a blockchain and has six components:

- **Block Version**: The version number of the software.
- **Merkle Tree Root Hash**: All transactions held in a block are aggregated in a Merkle tree by a cryptographic hash, which forms root hash of the Merkle tree.
- **Time stamp**: The time in seconds at which the block was generated or mined.
- **nBits**: The current difficulty (the effort required to mine a block) that was used to create the block.
- **Nonce**: The 'number only used once' (nonce) is a number that is attached to a hashed block to enforce security.
- **Parent Block Hash**: Forms the connection between the blocks thereby a chain of blocks, which also helps to identify the chronological order of the blocks.

5.1.2.1.1.2 Block Body

The block body holds all confirmed transactions made by the block and are verified by the miners. The list of transactions made by a block is represented in terms of a Merkle Tree. Figure 5.3 depicts the detailed structure of a block in a blockchain, and Figure 5.4 shows the chain of blocks where every block refers to its predecessor.

5.1.2.2 Miners

Miners are the ones that create new blocks on the chain using a method called mining. Every block in the chain has its unique nonce and hash. The block also points to the hash of the previous block in the chain. This makes the mining process difficult, especially on a long chain. Miners use specialized software to solve the exceedingly complicated mathematical problem in order to come up with a nonce that generates a recognized hash. Four billion viable nonce is the hash combinations that must be mined to determine the correct nonce, which is referred to as the golden nonce. The block is appended to the blockchain once the miner identifies the golden nonce.

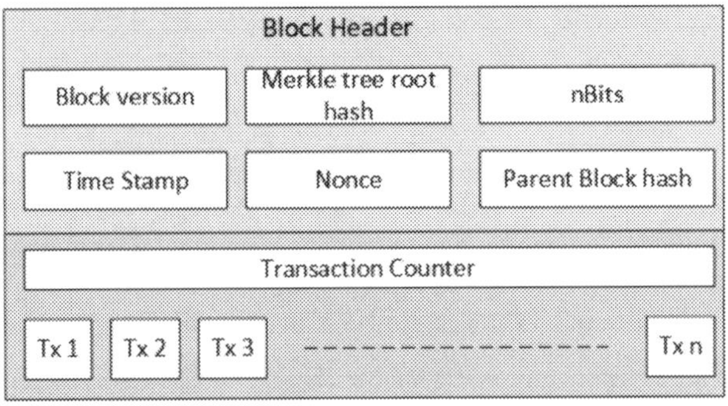

FIGURE 5.3 Block structure.

Any changes or modifications of a block in the blockchain involve re-mining of the block to be modified and all the trailing blocks. This makes the manipulation of blocks very tough in blockchain technology. However, safety-in-math is implemented to ensure security although the golden nonce demands a massive amount of computing power and time. When a block is successfully mined, the block is accepted by all of the nodes on the chain, and the miner is compensated in the form of cryptocurrency.

5.1.2.3 Nodes

The nodes connected to the chain in the blockchain form a P2P network. Nodes can be any electronic device that maintains copies of the blockchain and network functioning. The network should accept newly mined blocks for updating, trust and verification through algorithms. These actions in the ledger can be verified and viewed because of the transparent nature of the blockchain.

5.1.2.4 Types of Blockchain

In general, blockchain is used to perform transactions or exchange information through a secure network. Depending upon the applications, the organizations, businesses, consortium and individuals use different types of blockchains. The following part of this subsection explains the four different types of blockchain.

(*i*) *Public Blockchain*

A public blockchain is a permissionless and nonrestrictive distributed network that is the most widely used nowadays. This type of blockchain does not have anyone in-charge; hence, anyone can read, write and audit the blockchain. Some of the commonly used public blockchains are Bitcoin, Litecoin and Ethereum. Individuals can sign in to the blockchain platform to become an approved node of the blockchain network. Once a node becomes part of the public blockchain network, it can perform operations like mining, access records, validate transactions and perform the consensus protocol for an incoming block where the decision-making is done through

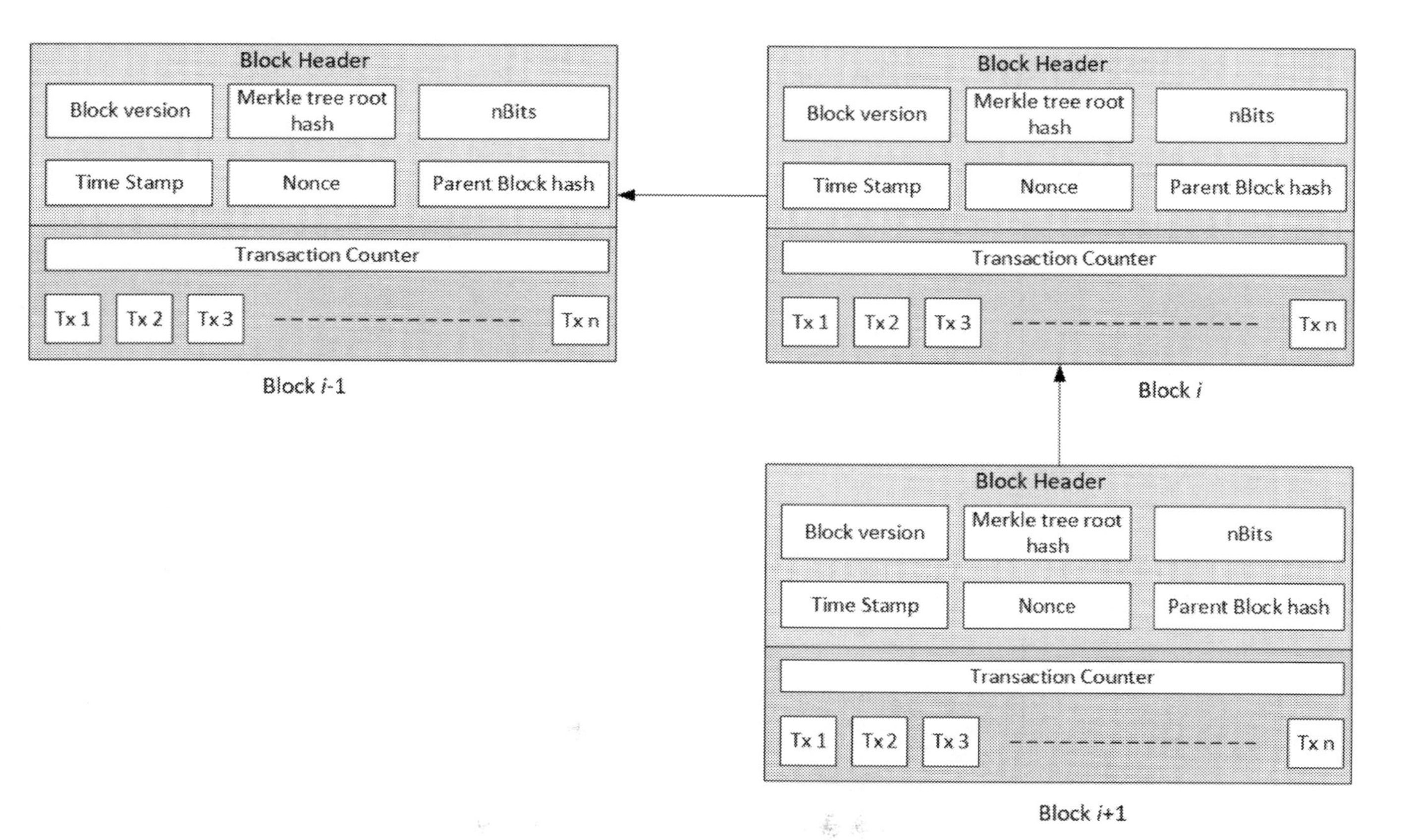

FIGURE 5.4: An example of a blockchain consisting of a continuous sequence of blocks.

consensus algorithms such as proof-of-work or proof-of-stake. Consensus algorithms refer to "a decision-making process for a group, where individuals of that group support the decision that works best for everyone."

(ii) *Private Blockchain*

Unlike a public blockchain, a private blockchain is a restrictive and permissioned blockchain that runs within an exclusive network, which acts as the private property of an organization or individual. It is a small and definite network. In this type of blockchain, only authorized persons can control the blockchain, perform reading/writing and provide access rights to nodes. Here, consensus or decision-making is reached based on the notions of those in-charge who determine mining rights. The private blockchain platforms include Bank chain, Ripple (XPR), IBM Hyperledger, Corda and Multichain.

(iii) *Consortium*

A consortium blockchain is a semi-decentralized blockchain where a blockchain network is managed by more than one organization. This blockchain is partially public and partially private and is the blend of both blockchains. The public and private nature of the blockchain is decided based on the consensus algorithm, where only some nodes are given rights to sanction transactions and manage the consensus process. Consortium blockchains are administered by a group of people who do not belong to any specific organization. The examples of consortium blockchains include Hyperledger, Quorum and Corda.

(iv) *Hybrid Blockchain*

A hybrid blockchain is the combination of both public and private blockchain which considers the features of both blockchains because it allows the node from a public blockchain system and a private blockchain system. In hybrid blockchain, the hashed data blocks are generated by the private blockchain network and are stored in the public blockchain without jeopardizing the data privacy. This kind of blockchain provides adaptable control over the blockchain, which also supports scalability and decentralization. Dragonchain is an example of a hybrid blockchain.

5.1.3 Significance of Blockchain

Blockchain is often seen as the technological boom of the modern era that has an enormous capability to disrupt and reshape plenty of commercial businesses. Since the launch of blockchain in 2008, various blockchain enthusiasts from all over the world came up with different applications. The notion of a distributed ledger without any central governing body has attracted many minds to explore this modern era technology and reap the benefits in the best possible way. As years progressed, there was a shift observed in the application domain of the blockchain toward a ubiquitous field of business applications. The following section explains how blockchain can change or disrupt some industries.

5.1.3.1 Banks and Other Financial Institutions

Banking benefits a lot by integrating blockchain into its business model compared with the other businesses. Financial institutions work in restricted times normally five days in a week up to 6 p.m. The large volume of transactions takes many days for verification and to transfer money from one account to the other. By integrating blockchain into banks, users can see their transactions processed in less than 10 minutes although the blockchain generation takes more time initially. Hence fund transfers between organizations are happening more swiftly and securely. With the integration of blockchain in banks, international money transactions become smoother and faster. Currently, it takes at least two days to transfer money from the United States of America to India; not only will such delays in money transfer be reduced to minutes but also the need to fill cumbersome and confusing forms will no longer be required.

5.1.3.2 Use of Cryptocurrency

Blockchain constitutes cryptocurrencies like Bitcoin and Ethereum. Currencies like the United States dollar are governed and verified by a central administration (Reiff, 2019), usually a bank or government. If a user's bank or the country currency collapses, the value of their currency may be in danger. Therefore, Bitcoin was born to overcome these scenarios. Bitcoin and other cryptocurrencies function without any central authority, which decreases risks and reduces the processing and transaction charges. Cryptocurrencies facilitate the countries with more stable currency and provide more extensive networks of individual and organization transactions across the globe.

Bitcoin and Ethereum are the most commonly used cryptocurrencies; however, the constant fluctuations in their prices make the cryptocurrencies a little risky for common people. Stablecoins are another type of cryptocurrency that is intended to be immune from market volatility, which makes them a more suitable form of payment than conventional cryptocurrency. Stability is created by pegging the worth of stablecoins to other 'stable' assets such as fiat currencies or gold. Tether, a stable cryptocurrency, has been created in such a way that it always holds an equivalent value of 1 US dollar which can be used to transfer money to anyone across the network.

5.1.3.3 Health Care Industry

Blockchain technology introduces a novel paradigm to transform the health care ecosystem and improve the security, privacy and interoperability of health data. This technology enables the Health Information Exchanges (HIE) to create electronic medical reports more efficiently and securely without use of intermediaries (Reiff, 2019). The medical report generated and signed could be encoded and saved on the blockchain with a secret key so that the records are only accessible by a specific individual, thereby guaranteeing privacy. The patients need not despair to bring all their medical records every time when they visit the doctors as the data would be stored on the blockchain and can be obtained by doctors upon the patients' authorization.

Prescription fraud is one of the most significant problems in the health care industry. The UN approximates that there are more than 29.5 million individuals around the globe with drug use disorders. In 2016, the US Centers for Disease Control and Prevention (CDC) recorded more than 42,000 deaths from opioid overdose in the United States. Blockchain provides a solution to avoid such frauds by storing the encoded prescriptions in the blockchain, which could be verified with the physician. This enables the doctors to explore the patients' prescription records along with their medical details.

5.1.3.4 Supply Chain Industry

Blockchain in SCM provides a swift and secure environment to handle supply chain processes like exchange, agreement, tracking process, etc. in self-executing supply contracts to automated cold chain management. The areas of SCM where blockchain technology can be implemented in the developing fields to enhance the business securely are listed following:

1. Automotive Supplier Payments

Blockchain enables money transfer throughout the globe without the need for conventional banking transactions, as purchases are carried out between the buyer and the seller, and it takes only minutes for secure business payment clearances. Australian vehicle manufacturer Tomcar utilizes Bitcoin to pay a few of its suppliers in Israel and Taiwan. On the other hand, the organization should also be careful to avoid clinging onto too much usage of Bitcoin, as Bitcoins are globally used and some international governments foresee it as a way for businesses to invest. Hence, companies that hold Bitcoins may be subject to taxation.

2. Traceability of Meat Production

The step-by-step goods status of production in companies can be maintained by blockchain. The data is permanent and immutable. The companies track each good starting from its origin. Global retailer Walmart utilizes blockchain (Infopulse, 2019) to trace the sales of pork in China. The system lets the organization view where every piece of meat originates from, each processing and storehouse steps in the supply chain and the stocks' sell-by date. Whenever necessary, the corporation can also view which batches of pork are affected and who purchased them, etc.

3. Electric Power Micro-Grids

Micro-grids have an increasingly important role in reducing carbon footprint, increasing the reliability of the power supply, maximizing energy utilization and reducing energy generation cost. Houses and offices with micro-grids may act as a prosumer. That is, they will not only be the consumer of energy but also be able to trade the surplus amount of energy (Yoo et al., 2017). To facilitate such energy trading and reliable operation of micro-grids, blockchain can play a significantly important role (Mollah et al., 2019). The Transactive Grid (O'Byrne, 2019) is a platform running on blockchain used to monitor and redistribute (trade) energy to a neighborhood micro-grid. This system automates the purchasing and sale of green energy to minimize the costs and plays a significant role in reducing pollution. The process uses the Ethereum blockchain platform, which is designed specifically for developing and executing smart contracts.

4. RFID-Driven Contract Bids and Execution

RFID tags are generally employed in the supply chain (O'Byrne, 2019) to collect data about the merchandise. The computers can read the tags automatically and then process them. Again, the question is why can't blockchain be used to store data in logistics? The plausible structure is designed this way. RFID tags for containers or pallets store the data and time in the shipment location. Logistics associates managing these applications can scan these tags and bid for a delivery contract where the associate offering the optimal price and service receives preference. Finally, a smart contract can also track the delivery performance and status.

5. Cold Chain Monitoring

Food and pharmaceutical goods often need specialized storage environments. Furthermore, industries take advantage of sharing warehouses and distribution hubs rather than paying for their own. Sensors on sensitive stock can record temperature, moisture, vibration and other environmental conditions and attributes. This information can be saved on a blockchain, which is permanent and immutable. If a warehouse condition diverges from the agreed protocols, each user of the blockchain will be notified about the present scenario. A smart contract can trigger a response to correct the situation. For example, depending on the extent of the variation, the action can be automatically carried out on the storage or inform the warehouse manager.

5.2 BLOCKCHAIN-BASED SMART DEVICE MANAGEMENT SYSTEM

IoT and blockchain technology both are disruptive technologies (Miraz, 2019), and they complement each other in creating a new era. Blockchain needs all participating nodes to provide consensus that is supported by IoT devices, whereas IoT devices require the security aspect of all the participating devices being protected, which is supported by blockchain features like transparency, privacy, immutability, digital identity and so on. In IoT, almost all devices are connected through the internet to maximize the revenue or benefits and minimize the resources, which leads to an effective decision-making intelligent system. The IoT devices are connected and communicating with each other. The data collected from various sensors and/or actuators embedded into hardware devices are shared through wireless technologies like ZigBee, radio-frequency identification device (RFID) and location-based technologies (Feki et al., 2013). These data are monitored, analyzed and trigger actions accordingly either automatically through actuators or manually depending upon the applications and scenarios. A variety of IoT devices are available that commune with the end users as a person like smart toys for children, smart watches for adults; smart appliances like smart TVs, smart bulbs, temperature sensors and thermostats that are smart enough to control the room temperature according to the number of people present in the room. IoT grows into a major driving force to bring this smartness into every part of the world in almost all fields. The diversity of IoT applications brings in a great challenge in terms of security during data generation, collection, transportation/communication, analysis, deployment, etc. The blockchain fusion with IoT technologies can provide security and make the system more versatile.

5.2.1 Architecture of Smart Device Management System

Many devices are connected from network to network (Evans, 2011). These devices are communicating with each other and also with other users and end devices for a variety of applications, making the system smart day by day. As per the dictionary, the meaning of SMART is able to understand quickly, that is intelligently, and act quickly in difficult situations. In the business world, SMART is an acronym that is specific, measurable, achievable, realistic and timely. The incorporation of smartness into devices introduces the smart device management system. It can be any system in which many devices are connected together to achieve goals. The current era is moving toward smartness in each and every day-to-day activity such as smart city, smart home and so on.

Smart device management includes provisioning and configuration of the devices, updating the firmware and software and automation. When the number and range of devices deployed in an IoT environment have to scale, then automation becomes an important feature that emphasizes the operation in a remote environment and simplifies the task, updates on multiple devices and addresses the security vulnerability.

Device management is the process of monitoring and maintaining all the devices from sensors, network components, processors and to the user interface system. Device management helps to automate the management of IoT devices throughout their life cycle and ensures the scalability, security and interoperability of IoT solutions. The data collected are aggregated in the processor and communicated to the end users for further action. The centralized processor here becomes a bottleneck and the single point of failure might crash the entire system. Hence the blockchain concept of decentralized network plays a major role in maintaining the log of all the participating devices, security, transparency and availability of the information at the right time to the right person in the right place. Nowadays each manufacturer uses their own software for the process of monitoring and maintaining the participating devices. The important aspect is to use a standard protocol for bridging the software used by different vendors. The protocol is the agreement, that is, the set of rules between the parties implementing the process of device management. It is the ability to communicate with the management platforms of other providers using nonstandard devices in combination with devices using standard protocols, avoiding vendor-locking.

Smart homes use technology such as computers, controls, information and communication and image display, all connected together to meet the automation requirements and provide effective control and management. The smart home [(Dorri et al.,

2017), (Gong et al., 2019)] meets the following requirements: (i) recommend to the occupants how to operate the home with the objective of minimizing energy consumption while not affecting the comfort of the occupants; (ii) monitor and communicate with a home remotely and check the activities at the home, thereby enabling decision-making in time and facilitating desired safety and comfort; and (iii) ensure the security and effective management of daily activities. The sensor meters are installed in each and every location to sense the information and send it to the processors, controller and actuators, where the real processing takes place and the final output is communicated to the display systems and mobile systems. Furthermore, intelligence is added into the processor for autonomous action and control, which is given to the administrator. The administrator maintains the different devices connected across the system and manages the scalability of the system according to the requirement and updates the programs whenever required. The safety of all devices has to be ensured and further care must be taken that the devices are not damaged because of mishandling or external environmental factors. Mainly all actions are taken based on the information received from the devices, hence the trustworthiness and reliability of the information is the foremost thing. If it is a single home management, then the people concerned will have control over the system because fewer people will be handling and the devices are restricted to one home alone. However, when the scalability increases like for a smart city, there are a few thousand houses, hospitals, schools, colleges, universities, banking sectors, government offices, private industries, public places and lot more. Then one can imagine the number of devices connected across all these sectors and the information and communication technologies required for effective decision-making. In terms of the security aspect, there is a huge void because of the centralized system and scalability issues. Hence, the incorporation of blockchain into the smart device management system will lead to a secure, smart system. Figure 5.5 depicts the overall view of the smart device management

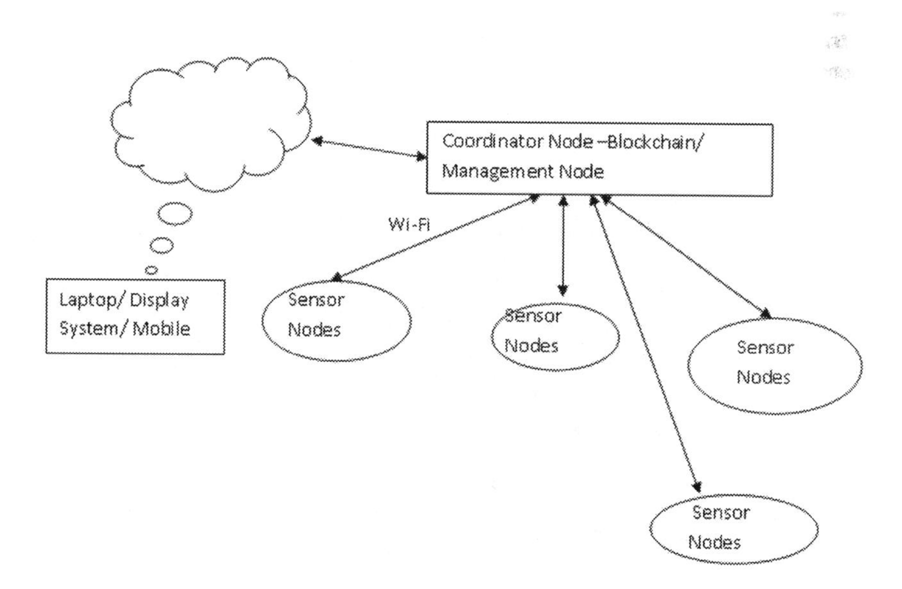

FIGURE 5.5 Block diagram of smart device management system.

system. The sensor nodes are connected to the coordinator node, which can be a management node or blockchain node that communicates either through Wi-Fi or directly. The management node updates the provisioning of the devices and checks the scalability devices whereas the blockchain node through smart contract ensures the transparency between the transaction happening between the sensor nodes and the end nodes. The node also takes care of the authenticity of the devices and access to the right person with the SHA function. Therefore, the security and reliability of the devices taking part in the transactions are governed by blockchain technology. The management node serves as an interface between the blockchain node, sensor nodes and the end devices. The management node is connected through the communication network to the end devices like display system or laptop or mobile devices.

5.2.2 IoT in Smart Devices

A thing that was made or adapted to fulfil a particular purpose is generally referred to as a device. IoT devices include mechanical, electrical, electronic hardware and software parts, wireless sensors, actuators and computer parts. These devices are connected to a particular object and can be operated through the Internet by exchanging data among the objects or people without the intervention of humans. The sensors in the devices detect and measure the information like pressure, temperature, humidity in electrical form then convert it into numerical form which is called data. The collected data from IoT devices such as smart appliances, smart TVs, transport, commercial security systems, traffic monitoring devices, weather tracking systems, etc. are utilized for different purposes ranging from monitoring until data analysis for future studies.

IoT devices can be broadly classified into three main groups, namely: consumer device, enterprise device and industrial device, which are meant to work in concert for people at home, in industry or in the enterprise, respectively. Consumer connected devices include toys, wearable gadgets, smart appliances, smart TVs, smart speakers, etc. Enterprise devices are the smart devices like smart cars, smart phones, smart locks, smart door bells, smart refrigerators, smart meters, smart tablets, smart bands, smart watches, smart key chains and others.

The IoT devices are connected to a server or cloud via the internet. The server or the cloud collects the data from the devices, performs analysis and computations on that data based on the application, stores and deletes the collected data. The controllers in the device are the key element to collect the data and periodically upload it to a server database that organizes the data based on the requirement. The IoT is the network of devices used in vehicles, home, building, retail, health care, safety and security, industries, etc. The network in IoT allows the things to connect, interact and exchange data. Because IoT data is unstructured, it can be easily stored in a public cloud. Most of the cloud providers offer low-cost scalable storage systems based on the technology of the storage object.

5.2.3 Smart Device Management Protocols

Message Queuing Telemetry Transport (MQTT) is one of the standard protocols in IoT and is a lightweight subscription-publish model transport protocol. Its variant,

MQTT-SN, is light, robust and overhead-less. This protocol runs over TCP/IP and supports Secure Socket Layer (SSL)/Transport Layer Security (TLS) to protect the information communicated, which is implemented in a variety of platforms. In order to cover this standard necessity, some additional protocols are introduced like OMA Lightweight Machine-to-Machine (OMA-LwM2M), which was developed by Open Mobile Alliance as a light, fast and structured protocol for low-capacity devices, and OMA-DM, a variant of LwM2M developed by Open Mobile Alliance, which is more oriented toward mobile applications and is more complex and structured than LwM2M. Broadband Form created a protocol, Technical Report-069 (TR-069), for remote management of customer premises equipment connected over the IP network, which is being widely used in telecommunication operators to provision their routers, etc. Constrained Application Protocol (CoAP) is a specialized internet application protocol designed for low power devices. The details of these protocols are given as follows:

(i) **MQTT**: It is a very suitable protocol for low power sensors. MQTT is based on the principle of publishing messages and subscribing to topics. Many clients connect to the broker and subscribe to the topics of interest. The clients can also publish messages to the topics. The broker and MQTT is a common interface for everything connected. No need to configure a topic, publishing on it is enough. On a hierarchy basis, topics are created with a slash (/) as a separator.

(ii) **OMA-LwM2M:** Irrespective of the device's hardware or the application, LwM2M manages the device and the interactions are based on the bootstrap interface, the client registration interface, the device management and service enablement interface and the information reporting interface. The bootstrap interface configures the devices for secure connectivity with device management, the client registration interface allows the devices to register and make their resources visible to web applications, hence allowing the applications to access and perform operations on resources using the device management and service enablement interface. This interface can read, write and execute actions; any changes in resources are observed and notifications are received when certain conditions are met with the help of the information reporting interface.

(iii) **TR-069:** It uses the Customer Premises Equipment (CPE) Wide Area Network (WAN) Management Protocol (CWMP) that provides support functions for autoconfiguration, software/firmware management, status and performance management and diagnostics. It is a bidirectional Simple Object Access protocol (SOAP) and Hypertext Transfer Protocol (HTTP)-based protocol and provides the interface between CPE and Auto Configuration Servers (ACS). It supports the large number of internet access devices like modems, routers, gateways and also the end user devices like set-top boxes and Voice over Internet Protocol (VoIP)—phones.

(iv) **CoAP:** It is a low overhead protocol designed to allow resource-constrained devices to communicate with other devices and with internet servers using a compact binary format. It uses the User Datagram Protocol (UDP), but has added TCP and TLS support to provide secure transmission and device

TABLE 5.1
Comparison of Various Device Management Protocols.

	MQTT	LwM2M	OMA-DM	TR-069
Type	Asynchronous messaging	Session	Session	Session
Overhead	—	Lighter	Heavier	Heavier
Footprint	Lighter	Lighter	Heavier	Heavier
Server load	Lighter	Lighter	Heavier	Heavier
Data model	Unstructured	Structured	Structured	Highly structured
Dynamic	No	Yes	Yes	Yes
Response time	Very Fast	Fast	Slower	Slower
Common use	Sensors, small/ constrained devices	Small gateways, sensors, small/ constrained devices	Mobile gateways	Fixed gateways
Advanced security	Basic	Advanced/certificate	Advanced/certificate	Advanced/certificate
Common reliability	Configurable	Configurable	High	High

authentication. The combination of CoAP and LwM2M lead to the management of low powered devices and wireless updates, keeping the devices secure.

The device management protocols MQTT, OMA-LwM2M, TR-069 and CoAP are discussed to efficiently manage the devices in an IoT environment.

Table 5.1 compares the device management protocols. The LwM2M uses a small code footprint and can be easily integrated across many IoT devices. It is configurable and secure, structured data model. It reduces the power consumption of IoT devices and effectively manages the devices. Smart IoT and blockchain are secured with the public key infrastructure (PKI) and the TLS/SSL. We have discussed so far the protocols used in IoT, the forthcoming section discusses the algorithms used in Blockchain.

5.2.4 Consensus Algorithm

All devices communicating with each other on a global scale and sharing a public ledger need a functional, efficient and secure consensus algorithm. The consensus algorithm is based on the concept of majority wins. All participating nodes are trustless ones, but based on the agreement between them, they trust the algorithm and validate the event/transactions depending upon a consensus algorithm. In the Bitcoin and Ethereum protocol, the consensus algorithm used is Proof of Work (PoW); other algorithms that came later are Proof of Stake (PoS), Proof of Activity (PoA), Proof of Burn (PoB), Proof of Capacity (PoC) and Proof of Elapsed time (PoE) (Nahas, 2020).

(i) **Proof of Work (PoW):** This algorithm states on what basis the block will be added to the chain of blocks, that is, blockchain. Based on the command executed, the block will be added. The block is added into the blockchain based on two functions: mining and verifying the mined block. The algorithm is designed in such a way that mining is difficult to execute, whereas verifying is an easy task. Mining of the block is calculating the hash value for the block of transactions. This hash value is similar to solving a cryptographic puzzle that has leading zero digits of the hash code calculated. The computational complexity increases depending upon the number of leading zero digits that exist. Once the hash value with leading zeros is calculated successfully by a miner, then the block gets added into the blockchain. Before getting added to the block, the verification process is done by all the participating nodes, comparing the hash function generated locally and received from the miner who solves the cryptographic puzzle. If no discrepancies in the hash value are found, the mined block is accepted. When many miners are participating in mining the blocks, the block received first will have the greater possibility to get added into the chain, and that miner will receive bitcoin reward and transaction fees.

The shortcoming with PoW based consensus is that whoever has more computing power has a greater chance to mine the blocks. This leads again to the concept of a centralized system for mining that is against the concept of blockchain technology. Additionally, it requires lots of electricity consumption because of the huge computational power. There is an inclination to set the mining pools in a country where electricity is cheaper; however, this results in favoring of one country or one group that has more resources. The number of transactions carried out in a second is less when compared with other credit card transactions. Because of this limitation PoW, Ethereum has slowly started moving to another consensus algorithm called PoS.

(ii) **PoS:** Proof of Stake is a consensus algorithm in which the name of the person validating the block is known as a validator and not a miner. The validator, in order to participate in the task of validating the transactions in a block, has to show their stake at hand, that is their cash value, like surety. The one who owns more stakes will have a greater chance of validating the block. For example, if a person has 500 coins and another one has 100 coins, then most likely the one who owns 500 coins will get a chance to validate. Validators are paid transaction fees for validating the blocks. No huge computational power is required and the transaction is also fast enough. But for the validated block to get added into the blockchain, different approaches are adopted in different protocol. In Tendermint, the validated block gets added into the blockchain based on majority voting, that is the maximum number of signatures signed by the participants. PoS was first adapted by Peercoin, later by Blackcoin and NXT. Now Ethereum is moving from PoW to PoS.

The problem with PoS is that when a person has more coins than others then there is an inclination toward them. Therefore, there was a thought of another consensus algorithm—the PoA.

(iii) **PoA:** Proof of Activity is an alternative incentive structure created for Bitcoin. In the Bitcoin structure, to avoid inflation, the bitcoin miners reward subsidy is halved on every 210,000 blocks approximately, for a period of 4 years. After that the miners will receive only the transaction fees. This might cause a security issue in which people will act on self-interest and spoil the system. So, an alternate hybrid approach, PoA, is introduced which is a combination of PoW and PoS. Mining takes place based on PoW but with no transactions in the block, which acts like a template. Then the system adapts PoS to validate the mined block once all the validators sign it. The selected validators are not available to complete the block, then the next block is selected and a new group of validators chosen, until the block gets maximum signs and gets added into the chain. The transaction fees are shared between the miners and the validators who signed the block.

The problem here is the high computational power because of PoW and the validator double signing the block because of PoS. So another consensus algorithm came into use, known as PoB. PoA is used only by Decred.

(iv) **PoB:** Unlike PoW, Proof of Burn does not require high computational resources; instead coins (burn) are sent to a non-retrieval address, that is, by committing the coins into a place of non-returnability, a miner gets the privilege to mine the system based on the process of random selection. The more coins one can burn, the more likely one is to get selected for mining of next block. It is an alternate to PoW (a PoW system without wastage of energy). The noticeable drawback of PoB is that the mining opportunity goes to those who are ready to burn more money. Slimcoin uses PoB in combination with PoW and PoS.

(v) **PoC:** According to Proof of Capacity, a mining opportunity is given to the miner that possesses the greatest amount of storage space among all the miners. A variation of PoC is Proof of Space and Proof of Storage. Burstcoin is an example that applies PoC in order to facilitate its blockchain operation. Similar to PoW, the PoC approach may lead to the concept of a centralized system for mining. That is to say, the miner with more capacity compared to others would be enjoying the mining opportunity.

(vi) **PoE:** Proof of Elapsed time is the idea of Chipmaker Intel. It works as PoW but with less consumption of electricity. Instead of solving cryptographic puzzles, it uses the Trusted Execution Environment (TEE) to ensure the blocks get mined in a random fashion. It is based on a guaranteed wait time in the TEE. Thousands of nodes will run efficiently on any Intel Processor that supports TEE. Therefore, this approach is again a single point trust-based approach relying on Intel.

5.2.5 Security Considerations for a Smart Device Management System

The security factors to be considered for the smart device management system are: confidentiality, authentication, availability, integrity, auditability and adaptability (Castor, 2017).

(i) **Confidentiality:** The data are collected from many sensor nodes and communicated through the network for further analysis and retrieval of information. The sensitive data/information should not be disclosed to an unauthorized person and should reach only the intended receiver. This property is ensured through the cryptographically linked blockchain.

(ii) **Authentication:** Only the intended or claimed identity can send or access the devices or information. The existing system has a centralized and federated identity system that is either centralized or semi-centralized, but with the incorporation of blockchain the decentralized identity system is available and ensures the right person accesses the system.

(iii) **Availability:** In order to facilitate successful operation, the availability of data/information is increasingly important. All the stored data/information should be available whenever an authorized person or device requires them. It is worth mentioning that it is very challenging to facilitate high availability of data/information in a centralized system. Taking the limitation of a single point of failure, Hadoop replicates data across multiple machines in a large cluster. Blockchain employs the distributed ledger concept, allowing data/information to be available in more than one node (all the transactions are available in every node participating on a distributed network), thereby increasing the availability of data/information.

(iv) **Integrity:** There is a vast amount of information from many devices communicating with each other and also with end nodes, and it should not be altered in any way from the source to the destination. This is assured in the blockchain technology, because it is an immutable ledger system that is tamper proof.

(v) **Auditability:** The information collected and the devices participating in the system have to be auditable; with the help of blockchain, this is ensured. Every peer stores a time stamp in the network, and transparency provides auditability and accountability.

(vi) **Adaptability:** Many devices are connected across various networks; therefore, their interoperability can be a cause for concern in some cases. The blockchain framework offers adaptability, because it is based on a distributed database, which is capable of working in a heterogeneous environment (the IoT environment can be easily adapted to the requirement).

5.3 SUPPLY CHAIN MANAGEMENT WITH BLOCKCHAIN

5.3.1 Challenges in Supply Chain and the Need for Blockchain

The supply chain plays a significantly important role in today's business from manufacturing industries, conventional stores to e-commerce business. As the name indicates, the supply chain is the chaining process between the suppliers and the buyers. The suppliers have to collaborate among multiple stakeholders to meet the demand of the buyers (Aste et al., 2017). Even a small shop has to coordinate among multiple parties for the purchase of the materials. In the day-to-day evolution of dynamic businesses, supply chain plays an increasingly important role. For example, a tea business factory owner has to coordinate with the tea estate owner for the tea leaves and with the logistics division for the movement of the tea leaves from the estate to the factory. Internally, the tea leaves have to be processed, packaged and finally delivered to the customers. There should be coordination between various stakeholders and the tracing of the final product, starting from procurement to delivery of the products. Figure 5.6 shows the e-commerce and the industry value chain (Laudon et al., 2016) of the different participating parties between the suppliers and customers. The suppliers have to deal with the manufacturers, distributors and transporters to distribute the products to be in right hands, either directly to the customers or to retailers. But in the traditional supply chain management, it becomes tedious to trace back the products and find the origin of the raw materials. The transparency between each stakeholder is not there in the traditional system as one has to rely on the other party in order for the business to be successful. If one party is untrustworthy, the entire system will collapse. That is, the visibility is not there and traceability is a big question mark. Because of this, the cost of the entire process is high and the process is inefficient. If any mishap happens to the product, tracing the root cause takes a long time. If it is across the border, the regulations differ between the countries and a lot of paperwork is involved, which will delay the system. With the help of blockchain, these problems can be made paperless, efficient and automated, which would lead to an increase in the speed of the system.

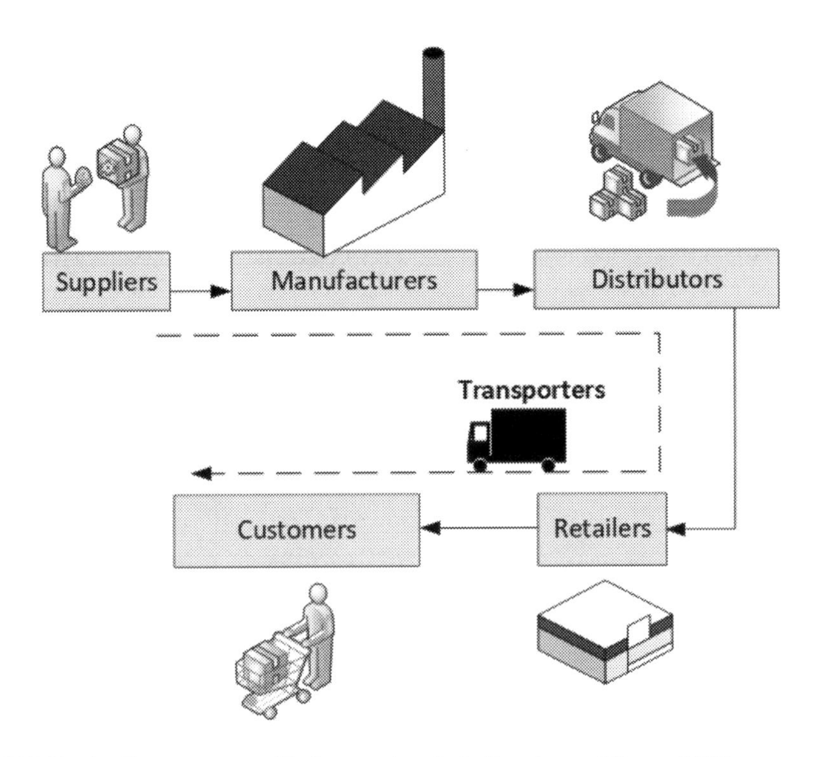

FIGURE 5.6 E-commerce and industry value chain (Laudon and Traver, 2016).

Hence the major issues in the growing environment of supply chain management are product visibility, traceability, cost effectiveness, trustworthiness and reliability.

Product visibility: Product visibility makes sure that the freshness of the product is known to all, including the customers. Customers can check for themselves what are the raw materials and where have the raw materials come from, when the product was been processed, how long it's been and the shelf life. This meets the demands of the customers in the modern era, where all the customer's tailor-made needs have to be satisfied. This also ensures the cleanliness for each customer and increases the loyalty between the consumer and the business people.

Traceability: Traceability is the process that helps to trace the whereabouts of the products involved. It helps the customers and the business people to check the origin of the products. In 2018, a lettuce-hosted *Escherichia coli* outbreak led to five deaths in the United States (US), and in October 2006, the US experienced a spinach outbreak from *E. coli* strain O157:H7 in which 199 people from 26 different states were infected (CDC, 2006). The Centers for Disease Control and Prevention (CDC) advised the people of the United States not to consume fresh spinach implicated in the outbreak and food containing spinach processed by Natural Selection Foods. The CDC also advised boiling spinach well before consuming it and not to eat it raw. This had a major impact on food consumables as people started inquiring for the freshness of the foods and the use by date. Business people have to be more aware of the product origin and in case the need arises, immediately state the origin of the products. When Walmart checked with the origin of the sliced mangoes in their store, it took them several weeks to find it in the traditional supply chain system (Nation, 2017).

Cost: Multiple parties are involved and all of them negotiate through intermediaries or agents, which is associated with an increase in cost, instead of dealing directly. The time taken for the product to reach the customer is high, which in turn increases the cost of the product. Hence, the intermediaries and the processing time result in the increasing cost of the products.

Trustworthiness: The suppliers, distributors and transport agencies all have to deal with the middleman and have to trust them. But if the middleman mishandles the system or changes the products, the business people will lose their customers' loyalty and their business will not be successful. Hence in any business or industry, trustworthiness is an important factor.

Reliable: A business or manufacturing industry strives to maintain a consistent performance of its product in order to earn a name, that is, establish a brand name for their organization. In a traditional supply chain, the business or manufacturing industry may experience negative consequences when the product is not delivered on time or the status of the product is not known as well as when the product does not meet the expected requirements. Furthermore, a processing delay by the intermediaries could lead to an increase in the price of products and decrease customers' satisfaction. Blockchain can pave the way for a reliable supply management system, thereby improving customers' satisfaction and reducing wastage of resources.

5.3.2 Benefits of Blockchain in Supply Chain

Blockchain has the features of a decentralized system that could remove the need for intermediaries. Once the transaction is stored in the block it is immutable and through the consensus protocol, the agreement between the un-trusted parties facilitates a trusted environment. The increase in the speed of the system by the elimination of the middleman and the reduction in cost leads to an efficient system. Hence the incorporation of blockchain into the supply chain system results in an automated, efficient, trusted and secure supply chain management system. It increases the reliability, reduces the cost, expedites the process, increases the transparency and enables traceability.

Transparency: When blockchain is embraced in a supply chain management system, the validated blockchain is available to all the participating parties (nodes) of the supply chain management system. All of the parties know the events taking place in each of the parties in a blockchain-based supply chain management system. Each and every happening will be updated in the blocks of the blockchain once the block of transactions is validated. Because the contracts between the parties are stored in the blocks and validated and included in the blockchain, neither of the parties can alter it and therefore have to act according to the contracts negotiated. This contributes the whereabouts of the happenings to all the nodes involved in the transactions and results in visibility to all participating parties.

Shared Ledger: The blockchain stores the negotiated contracts between the parties in agreement as a smart contract; it is a programmed action, which will be executed only when the conditions are met. Therefore, the smart contract eliminates the need for third parties when the transactions take place between the concerned people. As the need of an intermediary is eliminated in a blockchain-based supply chain, the reliability of the system is undoubtedly increased. Because blockchain is based on a shared ledger (decentralized), the visibility is ensured (all the happenings across multiple parties are known and automated). It also reduces the human error and miscommunication between the involved persons.

Cost and Speed: As mentioned previously, the middleman is eliminated and the actions to be carried out are presented as a software program known as a smart contract. This smart contract facilitates automatic execution of the program once the predefined conditions are met. This speeds up the process because the system can trigger an action without human intervention. It also ensures the security of the system because of the incorporation of these contracts into the blockchain as an immutable ledger. The cost is reduced significantly as there is no need to pay for each transaction

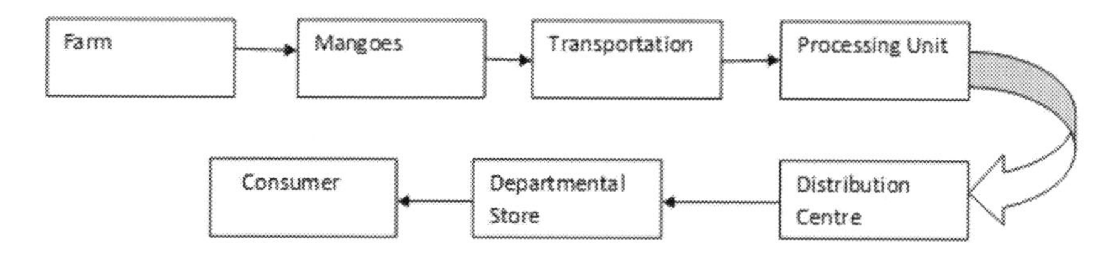

FIGURE 5.7 Supply chain of sliced mangoes.

between the intermediary parties. Incorporation of the smart contracts into the blocks of blockchain speeds up the process and reduces the cost of the overall system.

Traceability: Another main task satisfied by the incorporation of blockchain in the supply chain is the traceability of the system. The tracking of the raw materials from the farm or forest or the origin to the final destination like distributors, retailers or consumers can be easily carried out with the help of an RFID tag or QR code. This automatic tracking was implemented in 2018 as a pilot project by IBM and Walmart to track the sliced mangoes in the Walmart stores. It was found that within 2.2 secs (Nation, 2017) they were able to trace the farm of the mangoes using the blockchain technology, whereas a traditional system takes several weeks. Because the supply chain management is an interconnected and interrelated system, there should be coordination between all the involved parties. With the incorporation of blockchain and IoT into the supply chain management, it can be easily tracked and the coordination process becomes simpler because of automated and smart contracts.

Figure 5.7 shows the travel process of the raw mangoes to the sliced mangoes into the hands of the customers. Many parties are involved and hence documentation will be high according to the different geographical locations in a traditional system.

5.3.3 Automotive Supply Chain

The automated supply chain system makes it human free and enables it to complete the task efficiently, quickly and with less cost. The supply chain is a coordination between different parties; all the rules and regulations, financial transaction and the final delivery system have to be incorporated as a shared ledger. This speeds up the entire system, thereby reducing the cost involved. This is possible not only with blockchain, but the sensors also play an important role. With the IoT and blockchain, the supply chain in the modern era would be able to deliver products more efficiently with reduced cost.

The sensors, RFID tag or QR code, play a crucial role in automating the system. The data from the sensor is communicated through the network to the processor according to the requirement. For example, to know the details of the origin of mangoes, a QR code will be pasted on the wrapper. If it is scanned, the details of the farm where the mango came from, when it was processed, how long it has been after plucking and after processing can be obtained by a mobile phone. This feature increases the loyalty between the customer and the seller results in business profit by attracting more customers. The customers can rely on the vendors, and their demand for freshness is met. Many devices are communicating between each other; hence the data in transit has to be protected. These data are sent as transactions and the documents about the transportation and the agreement between the different participants are added as smart contracts. These transactions and smart contracts are blocks, and after validation by the miners, they are incorporated into the blockchain. This becomes immutable because of the cryptographically hashed property that secures the transactions and increases the reliability between the parties.

Once the contract, that is, the condition is met, the process moves to the next step automatically and makes the action visible to the participants involved in the system. Human error is reduced, the speed is increased according to the processor speed, and the time taken for the completion of the entire process is less. The overall production cost is reduced because of the automated system and the consumers get the benefit of lower prices for the products. The system ensures security, visibility, traceability, speed and reliability.

5.3.4 Drugs and Pharmaceuticals

To date, many industries have started using the blockchain technology in their supply chain management system. The drug and pharmaceutical industry has adopted the blockchain technology for sharing prescriptions, facilitating access to medical records, tracking and reporting of clinical trials and so on (Challener, 2019).

Figure 5.8 shows (Khezr et al., 2019) the blockchain-based supply chain management in the pharmaceutical industry. The pharmaceuticals research and development center does the research on the drugs; once it is validated, this result is added as a block into the blockchain. The drug under research is then sent to the manufacturing unit for production. The secure aspect is that it is owned by the company and they are the authenticated person to have the ownership of the drugs. The manufacturing unit

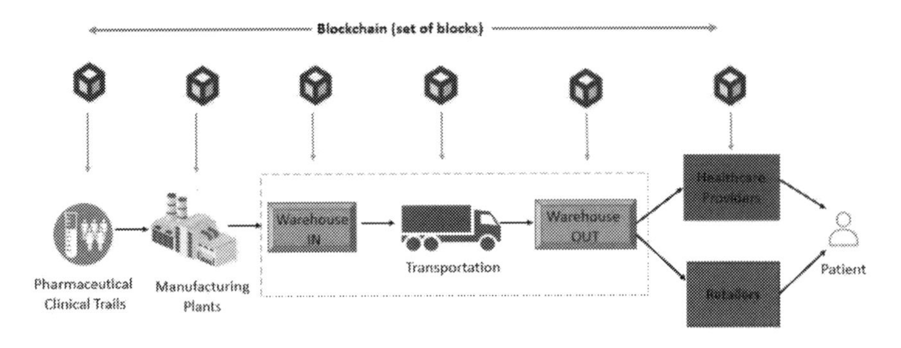

FIGURE 5.8 Blockchain-based supply chain management (Khezr et al., 2019).

has to develop the drug as per the specification given by the research center. These patented rights cannot be claimed by anyone except the owner.

No one can alter patent rights of ownership, since the patent rights of ownership (it is part of a shared immutable ledger) are incorporated into the blockchain. It is protected and transparency exists because it is shared with all the participants involved in the system.

Once the drug is manufactured, it is stored in a warehouse. After the approval of regulations as per the certification of drugs and the financial terms between the manufacturers and the drug stores, it is transported to the drug distributers. The certifications are stored as a smart contract in the blocks of the blockchain and hence it is automatically executed only when the conditions are met as per the regulations.

With known persons, the authorities may change the regulation and there will be a bias. Hence the system might be compromised. But because the smart contract is a program, no bias occurs and the regulations are followed exactly. It applies the same rules for all the involved participants. Without any intermediaries and bias, the drugs directly reach the wholesale market and retail markets, respectively.

Finally, it is made available to the patients. The patients using an RFID tag or the QR code that would be presented on the drug packaging would be able to retrieve the complete details of the drugs and hence they can be assured of their safety and health. The possible information that patients would be able to know includes the composition of the drug, manufacturer, the date the drug was manufactured, expiry date of the drug, side effects, precautions and the procedures to be followed for the intake of the drug. Because of the transparency that blockchain offers, we would witness in a foreseeable future competitiveness among the drug manufacturers to become more transparent and provide better quality products, which in turn would contribute to better health and establishing a trusted relationship between patients and health professionals.

The advantage of a blockchain-based supply chain in the pharmaceutical industry is that it is secure and trustworthy, response is fast and overall production cost lowered because of the noncentralized system. Transparency, provenance and a secure system are possible because of the automated supply chain management.

5.3.5 Agriculture

Blockchain technology was initiated in the banking sector for financial transactions with the Bitcoin protocol. However, the underlying blockchain technology was applied later in various sectors like the real estate industry, health care sector, law and so on. Recently, we have been witnessing more and more applications of blockchain in the agriculture industry. Blockchain can benefit small farmers to support their financial needs by sharing resources from land to tractors (Datta, 2019). ICT can play a significant and important role in increasing agricultural yields (Shams et al., 2020). When blockchain is integrated with IoT in agriculture, it can be surmised that the agricultural industry will make a big leap forward. Blockchain is a disruptive technology that benefits the consumers and producers. For consumers, it ensures their safety and healthy food consumption. For producers, it makes a competitive environment leading to product quality to remain in the market and make profit. A healthy environment becomes prevalent with trusted, secure, transparent and provenance of the food among all the stakeholders.

Figure 5.9 depicts the process of the supply chain in agriculture. The crops are cultivated at the farm, and the sensors are placed on the field in order to obtain farming-associated data in real-time. The data from the sensors are collected—it can be temperature, moisture, humidity of the soil—and can be used to set the appropriate operation of the farm so as to maximize production while reducing cost. Additionally, such data can be used further to understand the quality of the crop, the suitable crop for the weather conditions and also predict the yield of the crops by analyzing previous data collected. Once the data collected from the sensors are processed, next the data needs to be analyzed to gain insights (e.g., prediction of yield, type of crops to be sown and the required water level for irrigation in a season). Such insights can be obtained through applying machine learning tools in the preprocessed data, as depicted in Figure 5.10. This information is stored in blocks and after validation by the miners added into the blockchain. It is then shared among the participating farmers in the network; hence all gain the knowledge of the seeds to be used for cultivation in their soil texture, required water content, and the attainable production yield. Then the crops, fruits and vegetables, are transported through an IoT-enabled vehicle (Takyar). This vehicle has sensors to measure the temperature, moisture level, humidity and pH level of the environment. These data are processed to preserve the food items during transportation and their location is tracked through the global positioning system (GPS). The travel duration of the crops is known, which is important because the fruits and vegetables are perishable. The products of the farm reach the manufacturing unit, where the required processing of the crops, fruits and vegetables, is carried out. Once processed, the details of the procedure are embedded in the QR code and pasted on. Then the processed food is again transported by the IoT-enabled vehicles to the distribution center. From the distribution center, the processed food is sent to the wholesale markets and retails markets accordingly. Finally, the consumers can buy the processed food from any of the markets. The QR code will

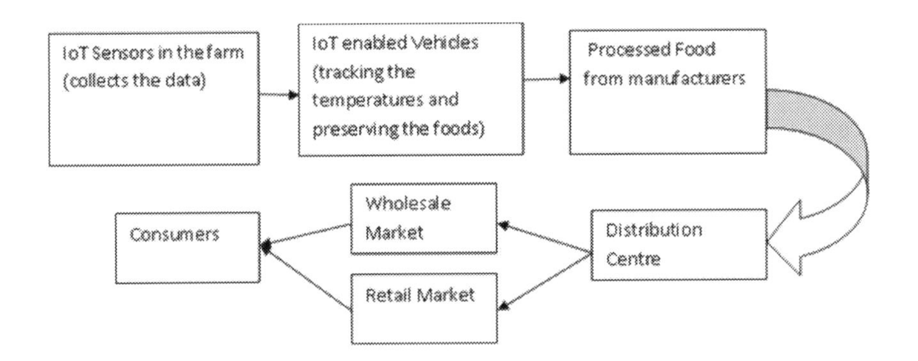

FIGURE: 5.9 Supply chain in agriculture.

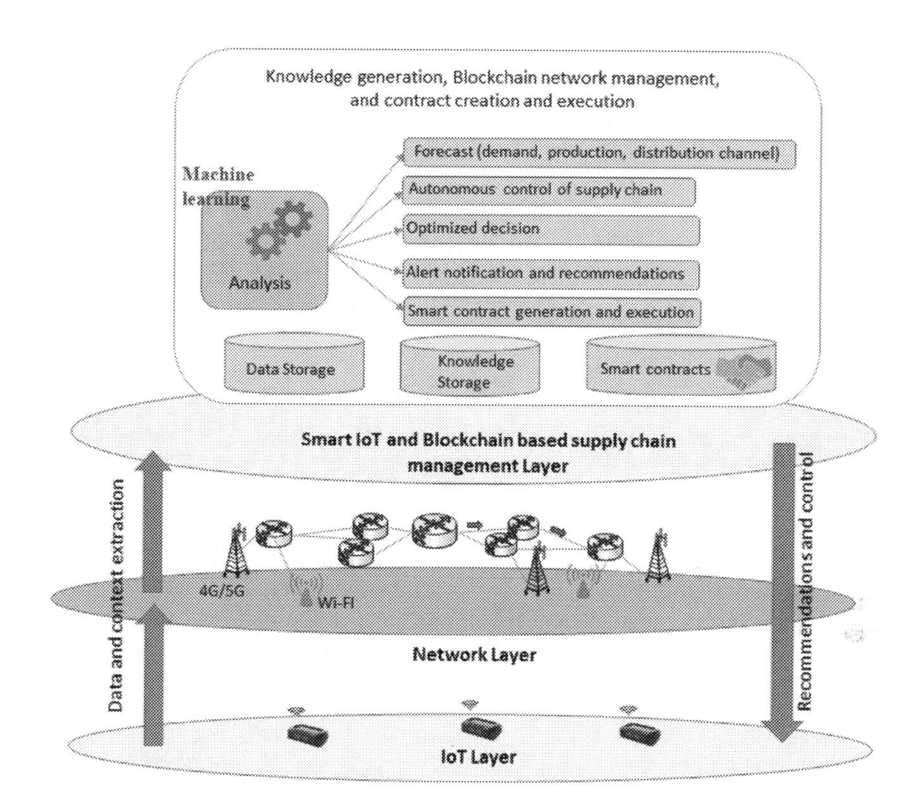

FIGURE 5.10 IoT to cloud: data gather, knowledge creation, control and smart contract management.

be available on the processed food and when scanned gives the required details. Because all the details are embedded into the block, it becomes more protected and distributed, which creates visibility among all the participants involved in the process. Consequently, this makes the entire food supply chain healthy and secure among the parties in the network. Furthermore, because blockchain facilitates direct dealing with different parties without any intermediaries, the benefit goes to all parties from farmers to the customers, including saving time and cost. This IoT-enabled agriculture system with blockchain technology leads to a healthy and safe environment, which the industry should strive to attain.

Figure 5.10 narrates how data is extracted from the IoT layer through the network layer and finally processed in the cloud-based supply chain management layer. The cloud may facilitate a platform where data, knowledge and smart contracts can be stored. Producing recommendations and control (trigger an action) when the conditions are met in smart contracts would be another important role of the supply chain management layer depicted in Figure 5.10. Such high level of intelligence obtained through data processing using machine learning algorithms would make the supply chain management autonomous in the foreseeable future.

5.3.6 Deployment of Blockchain in the Supply Chain

The various phases of the supply chain management system are raw materials, transportation, processing, delivering, storing and purchasing.

Raw Materials: In a food processing industry, the freshness of the products plays a primary role. With the help of blockchain, the date on which the fruits or vegetables were reaped or cereals were harvested would be known because the time, date, and name of the farm would be incorporated into the blockchain.

Transportation: Goods are transported from the place of procurement to the processing/manufacturing location and then the finished goods are transported to the stores/consumers. This transport of raw materials or final product is important in the case of perishable products. With the incorporation of the blockchain, the quality of the goods can be assured. With the combination of IoT and blockchain, a warning system could be developed to identify the perishable products and the timeline for it to be fresh enough.

Processing: Ingredients used in the food products, the process carried out in the manufacturing location and any mishap in the process can be easily known to the involved participants; delays can be avoided in case of emergency maintenance; if an employee has left the organization, the details of the process carried out is added in the blockchain and hence easy reference is possible.

Delivering: Delivering goods on time is important. Things not available at the right time to the right person are wasted. With incorporation of the blockchain, delay is avoided and things will be in place at the proper time.

Storing: Storing the goods and products is another critical task in which the expiration dates of the things should be known and an alert should be given when the time is nearing. The warehouse operators can act accordingly.

Purchasing: Customer loyalty is important and that loyalty could be built by incorporation of blockchain and regular customers can be retained. Services can be provided and the quality of products ensured.

5.4 CONCLUSION

Blockchain technology is a disruptive technology that supports a decentralized, cryptographically hashed, distributed environment. The transactions stored in the block once validated and added into the blockchain cannot be tampered with and hence blockchain plays a major role in protecting the information. It also tracks the information stored in it pseudo-anonymously. A smart supply chain system provides an automated distribution of materials between the suppliers and consumers. The incorporation of blockchain in the supply chain enables the tracking of the origin of the raw materials within seconds compared with several weeks in a traditional supply chain system. Multiple parties are involved in the supply chain across the globe; hence the transparency and integrity are difficult in a traditional system. However, with incorporation of blockchain, the complex tasks become easy because of the distributed shared ledger, which is visible to all the participants involved in the network. The detailed record of the information about the suppliers, processing units and the customers are available to anyone in the network without logging in for a request to the centralized server, because it is a decentralized peer-to-peer network.

When blockchain is embraced in the supply chain, the customer would get all the details of the products purchased like the ingredients or raw materials of the product, duration of the shipment of the product from farm or mine to the manufacturing unit, and also after processing or manufacturing into the distribution center. It also provides the details of the manufacturers, the life of the product and the status of the product, whether they are in good condition or not. This means a lot to the customers, ensuring the quality of the product and that it is worth its cost. In the perspective of the suppliers, they get the details of the accountability of the product including the optimization of resources and hence reduction in the cost of goods, that is, the manufacturing cost would be less because there is no middleman and goods are available in time for processing. Because the costs of the goods are less, they can compete in the market with lower selling prices without compromising the quality. The exact shipments of the raw materials are known and hence the processing functions can be planned. Thus, transparency of the goods helps all people participating in the network and increases the efficiency of the system.

The terms and conditions (agreement) between the supplier and the buyer are written as a contract. It is a software program embedded in the block that eliminates the need for intermediaries to observe any contract. No intermediaries are involved; hence the execution is fast and only when the agreement is satisfied is the process executed. The next step is taken accordingly and all the parties in the network are satisfied. The actions are carried out without human intervention so any cheating or fraud can happen and trustworthiness is ensured between the non-trusted parties in the system.

Blockchain will bolster the security of the information. All the information is stored in the block with the time stamp and a hash function; the data are protected and the identity is not revealed. The previous hash function is added to every block and validated with the help of miners based on the consensus algorithm.

Smart supply is the need of the hour with the growing population and globalization phenomenon. The minimization of resources, profitability of the business, transparency, tamper resistance, privacy and security of the information is possible because of the incorporation of blockchain in the supply chain; otherwise, it might have been too complex and not effective. But blockchain has t its limitations in scalability and is in the beginning stage. Once the drawbacks are eliminated and it is implemented fully fledged in the smart supply chain, it might lead to a very effective system.

REFERENCES

Aste, T., Tasca, P., and Di Matteo, T. 2017. Blockchain Technologies: The Foreseeable Impact on Society and Industry. *Computer* 50:18–28.

Castor, A. 2017. A (Short) Guide to Blockchain Consensus Protocols. www.coindesk.com/short-guide-blockchain-consensus-protocols (accessed May 21, 2020).

CDC. 2006. Multistate Outbreak of E. coli O157:H7 Infections Linked to Fresh Spinach (FINAL UPDATE). www.cdc.gov/ecoli/2006/spinach-10-2006.html (accessed May 21, 2020).

Challener, C.A. 2019. Why the Industry Is Moving toward Blockchain Technology. www.pharmasalmanac.com/articles/why-the-industry-is-moving-toward-blockchain-technology#:~:text=There%20are%20many%20potential%20uses,provider%20credentialing%2C%20quality%2Dof%2D (accessed June 10, 2020).
Datta, M. 2019. Smart Farming is the Future of Agriculture. www.geospatialworld.net/blogs/smart-farming-is-the-future-of-agriculture/ (accessed May 21, 2020).
Dorri, A., Kanhere, S. S., Jurdak, R., and Gauravaram, P. 2017. Blockchain for IoT Security and Privacy: The Case Study of a Smart Home. Proceedings of the IEEE International Conference on Pervasive Computing and Communications Workshops (PerCom 2017). Kona, HI, USA: IEEE. 618–623.
Evans, D. 2011. *The Internet of Things: How the Next Evolution of the Internet is Changing Everything*. San Jose, CA: Cisco IBSG. www.cisco.com/web/about/ac79/docs/innov/IoT_IBSG_0411FI NAL.pdf (accessed January 21, 2015).
Feki, M.A., Kawsar, F., Boussard, M., and Trappeniers, L. 2013. The Internet of Things: The Next Technological Revolution. *Computer* 46:24–25.
Gong, S., Tcydenova, E., Jo, J., Lee, Y., and Park, J.H. 2019. Blockchain-Based Secure Device Management Framework for an Internet of Things Network in a Smart City. *Sustainability* 11:1–17.
Infopulse. 2019. Blockchain in Supply Chain Management: Key Use Cases and Benefits. https://medium.com/@infopulseglobal_9037/blockchain-in-supply-chain-management-key-use-cases-and-benefits-6c6b7fd43094 (accessed June 22, 2020).
Khezr, S., Moniruzzaman, Md., Yassine, A., and Benlamri, R. 2019. Blockchain Technology in Healthcare: A Comprehensive Review and Directions for Future Research. *Applied Sciences* 9:1–28.
Laudon, K.C., and Traver, C.G. 2016. *E-commerce: Business, Technology, Society*. 12th Edition. London: Pearson Education, Ltd.
Miraz, M.H. 2019. Blockchain of Things (BCoT): The Fusion of Blockchain and IoT Technologies. Advanced Applications of Blockchain Technology. In *Advanced Applications of Blockchain Technology. Studies in Big Data*, eds. S. Kim and G. Deka, 60:141–159. Springer, Singapore.
Mollah, M.B., Zhao, J., Niyato, D., Lam, K-Y., Zhang, X., Ghias, A.M.Y.M., Koh, L.H., and Yang, L. 2019. Blockchain for Future Smart Grid: A Comprehensive Survey. *IEEE Internet of Things Journal* 2020:1–26. https://arxiv.org/pdf/1911.03298.pdf.
Nahas, M. 2020. Lightweight M2M: Device Management for Low Power Devices. www.telit.com/blog/what-is-lightweight-m2m-lwm2m/ (accessed May 21, 2020).
Nation, J. 2017. Walmart Tests Food Safety with Blockchain Traceability. *ETHnews* [Online]. www.ethnews.com/walmarttests-food-safety-with-blockchaintraceability (accessed June 10, 2020).
O'Byrne, R. 2019. How Blockchain Can Transform the Supply. www.logisticsbureau.com/how-blockchain-can-transform-the-supply-chain/ (accessed June 22, 2020).
Reiff, N. 2019. Guide to Blockchain. www.investopedia.com/terms/b/blockchain.asp#blockchains-practical-application (accessed June 22, 2020).
Shams, S., Newaz, S.H.S., and Karri, R.R. 2020. Information and Communication Technology for Small-Scale Farmers: Challenges and Opportunities. In *Smart Village Technology*, eds. S. Patnaik, S. Sen, and M. Mahmoud. Modeling and Optimization in Science and Technologies, 17. Springer, Cham.
Takyar, A. Blockchain in Agriculture—Improving Agricultural Techniques. www.leewayhertz.com/blockchain-in-agriculture/ (accessed May 21, 2020).
Yoo, Y.S., Hwang, T., Kang, S., Newaz, S.H.S., Lee, I.W., and Choi, J.K. 2017. Peer-to-Peer based Energy Trading System for Heterogeneous Small-scale DERs. ICTC 2017, Jeju, S. Korea. October 18–20.

6 Blockchain-Based Health Care Applications

M.R. Manu, Namya Musthafa, Divya Menon, Sindhu S and Soumya Varma

CONTENTS

6.1 BLOCKCHAIN BASED HEALTH CARE APPLICATIONS-CHALLENGES AND OPPORTUNITIES

The blockchain technology has a variety of skills such as interoperability, security and privacy that would be beneficial for patients by involving an efficient electronic mechanism. In health care, the electronic medical record improves efficiency and yields better health outcomes for patients when the blockchain methodology is used. The blockchain provides secure, shared immutable records about peer digital transactions. The main advantage of using the blockchain methodology is that it provides a non-centralized data management system. The cryptographic algorithm provided by the blockchain architecture make it more secure. Table 6.1 provides more details about opportunities and threats [16].

TABLE 6.1
Opportunities and Threats in Blockchain.

Opportunities	Threats
Reduce transactions	Offline transaction threat
Reduce fraud	Cross border threat
Bring transparency	Scalability
Protect from hacking	High consumption of power
Low cost and no third-party organization	Authorization

6.1.1 Latest Healthcare System

The patients' health is the essence of health care industry. The most significant challenge in health care is providing patients with superior level, effective treatments; the patient's mental care is important and plays a vital role in it [10]. The main problem in the quality of health care is the gap between the payer and the provider. Another issue is the misuse of the patient's data, which reflects the quality of health care management and patients will face critical problems. This has been depicted in Figure 6.1 [9].

The outdated systems and equipment used in the health care industry ruins the authenticity and accuracy of treatment [2]. This will lead to time-consuming diagnosis and the cost factor of the patients automatically increases. So, the health care system should provide advanced as well as smooth functioning and be reliable and transparent.

6.1.2 Medical Cyber Physical System

The integration of advanced analytic computation with a physical processing system is referred to as a cyber-physical system. If the system is focused on patient-centric medical integration, it is called a medical cyber-physical system and it is denoted in Figure 6.2. It is a smart intelligent health care monitoring system that will autonomously diagnose the disease without involving a doctor [9]. The rapid growth of information technology and the advancement in analytics toward the manufacturing of medical equipment for diagnostic purpose was a milestone in the medical industry. A cyber-physical system facilitates the huge medical data into useful patient-centric information.

The advanced use of cyber-physical systems will recognize medical information that can be invisible to the human eyes and minimize the medical error rates. The faster processing of medical diagnostic information is useful for timely recognition of chronic disease. The system consists of different layers such as data collection, data management and application which provides better operational efficiency in processing [1]. The data privacy of the medical data is ensured by different cryptographic techniques such as AES, Elliptical Cryptography and the Diffie-Hellman key exchange protocol. Predictive diagnostic systems are used to analyze the massive

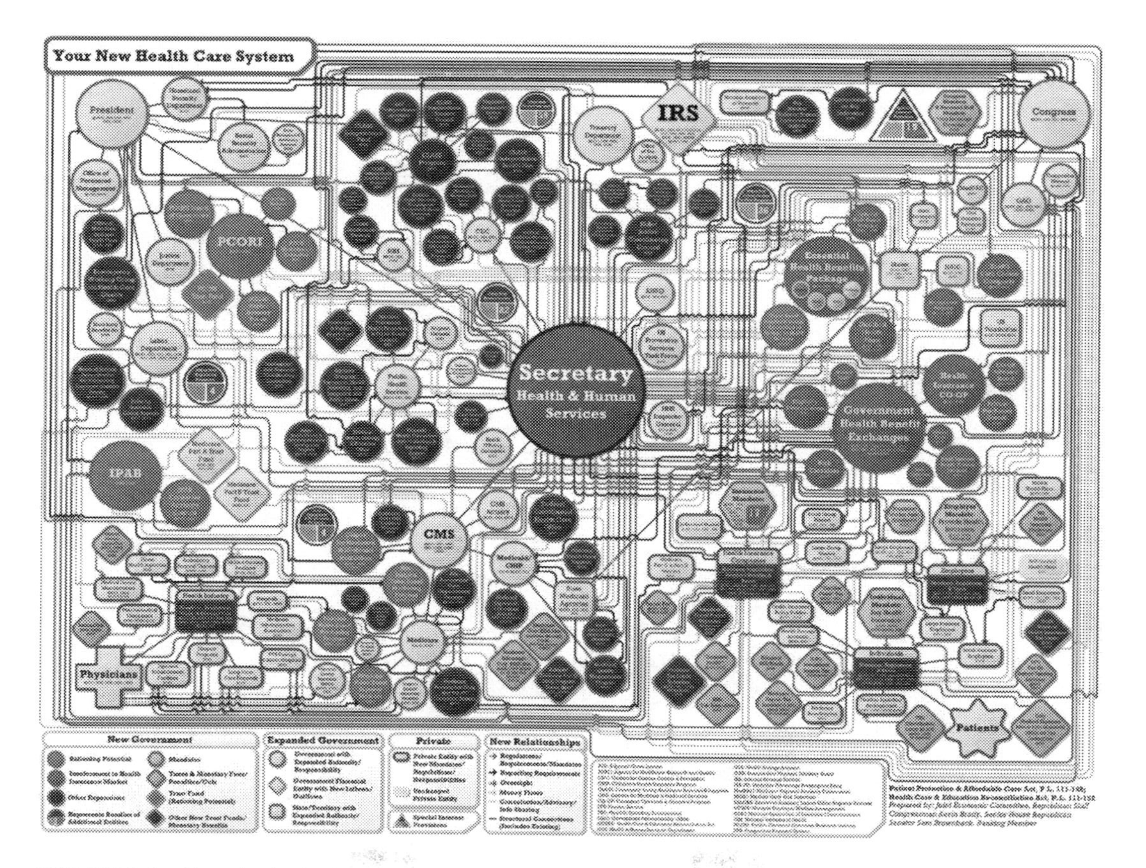

FIGURE 6.1 Complexity in the current health care industry.

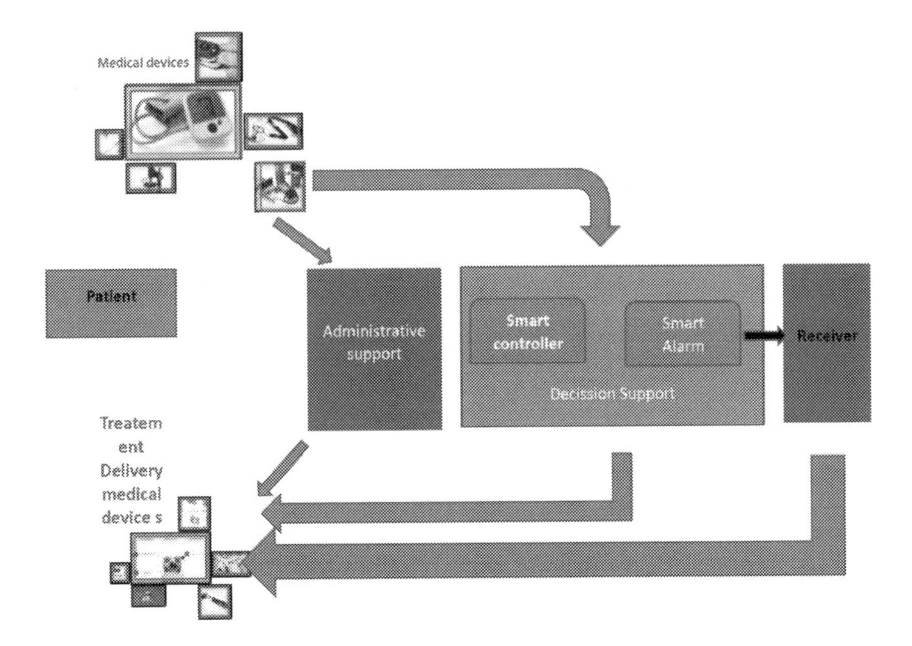

FIGURE 6.2 An overview of a cyber-physical system.

information in health care to diagnose various diseases [5]. The internet of things applications also provides a definite model for the medical cyber-physical system.

6.1.3 Medical Bigdata Mining and Processing in E-Healthcare with Blockchain Security

The extraction of huge complex data sets from a large repository to recognize a variety of hidden patterns and form a relationship between them is referred to as big data mining. The knowledge repository also includes heterogeneous medical health records, and it will extract the information based upon pattern analysis and generate a graphical presentation from the available data source. This is helpful for clinicians to visualize and understand quickly and make the right decision. This process, referred to as medical big data mining and processing, is illustrated in Figure 6.3 and has shown an indication that results in a drastic change in the health care industry, as the rapid growth of data with healthcare information was a major challenge in this field. The big data technology involves a recovery solution for this challenge that will store and process large complex data unveiled by a traditional data processing system. For processing the heterogeneous medical records, various data mining algorithms are available. The data mining algorithms are mainly classified into two categories: descriptive (unsupervised learning) and predictive (supervised learning) [11]. The descriptive data mining method involves clustering the medical information based on their similarities whereas predictive infers predictive rules based upon prior training medical records. The major role played by the supervised learning

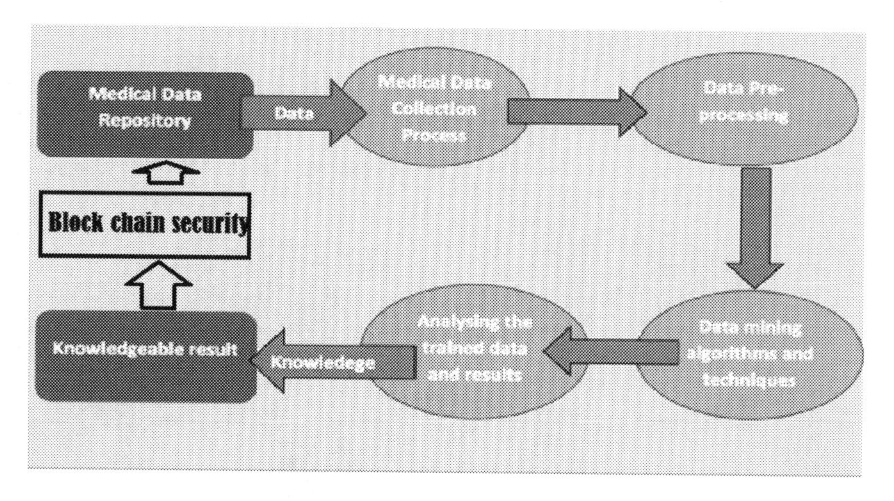

FIGURE 6.3 Medical big data mining process.

algorithm used in prognosis and diagnosis in e-health care involves an artificial neural network (ANN), deep learning method, genetic algorithm and various other techniques.

Big data mining techniques have plenty of application in e-health care such as predictive modeling, diagnostics analytics and disease and safety surveillance in the medical industry. These techniques will uncover information including hidden patterns and unknown correlations and improve operational efficiency providing competitive advantages over traditional health care methods. This techniques also equipped on Electronic health records, food allergy prediction, web and social data enabled to identify the optimal practical solutions for medical experts for diagnosing disease faster than the traditional methods. The structures and unstructured information from the medical database can be processed using Hadoop, Spark system with tensor flow or Kera packages. The cryptographic security measures involved in the secure hashing method in blockchain will be a pivotal function for medical big data processing in order to secure the patients' information and privacy

6.2 DIGITAL IDENTITY MANAGEMENT SYSTEM

Digital identity can be defined as the information on an entry used by the computer system to constitute an external agent. The particular agent may be either a person, device, organization or an application. The information assembled in a digital identity allows the assessment and authentication of any user interacting with a web-based business system without involving any human assistance. Such a system allows our access to computers and services to be automated and helps the computers to consolidate relationships. Recently this kind of identity system has been noticed to be in widespread use and hence the information to represent the people in a trustworthy digital format in computer systems. Nowadays we often use digital identity in certain ways that requires the data stored in the computer system to be linked to

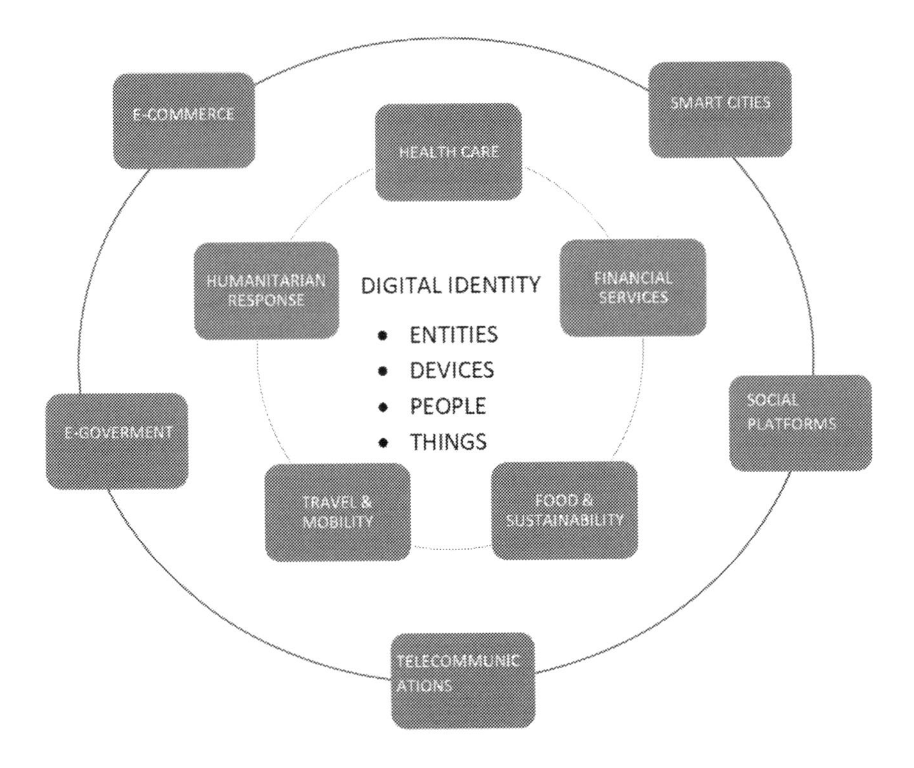

FIGURE 6.4 Applications of digital identity.

corresponding civil or national identities. In some aspects, "digital identity" plays a role in the entire collection of information engendered or produced by a person's online activities. This information can be username and passwords, search activities, date of birth, social security number, e-shopping history and whatever information is publicly available and not anonymized. With self-sovereign identity (SSI), an individual identity holder can fully create and control their credentials without seeking the permission of a centralized authority, and it gives control over how their personal data is shared and used. This can be used in digital identity to acquire the right to create and control unique identifiers along with some setup to store identity data.

Figure 6.4 shows some of the core applications of the digital identity. With the help of digital identity, we can leverage the technology at our fingertips. Digital identity has recently gathered great attention in broad areas. From government services to organizational operations, there are real time applications.

Identity management or identity and access management involves all the processes and technologies within any organization that are used to identify, authenticate and authorize someone who wants to access services or systems in that particular organization or other associated ones [30]. To understand this concept, we can consider some examples, such as employees accessing software or hardware inside the company, or in educational organizations where the administration members have a particular access space whereas the teachers and students are assigned different access portals.

In a governmental portal, there is the issuing and verification of birth certificates, national id cards, passports or driver's licenses where the users are allowed to not only prove their identity but also access services from the government and other organizations.

6.2.1 Issues Faced by Current Identity Management Systems

A digital identity is a revolutionary approach that reduces the level of administration and increases the speed of processes within organizations [25]. This allows for a greater interoperability between departments and other institutions. But a problem arises when this digital identity is stored on a centralized server, because it becomes likely target of hackers; a solution is required, as shown in Figure 6.5. Unfortunately, we are still depending on centralized servers in many organizations. Hence most of the identity management systems are weak and outdated.

The most important characteristics of identities are portability and verifiability. That is, irrespective of the location and time, the identity can be digitized and can be verified. But the identity should also possess privacy and security.

Upon focusing on the use of digital identity in health care, we may come across some issues. An example of a digital identity management ecosystem is shown in Figure 6.6. We know that half of the world's population can't access quality health care. The lack of interoperability between actors in the health care space such as hospitals, clinics, insurance companies, doctors, pharmacies, etc. leads to inefficient health care and delayed care and frustration for patients. The health care industry remains one of the most highly targeted for cyberattacks, and as a result there were huge number of breaches in recent years. Healthcare IT professionals are trying hard to attain security and ensure system availability while supporting the complex workflows of their providers [18]. The rapid jump to health care's digital transformation along with the rapid growth of service locations and the integration of cloud apps and services creates an inescapable challenge.

There are some common myths persistent in the health care field dealing with digital identity management. Such as people misjudge the health care ID as a trusted

FIGURE 6.5 Problem and solution overview.

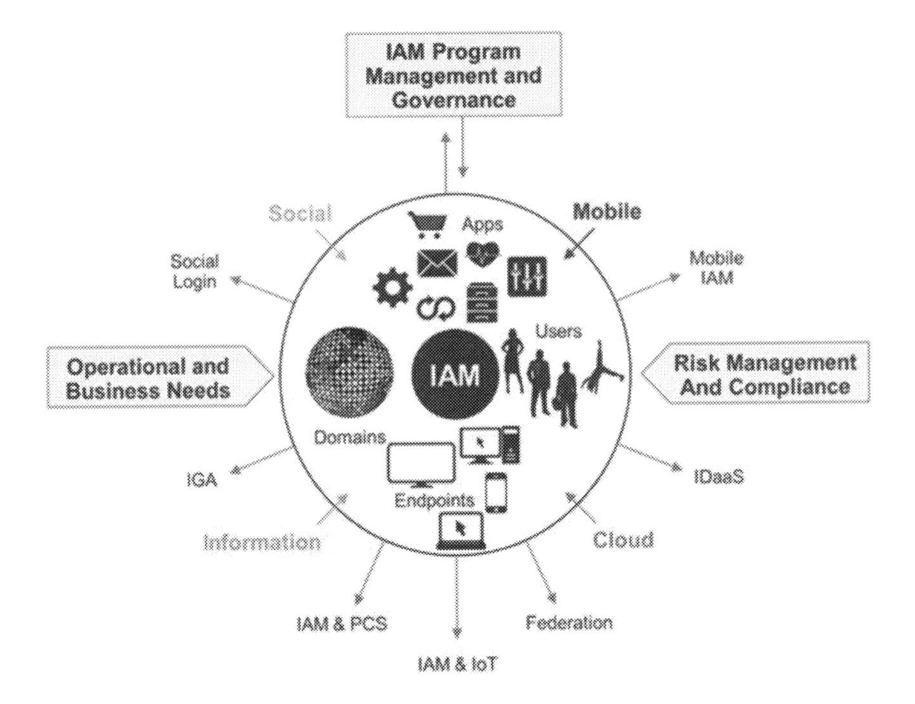

FIGURE 6.6 Example of identity management ecosystem.

digital ID. But they are a relevant Id that helps us to access all the services in the hospital. Figure 6.7 gives a simple idea of how the health care ID plays its role [19].

A patient accessing the system for the first time has to register with their basic information to get a health care ID [23]. As a result, they will have an electronic health record (EHR) in the health care database. Each EHR will be assigned to a unique ID that plays the role of our health care ID. This ID will help us in all hospital-based services and helps to access the EHR, which has up-to-date health records. Then how does health care become an easy target of hackers? The reason is that there insurance companies that give aid for medical purposes. Sometimes the hackers misuse the patient ID to acquire this amount in fraudulent ways. Hence it is important to make these data secure and protected.

6.2.2 Cryptography in Identity Management System

Providing security to such ID are not at all a secondary job. Whenever we are requested to prove anything about our identity, such as our name, address or unique IDs, there is a procedure of authentication. The data we are claiming about ourselves are examined and verified to be true or false by a verifying entity. This process is usually done through the verification of relevant identifying documents.

This complex procedure of identity verification and authentication make privacy concerns arise. Will the authority be able to access the remaining information

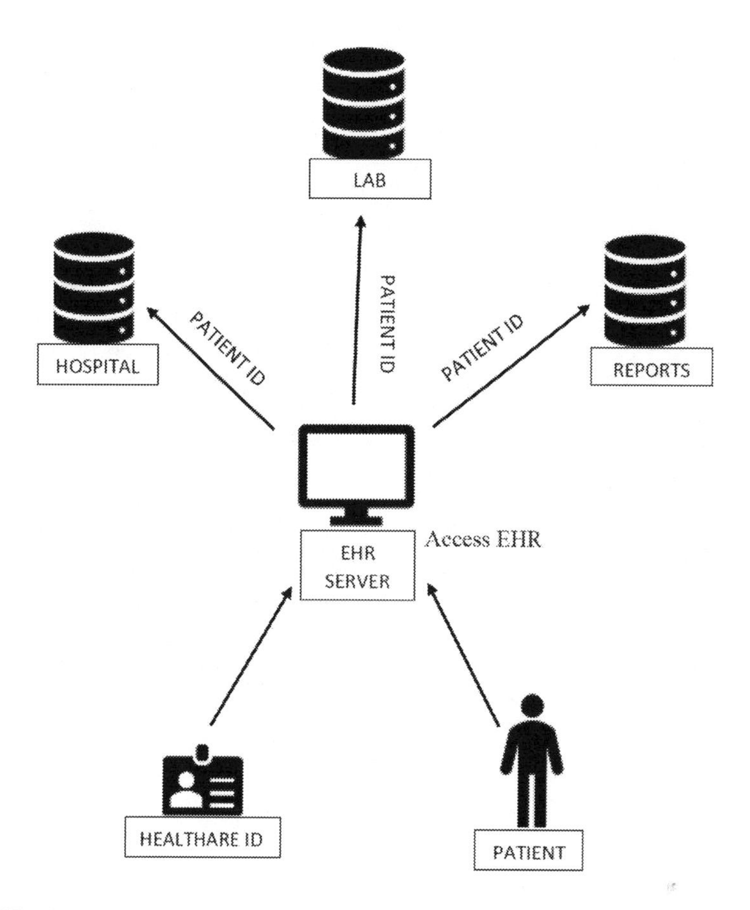

FIGURE 6.7 Role of patient ID.

contained in my document while they are looking at it to verify that information? Does an entity that requests proof of my age need to know the day and month I was born? And the concerns keep growing [8].

Blockchain has become the backbone to almost all of the new generation fields that require security and privacy. In an identity management with blockchain scenario, the person will prove their personal details only by fulfilling certain requested requirements without revealing the actual details.

6.2.3 A Blockchain-Based Identity Management System

In identity management, blockchain enables everyone in the network to have the same source of trustworthy data that credentials are valid and who attested to the validity of the data inside the credential without showcasing the actual data.

The three entities in identity management with blockchain are owners, issuers and verifiers.

The corresponding roles of each entity are identity owners, identity issuers and identity verifiers.

The identity issuer is a trusted party, such as local government, and can provide personal credentials for an identity owner, that is, a legible user. By issuing a credential, the identity issuer confirms the validity of the personal data in that credential. The identity owner now can make use of those credentials in their personal identity wallet and use them later to prove statements about his or her identity to a third party such as the verifier who verifies the credentials at the time-of-service provision.

A credential can be a set of multiple identity attributes like name, age, date of birth and so on. The usefulness and reliability of a credential fully depends on the reputation/trustworthiness of the issuer [28].

The most important characteristics of blockchain is that the verifying parties do not necessarily need to check the validity of the actual data in the proof provided. But rather blockchain can be used to check the validity of the attestation and the attesting party, like the government, from which they can determine whether to validate the proof to establish the trust between the parties by guaranteeing the authenticity of the data and attestations, without actually storing any personal data on the blockchain. This is really important as a distributed ledger is immutable. Hence anything that is put on the ledger can never be altered or removed. So, we have to keep a keen awareness on not placing any personal data on the ledger. But we can put the references and the associated attestation of a user's verified credential on the ledger [30]. A pseudonymization technique is used to ensure privacy through non-correlation principles. Here are some of the things that can be stored in the ledger instead of actual private information:

- Public decentralized identifiers (public DIDs) and associated DID descriptor objects (DDOs) with verification keys and endpoints
- Schemas
- Credential definitions
- Revocation registries
- Proofs of consent for data sharing

DIDs are a newly introduced type of unique identifier for verifying digital identities and are entirely controlled by the identity owner. Therefore, they are independent from centralized registries, authorities or identity providers. Some of the properties that DIDs should have are described as the following:

- DIDs should be permanent so that they will be nonreassignable. When alterations are made, there might be higher risk of intrusion. This reduces privacy and security.
- DIDs should be resolvable, that is, each DID resolves to a DID Document that states the "public keys, authentication protocols and service end points necessary to initiate trustworthy interactions with the identified entity". Through the DID Document, an entity will understand how to use that DID.
- DIDs should be cryptographically verifiable. The DID owner can prove their ownership of the DID by using cryptographic keys. The public key

contained in the DID Document can also be used to attest to the authenticity of the issuing authority's signature associated with a credential.
- DIDs should be decentralized. DIDs do not depend on a central authority as most of the current systems do. Distributed ledger technology ensures trust as it allows everyone to have the same source of truth about the data in the credentials.

6.3 FINANCIALS, INSURANCE AND RECORDS

Blockchain technology is a huge leap forward that will bring about significant efficiency gains, cost savings, transparency, faster payouts and fraud mitigation while providing permission for data to be shared in real-time between various parties in a trusted and traceable manner. Blockchain can even enable new insurance practices to enhance service experiences and build better products and markets [26].

Insurance companies operate in a highly competitive and vulnerable environment in which both retail and corporate customers expect the best value for money and the best online experience without compromising the security measures [26]. Blockchain technology represents an important platform for positive change and growth in the insurance industry. From Figure 6.8, we can understand the relationship between the insurance provider, verification authority and user.

Enterprise Ethereum refers to an outlined set of guidelines and technical specifications to accelerate the adoption of blockchain technology among enterprises. The specifications provide businesses with the power to leverage both Ethereum-based private chains and therefore the public main net. By applying Ethereum's smart contracts and decentralized applications, insurance is often conducted over blockchain accounts, introducing more automation and tamperproof audit trails [21]. Notably, the low cost of smart contracts and their transactions means that many products can be rendered more competitive for penetration of underinsured markets in the developing world. Blockchain Based Smart Contracts for Healthcare is used from Ethereum to create smart representations of medical records that are stored on the network in individual nodes; an illustration is shown in Figure 6.9. The contracts are built to contain record ownership metadata, permissions and data integrity. Smart contracts have been designed to handle different medical workflows and then manage data access permission between different entities in the health care ecosystem.

Blockchain can be applied throughout the insurance industry and across many lines of business, such as:

- Registries of high-value items and warranties
- Know-your-customer (KYC) and anti-money laundering (AML) procedures
- Parametric (index-based) products
- Reinsurance practices
- Claims handling
- Distribution methods
- Peer-to-peer (P2P) models

Blockchain can create an immutable and trustworthy record of products' provenance for the benefit of all stakeholders. Not only that, blockchain tracks products'

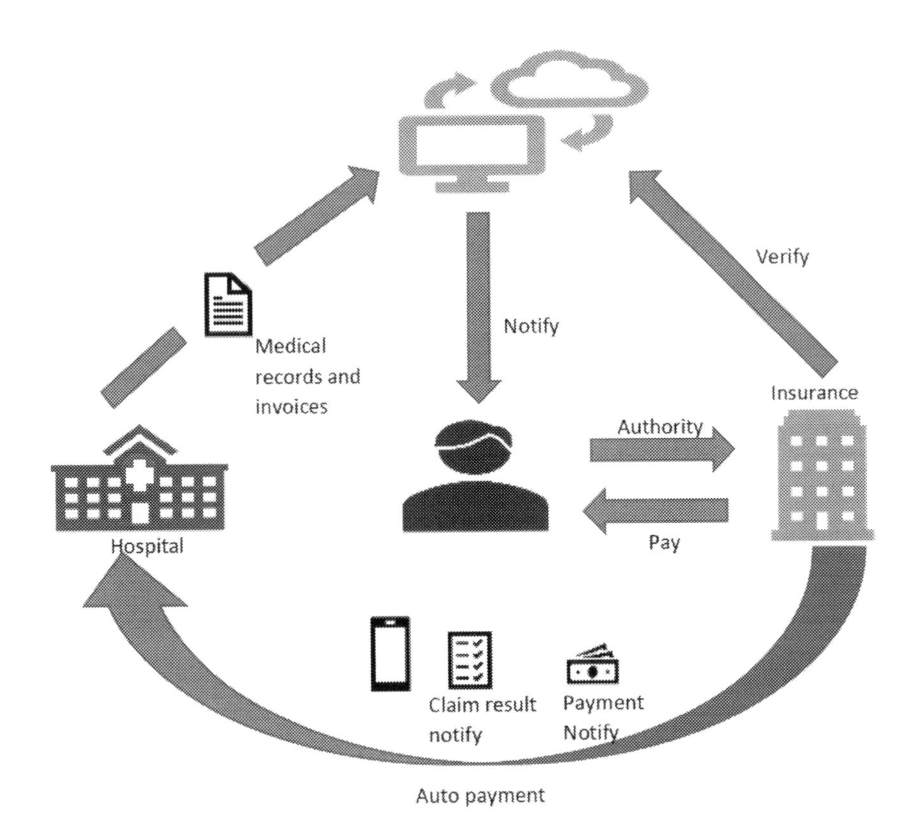

FIGURE 6.8 Online health care insurance workflow.

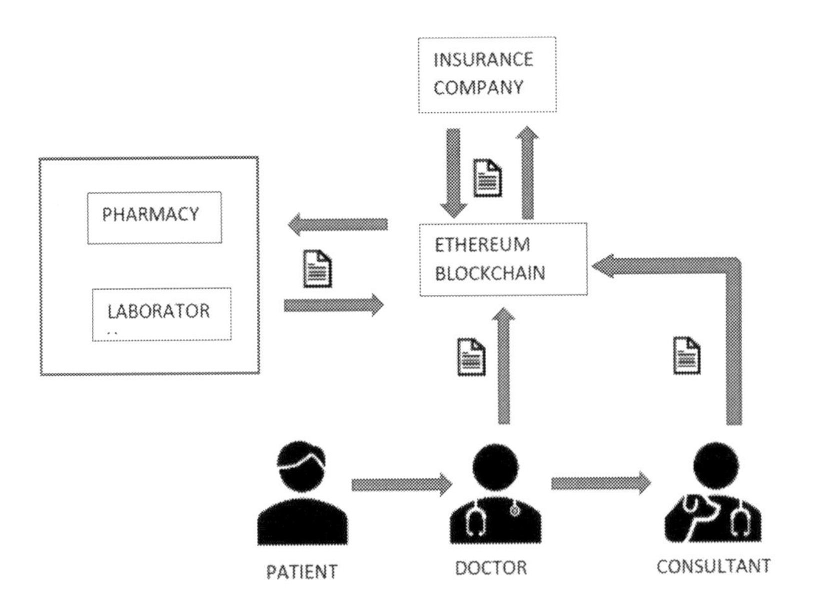

FIGURE 6.9 Ethereum blockchain in health care insurance.

ownership and claims in real-time and even across borders. Enhancing industry-wide efforts to mitigate claims fraud by superior data and data-sharing is also done. All these help blockchain to impact registries of high-value items and warranties.

In the insurance sector, blockchain automates most or all parts of parametric insurance. A policy's logic is embedded in a smart contract and lets an oracle (digital feed) trigger execution upon a predefined loss event and also settles and clears all transactions without manual intervention.

Blockchain does all such activities like automating risk modelling, audits and compliance checks as well as binding towers of risk and treaties on a single time-stamped smart contract. It allows primary insurers, reinsurers, brokers and regulators to share data securely in real-time. A trusted, tamperproof, industry-wide record of claims is being created and claims fraud is reduced by eliminating data silos. Customers are provided greater control over their own data, including access rights along with access to numerous carriers on the same platform that allows them to manage various policies on the same platform. Blockchain is used to coordinate the actions of multiple parties at minimal cost on an online platform. Even transactions for paying premiums or claims have become faster, easier and cheap [26].

REFERENCES

1. Chen, Zhangwei, Ziyong Li, Junjie Li, Chengbo Liu, Changshi Lao, Yuelong Fu, Changyong Liu, Yang Li, Pei Wang, and Yi He. "3D printing of ceramics: A review." Journal of the European Ceramic Society 39, no. 4 (2019): 661–687.
2. Campa, Riccardo, Konrad Szocik, and Martin Braddock. "Why space colonization will be fully automated." Technological Forecasting and Social Change 143 (2019): 162–171.
3. Faisal, Asif, Md Kamruzzaman, Tan Yigitcanlar, and Graham Currie. "Understanding autonomous vehicles." Journal of Transport and Land Use 12, no. 1 (2019): 45–72.
4. Infield, David, and Leon Freris. Renewable energy in power systems. John Wiley & Sons (2020), ISBN: 978-1-118-64993-0, February 2020.
5. Pirandola, Stefano, Ulrik L. Andersen, Leonardo Banchi, Mario Berta, Darius Bunandar, Roger Colbeck, and Dirk Englund et al. "Advances in quantum cryptography." arXiv preprint arXiv:1906.01645 (2019).
6. Hou, Jianwei, Leilei Qu, and Wenchang Shi. "A survey on internet of things security from data perspectives." Computer Networks 148 (2019): 295–306.
7. Hassan, W.H. "Current research on Internet of Things (IoT) security: A survey." Computer Networks 148 (2019): 283–294.
8. Li, Jin, Yanyu Huang, Yu Wei, Siyi Lv, Zheli Liu, Changyu Dong, and Wenjing Lou. "Searchable symmetric encryption with forward search privacy." IEEE Transactions on Dependable and Secure Computing (2019). **DOI:** 10.1109/TDSC.2019.2894411, ISSN: 1545-5971.
9. Mistry, Ishan, Sudeep Tanwar, Sudhanshu Tyagi, and Neeraj Kumar. "Blockchain for 5G-enabled IoT for industrial automation: A systematic review, solutions, and challenges." Mechanical Systems and Signal Processing 135 (2020): 106382.
10. Patel, Vishwani, Fenil Khatiwala, Kaushal Shah, and Yashi Choksi. "A review on blockchain technology: Components, issues and challenges." In ICDSMLA 2019, pp. 1257–1262. Springer, Singapore (2020).
11. Wazid, Mohammad, Ashok Kumar Das, Sachin Shetty, and Minho Jo. "A tutorial and future research for building a blockchain-based secure communication scheme for internet of intelligent things." IEEE Access 8 (2020): 88700–88716.

12. Tseng, Lewis, Liwen Wong, Safa Otoum, Moayad Aloqaily, and Jalel Ben Othman. "Blockchain for managing heterogeneous internet of things: A perspective architecture." IEEE Network 34, no. 1 (2020): 16–23.
13. Jain, Anshul, and Tanya Singh. "Security challenges and solutions of IoT ecosystem." In Information and Communication Technology for Sustainable Development, pp. 259–270. Springer, Singapore (2020).
14. https://tykn.tech/identity-management-blockchain/
15. https://www2.deloitte.com/us/en/pages/public-sector/articles/blockchain-opportunities-for-health-care.html
16. https://hackernoon.com/blockchain-in-healthcare-opportunities-challenges-and-applications-d6b286da6e1f
17. https://www2.deloitte.com/us/en/pages/life-sciences-and-health-care/articles/blockchain-in-insurance.html
18. https://healthcareweekly.com/blockchain-in-healthcare-guide/
19. www.ibm.com/blogs/blockchain/category/blockchain-in-healthcare/
20. www.ibm.com/blogs/blockchain/2020/09/growing-the-veterinary-learning-credential-network-with-blockchain/
21. www.ibm.com/blogs/blockchain/2020/08/disrupting-veterinary-medicine-and-credentialing-through-blockchain/
22. www.ibm.com/blogs/blockchain/2020/06/blockchain-newsletter-for-june-covid-19-spurs-blockchain-innovation/
23. www.ibm.com/blogs/blockchain/2020/07/blockchain-newsletter-for-july-are-digital-health-passports-in-our-future/
24. https://blockchainhealthcaretoday.com/index.php/journal
25. https://tykn.tech/identity-management-blockchain/
26. https://consensys.net/blockchain-use-cases/finance/insurance/
27. https://trialsjournal.biomedcentral.com/articles/10.1186/s13063-017-2035-z
28. https://www2.deloitte.com/us/en/pages/consulting/articles/blockchain-in-clinical-trials-research-patient-data-donation.html
29. www.healthcareitnews.com/news/blockchain-use-case-healthcare-supply-chain-0
30. How it Works | MyDigiLife—we value your IDENTITY—an life ecosystem on Blockchain (assetchain.in)

7 Developing Sustainable Solutions for Waste Management in Smart Cities Using Blockchain

Jeyamala Chandrasekaran and A.M. Abirami

CONTENTS

7.1 INTRODUCTION

The demand for basic amenities such as clean environment, water, energy and infrastructure are increasing at a rapid rate because of ever increasing population. Citizens from rural areas are moving to urban areas for easy access to these basic amenities and for better employment opportunities, health care and other facilities. Around 1.5 million people are added to the global urban population per week. As a result, the demand for city spaces and infrastructure has to expand to accommodate the increasing urbanization needs. Urban challenges like inadequate water supply, inadequate power supply, poor services, high cost of living, increased pollution and increased traffic jams have forced governments to extend their vision on smart cities. Smart cities aim to improve operational efficiencies and provide better services to their citizens. The objective of smart cities is to develop the core infrastructure of a city to promoting decent and quality life for its citizens. Smart cities aim to provide a clean and sustainable environment with a prime focus on developing smart solutions. The smartness layer is added incrementally. There is no standard definition of smart cities because it varies from state to state and from country to country. The definition of smart city in India does not hold true for the smart city definition in Europe. However, the smart city in any urban ecosystem is focused on institutional, physical, socioeconomic and economic infrastructure. The prime focus is on the development of sustainable solutions for the betterment of human lives. The strategic plan for the development of smart cities includes a pan-city initiative in which at least one smart solution is applied city-wide, develop areas in a step-by-step manner, retrofitting, redevelopment and greenfield. The core infrastructure elements of a smart city include

- Adequate water supply
- Assured electricity supply
- Sanitation, including solid waste management
- Efficient urban mobility and public transport
- Affordable housing, especially for the poor
- Robust IT connectivity and digitalization
- Good governance, especially e-governance and citizen participation
- Sustainable environment
- Safety and security of citizens, particularly women, children and the elderly
- Health and education.

The mission of developing smart cities is to improve the standard of life of its citizens by promoting local area development and harnessing technology. In India, the

Smart Cities Mission was launched on 25 June 2015 with the aim of creating 100 smart cities. The objective, as defined by the Ministry of Housing and Urban Affairs (MoHUA), is 'to promote sustainable and inclusive cities that provide core infrastructure and give a decent quality of life to its citizens, a clean and sustainable environment and application of "Smart" Solutions'. Appropriate technologies are to be identified to produce smart outcomes. The major focus areas in development of smart solutions are

- E-governance and citizen service
- Urban mobility
- Energy management
- Water management
- Waste management
- Economy and employment
- Health
- Education
- Air quality
- Sanitation
- Transportation and mobility
- Open spaces
- Housing and inclusiveness
- Safety
- Intelligent government services

Blockchain technology seems to be a promising direction for many of the domains listed in smart cities. Blockchain can provide secure, transparent, efficient and resilient services in building a smart city, thereby increasing the productivity and economic growth of a nation. This chapter presents a detailed investigation on application of blockchain technology to waste management in smart cities/

7.2 WASTE MANAGEMENT SYSTEM

Waste management, otherwise called waste disposal, deals with procedures and policies for managing waste starting from generation to final disposal. Various subdomains of waste management are represented in Figure 7.1.

Waste management is a broad domain that deals with the generation, characterization, minimization, collection, separation, treatment and disposal of all types of wastes [1]. Waste can be solid, liquid and gas generated out of industrial, biological and household activities. The waste generated day by day is increasing in volume as well as toxicity. The World Bank estimates that the amount of municipal solid waste created annually will almost double by 2025. The waste generated has adverse effects on human health, climate and the environment. The increase in the volume of waste generated is attributed to the increased rate of urbanization and population in many of the developing countries. Various stages in waste management are represented in Figure 7.2.

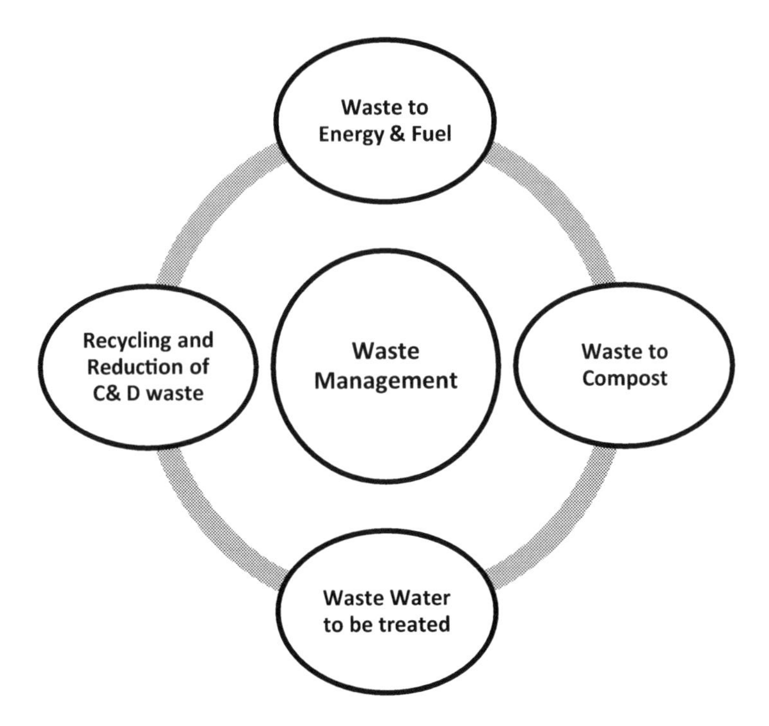

FIGURE 7.1 Sub domains of waste management.

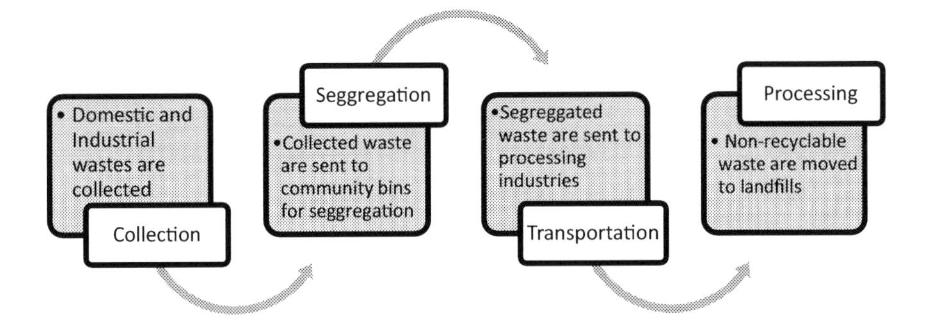

FIGURE 7.2 Stages in waste management.

Major challenges in the present-day waste management include:

- Increasing rate of waste generation

There has been a tremendous increase in the amount of waste generated. The World Bank report states that in the United States, around 250 million tons of wastes are generated annually. The per-capita waste generation is about 4.6 pounds per day.

The quantity of waste generated has tripled over the last 30 years. The per-capita waste disposal is expected to rise to 1.5 kg per day by 2025.

- Finding new disposal sites
- Landfill issue

Harmful greenhouse gases are generated in landfills. Nonrecyclable waste often contains toxic materials that can precipitate and leach out of the soil. The penetration in the ground eventually makes the groundwater to a hazardous. The natural aerobic process of decomposition is also affected leading to the production of highly inflammable greenhouse gases. These gases are highly lethal to living organisms

- Improper waste disposal and waste management methods
- Aspirational recycling

Before 2018, China processed more than half of the world's export of plastic, paper and metals. The recent regulations of China cut down 24 different categories of solid waste like scrap, plastic and mixed paper. This has resulted in an enormous amount of input, which cannot be handled by the recycling companies. As a result, many of the recyclable wastes are ending up the landfill.

- Lack of accountability
- Difficulty in tracking of e-waste
- Fraud and manipulation

The waste management sector has turned out to be a profit-making venture in recent years. Every entity involved in waste management has become a victim of corruption. A systematic looting takes place in all stages because the severity of the issue has not been understood. The current waste management systems do not have strict verification mechanisms to examine the reports submitted by the contractors on the quantity of wastes disposed. The reports are easily manipulated and payment frauds have increased.

In order to address the preceding challenges, the government has to maintain and administer a single platform that will monitor and track all waste management activities. Blockchain seems to be a viable solution in establishing a highly decentralized system that can track all events of waste management starting from waste generation to waste disposal. Blockchain, sometimes referred to as distributed ledger technology (DLT), makes the history of any digital asset unalterable and transparent through the use of decentralization and cryptographic hashing. The transparency in the blockchain enhances the accountability and easy tracking. By using blockchain, a new waste management application can be created that can trace the quantity of waste generated to when, where and how the wastes are disposed. The complete waste management process can be automated, time stamped and made immutable thereby preventing frauds and manipulation of records.

7.3 BLOCKCHAIN TECHNOLOGY

A blockchain is a peer-to-peer distributed ledger that is cryptographically secure, append-only, immutable and updatable only via consensus or agreement among peers. The concept of blockchain is purely based on decentralized and peer-to-peer systems. The record of all the transactions that happened is maintained in a peer-to-peer network. The transactions are stored in a number of interconnected systems. The technology has no single point of vulnerability as it is distributed and maintained across multiple nodes. A set of transactions are stored in a block. Each block is connected to the previous block through cryptographic protocols. Changes cannot be made to the earlier transactions, thus enabling tamper proof detection. The key innovation in blockchain is that all transactions can be validated and verified without the need for a trusted third party. The following characteristics of blockchain technology have made it feasible for application in waste management.

7.3.1 Decentralized System

The entire blockchain is built on the concept of decentralization. There is no single owner in the network. Every entity in the network maintains a copy of the ledger and every block. The use of trusted authorities and middlemen has been completely eliminated.

7.3.2 Transparency and Privacy

As every node in the blockchain network maintains a copy of the transactions and ledger, the transparency of the system has been greatly increased. The use of strong hashing algorithms makes the transactions in the blockchain tamper proof. Blockchain also provides the necessary privacy. The required data can be published and the rest of the data can remain private as decided by the users in the network.

7.3.3 Security

Blockchain uses strong cryptographic algorithms to provide security. The hash of every block is stored in the next block and thereby interconnects all the blocks in a network to form a chain. If a hacker tries to hack any one of the blocks, he has to hack all the blocks to gain access into the system, which is proved to be computationally infeasible.

7.3.4 Shared Ledger

Blockchain is said to be immutable and is an append-only distributed system, thereby eliminating the single point of failure.

7.3.5 Consensus

The transaction committed gets added to the blockchain only if all the parties verify its correctness, thus providing transparency.

7.3.6 Provenance

The entire history of an asset or a transaction is maintained in the blockchain.

7.3.7 Finality

Once a transaction is added to the blockchain, it can never be reverted.

The facilities offered by using blockchain for waste management are depicted in Figure 7.3.

7.4 ADVANTAGES IN USING BLOCKCHAIN FOR WASTE MANAGEMENT

Land size is large and there are more industries in the countries like America and Germany [2]. They follow complex waste management by accumulating the general solid wastes in different color-coded bins for nearly two weeks; then they may be sent for recycling or filled in open space dump yards. However, this luxury solution is not feasible with smaller countries or larger populations. When most of the land is filled, it may be covered with clay and plastics which in turn reduces oxygen-driven decomposition. It results in the increased generation of methane, which is 21-times more harmful than carbon dioxide. It creates a 'greenhouse effect', hence 'global warming'. Centralized deposit or landfills of waste impacts the environment by generating hazardous or harmful materials and pollutes either ground level resources or the atmosphere. Although adequate measures are taken by every country to reduce the quantity of waste produced, there exists an urge to manage them effectively and efficiently. Waste production is not equal across all houses or industries. Waste transportation needs to be managed accordingly from abundance to scarcity. Public–private partnerships can play a vital role in sharing the rides from waste collectors to recyclers. Waste tokens may be generated and rewarded to all parties involved in this process, which in turn creates a natural asset marketplace by enabling all stakeholders to participate in the climate value chain.

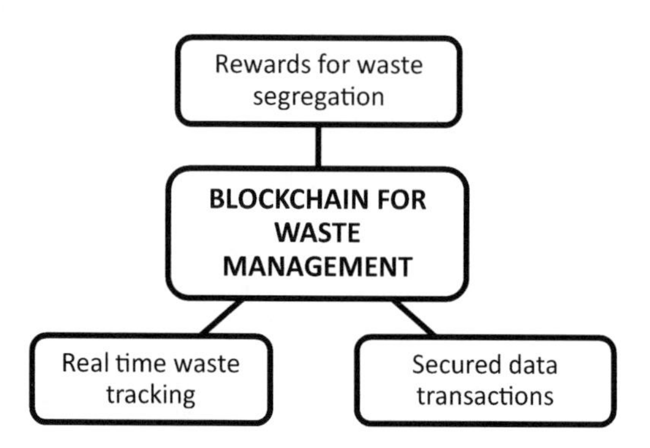

FIGURE 7.3 Blockchain-based waste management system.

IBM started applying blockchain principles and techniques to achieve some of the United Nation's Sustainable Development Goals [3]. IBM considers that the problems and challenges faced by the waste management and recycling industries can be alleviated by the blockchain technology. Use of blockchain technologies would not reduce the use of plastics by the citizens but move these wastes to recycling industries in an easier way. The new system would categorize the types of wastes by scanning the QR codes on the trash bags and identify the suitable processing alternatives where the wastes need to be sent. When a recycling company has a huge pileup of scrap materials, it can find its processing alternatives in the blockchain network and reach out to them. Thus, blockchain technology reduces one's burden and avoids landfilling by creating efficient communication between the parties involved in the recycling problem.

As of now, the accountability for handling wastes lies with the producers. The consumers are clearly given instructions in their packages. However, it is hard to track whether both consumers and producers follow the rules and regulations while disposing of wastes. If blockchain techniques are put in place for handling the waste supply chain, the tracking of items from production to disposal is easily traceable, which in turn results in the increased responsibility and accountability among consumers and producers. Thus, the blockchain technology provides efficient transaction and communication and helps in simplified waste supply chain management.

Population growth and the increased rate of urbanization in developing countries like India and Nigeria result in huge problems in managing solid wastes. It becomes crucial and the introduction of modern technology is needed for handling this issue. Blockchain technology with IoT sensors and RFID taggers would provide a sustainable solution to solid waste management. Also, it helps in real-time tracking of vehicles and the generation of data. With the use of IoT sensors, details like the amount of waste collected, amount of waste that can be recycled, amount of waste that is hazardous and nonbiodegradable can be obtained from the vehicle and these data can be updated in the blockchain network.

Generally, the recycling of waste is outsourced to third parties. By the use of the blockchain platform, all the stakeholders involved are traceable and waste management activities would be transparent. Issues like fraud and manipulation of data, loss of information, etc. can be handled by ensuring authenticity and data integrity [4].

Bitcoin, Ethereum and blockchain are different decentralized platforms where digital money exchange, storage and validation of data are feasible. Blockchain technology has many success stories as it maintains and manages data in the distributed environment. This non-centralized storage of data avoids the risk of data corruption and ensures its integrity. As of today, the uses of bitcoins become more relevant in social aspects. Application of blockchain technology in waste management has been adopted in the municipality of Sao Paulo state, Brazil [5]. It replaced the paper-based Green coin system with bitcoins involving all the stakeholders like volunteers, store owners and public agents.

Waste Token is a cryptocurrency that helps the stakeholders gain profit through transaction fees between sellers and buyers. These currencies are paid in the form of a digital wallet for the stated time period. Integrating blockchain technology and cryptocurrency establishes a new marketplace globally for environmental sustainability services and technologies [6]. This business model needs no investment and government rules and policies also would help in setting up these transactions. Proper and smart waste management has many advantages: carbon mitigation, climate neutral fuels, waste for energy and plastics accountability [7].

Cryptocurrencies like bitcoin are playing a vital role in waste management and the recycling process. Some producers are not keen whether the scrap materials reach the right destination for recycling or reusing purposes. The blockchain transformation enables easier maintenance of a digital ledger for tracking the materials, making the data available to the public and ensuring regulatory compliance [8]. Marine Transport International, a New Jersey-based freight forwarder, implemented a pilot study successfully [9]. Their blockchain-enabled tracking system links recycling suppliers, port operators and ocean carriers. Rubicon, an Atlanta-based startup company, provides cloud services for recycling and waste management.

Several countries like Canada, Columbia and the Philippines initiated a strategy to reduce the use of plastics; they started Plastic Bank, a global venture company for recycling plastic wastes [4]. The initiative says that the people who bring plastic wastes to the recycling center are given blockchain-secured digital tokens that can be used as food coupons or in phone charging units. Also, they can track how their deposits are converted into recyclable products.

As of today, there exists difficulty in tracking the workflow such as waste collection, waste transport and waste disposal or recycle. The studies on the present waste management system show that the loopholes in the process create contamination in the oceans, outbreak of epidemics, and global warming [10]. Blockchain's collaborative approach provides a unified solution to the country and public–private sectors' partnership. Hashcash in the Blockchain platform tries to streamline the processes and the stakeholders involved in it. It covers all three aspects like monitoring, analysis and management related to waste disposal and recycling processes. The use of smart contracts and digital ledgers avoids misuse of the data and ensures traceability from the collection to disposal or recycle.

The UK government has taken the initiatives among startups to make use of blockchain and machine learning techniques in the process of waste management [11]. The government feels that the absence of digital records on waste induces the irregularities in the system. The government has identified a few projects that uses IoT sensors and blockchain technology to enhance traceability. These projects work with the objectives of Zero Waste to Land, thereby increasing circular economy by recycling or reusing the waste.

7.5 CASE STUDIES ON BLOCKCHAIN FOR WASTE MANAGEMENT

7.5.1 Blockchain for Plastic Waste Collection and Reuse

Plastic Bank is an application for waste management supported by IBM technology. It uses blockchain technology to provide rewards for the exchange of plastic waste for goods in a secure manner. Blockchain is used to track the entire cycle of recycled plastic from collection, credit and compensation through delivery to companies for re-use. The technology has succeeded in developing trust among customers, plastic collectors and corporations without any centralized authority. Plastic Bank was opened in Haiti in 2014. Plastic Bank has extended to over 30 locations and involves 2000 community members. Plastic Bank has a record of collection of 3 million Kg of plastic which includes a collection of 1,205,396 Kg in 2019 alone. Plastic Bank supports an educated population with its environmental and social initiatives. The needs of the community are identified by having a valuable conversation and engagement with plastic collectors. This is facilitated by the Plastic Bank collection model. The model does not provide a one-size-fits-all solution but provides differentiated services based on the community needs. Although the plastic collectors could not read or write, literacy programs were organized by the Plastic Bank. Plastic collectors were provided training on using mobile apps and on using internet transactions. Thus, Plastic Banks serve not only as recycling depots but also as community centers. Collection of plastic waste is incentivized by providing a consistent, above-market rate for plastic waste. Many plastic collectors have escaped from poverty and now have the options to exchange the collected plastic for school tuition and health insurance. Blockchain technology has not only prevented the plastics from entering the oceans but also has created opportunities for generating income, goods and services.

7.5.2 Blockchain for Waste Management in Cargo Ships

A Slovenia-based company, called Carbon Offset Initiative (COI), implemented blockchain technology for monitoring and analysis of waste left by large cargo ships [12]. Applications of blockchain technology such as payment medium and quicker and easier communication enables it to be used in all domains. The technology is being used in ecology and sustainability sectors. The maintenance of data like routing information, fuel levels, spillages, etc. in the maritime industry is done manually, which in turn results in poor quality in waste management services.

The blockchain-based solution uses the Ethereum protocol and smart contracts to have secure and verifiable transactions and payments. All parameters are monitored and stored in immutable, publicly accessible storage and there would not be any data malfunctions. All stakeholders from ship operators to recycling centers would benefit from adopting this technology. Different advantages include cutting fuel costs and oil incineration and settlement between the parties. Various types of sensors and cameras installed in the vessels help in collection of data like temperature, humidity, fuel consumption and routes and get recorded in cloud data storage.

7.5.3 Artificial Intelligence-Based Blockchain for Recyclable Solid Waste

Klean Industries Inc., an industry in Vancouver, Canada, has implemented KleanLoop, a blockchain-based platform for the waste-to-energy process [13]. Having learnt its success rate during its beta testing, the KleanLoop would be rolled out and start collecting essential information on waste and recyclables from all the stakeholders. KleanLoop aims to provide a transparent and secure marketplace between the parties involved with a suitable rewarding system by KleanCoin, the digital token. This application tracks the waste in all of its states like creation, transportation, recycling, repurposing, conversion and reusing. This platform makes all the data and transactions publicly available to all stakeholders in the distributed ledger. In order to provide compliant and sustainable solutions, the data must be accurate. This blockchain platform ensures that all industries are compliant with the country's regulations and policies. KleanLoop also uses artificial intelligence (AI), machine learning and facial recognition techniques to have an efficient waste management process. These technologies enable faster processing of massive data, provide market insights and strengthen the ecosystem.

7.5.4 Blockchain for Incentivizing the Effective Use of Rural Wastes

This case study considered the Yitong system in Changzhi city, China, for recycling of rural wastes into the production of energy. With the help of the government, the company collects agricultural wastes like crop residue and rural solid wastes and converts it into energy, byproducts like animal feedstock and fertilizers. However, in the field study, the team observed that mostly women were involved in the cleaning and separating process using their hands; frequent rain damages the waste by making it wet; transporting wastes to dump yards is also not feasible as the roads are narrow; and farmers do not have adequate motivation in waste conversion to energy. These challenges make the wastes accumulate near the farms or they may be burnt by farmers in the fields, which hinders the local ecosystem. It causes greater impact and has a hazardous effect on the environment.

With blockchain technology, it becomes easier to build and manage data in the distributed environment. This blockchain model uses QR codes, phone apps and smart bins to have better efficiency, considering the incentives, economic affordability and feasibility. Digital coupons and cryptocurrencies are part of the incentives to farmers, and it establishes long-term trusting relationships between farmers and company managers [14].

Agricultural waste and nonbiodegradable waste are separated and collected in smart bins. The electrical energy and fertilizer produced are given back to the farmers, which in turn reduces the expenditure of the farmers. This smart waste management increases local employment and improves the local ecosystem.

7.5.5 Blockchain for Waste Management in the Oil and Gas Industry

Waste water and solid waste management is one of the challenging issues in the oil and gas industry. Proper disposal is needed for solid waste like drilling mud and drill cuttings [15]. Blockchain technology is useful here as an efficient database management platform. It provides optimal management strategies for large databases and transactions. It is difficult to obtain details like available source water and options for waste disposal and management for the operators in the oil and gas industry. The use of blockchain technology would avoid errors in reporting and enable easy monitoring. In this industry, there is a need for tracking source water transactions, waste water generation, shipments of waste water, waste water disposal quantities and pollutant concentrations of material sent for disposal.

Blockchain technology uses a shared and synchronized digital ledger to track the transactions between the parties. It facilitates accurate assessment of transactions in all blocks. Data in one block cannot be altered without altering it in the subsequent blocks, so that the data integrity is ensured. Implementation of this technology would reduce the tax amount, auditing cost and promote a positive attitude of the operators as they would be rewarded with incentives.

7.5.6 Blockchain for Waste Management in Railway Stations

The French railway management system effectively applied blockchain technology, developed by SNCF subsidiary Arep [16], for waste management in the stations,. Each station bin is assigned a block; the bin uses Bluetooth to collect data like the type of waste and quantity. The technology also assists in how the waste moves around, whether it goes to the recycling industry or for land disposal. This technology helps the government determine whether each trading company produces waste within the limit recommended by the European Emission Trading System for carbon quota. This blockchain-enabled tracking system ensures responsibility and accountability between the producers and the consumers. As each product would have QR codes embedded in it, when it is transferred to another person, new block is created for this transaction. Severe rules like heavy penalties would be given to the party who throws the item freely to the land or beach. Thus, blockchain would track the flow of goods from its production to disposal. However, there exist practical difficulties and challenges while imposing these strict regulations to the public.

7.5.7 Blockchain for Tire Recycling

Due to wear or irreparable damages, tires used on vehicles become no longer usable. The process of recycling of waste tires is called as tire recycling. Waste tires are generated in large volumes and include components that are ecologically harmful. Waste tires occupy a significant volume in landfills as they are highly durable and nonbiodegradable. Recycled tires can be used for fuels, in the cement manufacturing industry, in making rubber and in civil engineering projects. Recycling of tires can be optimized with the application of blockchain technology. Potential applications of blockchain include incentivizing recycling and waste sorting. The companies are aiming to use blockchain to incentivize better recycling and waste sorting. Application of blockchain ensures that the recyclable wastes do not end up in landfills. Because of digital tracking, fraudulent activities are prevented and deeper analysis of supply chains are made possible.

EMJAC is a blockchain solution company with a focus on employing state-of-the-art blockchain technology to provide transparency and traceability across the logistics flow of the recycled tire supply chain. EMJAC is built on the top of Ethereum blockchain using TRU technology. The transparency and traceability of tire recycling are enhanced by blockchain. It also ensures that the trapped energy and carbon in waste tires are fully recovered and reused. According to EMJAC, 100% of waste tires will be recycled into four valuable commodities, namely 45% refined diesel, 35% carbon black, 10% steel wires and 10% synthetic gas. EMJAC is committed to bringing this prototype of efficient tire recycling into a fully working product.

7.5.8 Blockchain for Waste Electrical and Electronic Equipment

Waste Electrical and Electronic Equipment (WEEE) includes old computers, televisions, laptops, refrigerators, air conditioners etc. The volume of such electrical and electronic waste is around 10 million tons per year. The WEEE export ban has been introduced because of harmful effects caused in local populations. Also, the devices contain raw materials that will become scarce in the near future.

In recycling of monitors in Italy, the tubular mercury lamp present in the monitor has to be removed. The identification of such monitors is a great challenge and involves human inspection. The second challenge is ensuring privacy to all of the stakeholders like component manufacturers, consumers, collection services and recycling companies. Manufacturers do not want to publish their trade secrets and consumers want to protect their possessions. Circularise has come up with a blockchain based solution wherein all the transactions between parties are secured (a block) and linked (the chain) to previous and subsequent transactions made by those parties [17]. No central database is maintained, and all participants share the common register. As all transactions are encrypted, the participants have privileges for viewing their own transactions only. Also, because of its cryptographical forte, these transactions cannot be changed. Encryption ensures that each participant sees only their own transactions, and those transactions cannot be changed. To maintain trade secrets, zero knowledge proofs are used. Certainty is ensured by simple questions but without revealing the secret. The longer it takes to search in the network, the more certain the answer.

7.5.9 AI-Based Blockchain for Improving the Circular Economy of Plastics

Plastic goods manufacturers prefer virgin polymers based on petrochemical feedstock and not recycled plastics [18]. This is because the quality of recycled plastics and the availability of plastics are not guaranteed. This chapter proposed a technique to separate plastics based on its types and enhance the reliability of information about recyclable plastics. The technique uses blockchain smart contracts and multisensor data-fusion algorithms with AI. In this model, the blockchain technology is effectively used for exchanging and validating information about supply, demand, specifications, bidding and offer prices. It results in improved resource efficiency and a novel profitable model for the circular economy of plastic waste.

Two types of smart contracts are maintained: one between the supplier of plastics segregator and a prospective buyer, the recycler; another one between the recycled plastic feedstock and the plastics good manufacturers. Two types of data, public and private, are maintained. The public data is open to evaluators for approval and the private data is available to sellers and buyers involved in the contract or transaction.

Private data includes price, segregation technique and origin of the waste. Buyers can validate the quality of materials and approve its quality compliance. These data would be made public based on the statements specified in the contracts, which are digitally signed by the parties involved with a time stamp.

7.5.10 Blockchain for Micromanaging Wastes from Households and Industries

Swachhcoin [9] is a blockchain-based technology for performing microlevel managing of wastes generated from households and industries. Swachhcoin aims to convert them into useful products that are of higher economic value. Waste management cannot be done by a group of people. It involves the willingness and contribution of a large group of the population. People are encouraged for proper waste disposal by monetary incentivization. Because of the intermixing of wastes, the economic value of the outputs obtained from the accumulated wastes has been significantly reduced. Swachhcoin uses SwBINS to eliminate the problem in segregation by using technologies like AI, big data, IoT and blockchain. Swachhcoin tries to eliminate the gap between the amount of waste generated and the amount of waste recycled. The tools and technology used in Swachhcoin are as follows:

7.5.10.1 SwATA

SwATA refers to the big data and its customization in the waste management industry. Frequent operations involved in the waste management industry are optimized to improve efficiency. Data collected in various stages of waste management are heterogeneous and are in different formats. The management of big data results in optimization of routes, effective maintenance of schedules and easy assessment of reports. Use of NoSQL provides support against processing unstructured inflowing data and protection against faulty prescriptive analytics. Swachhcoin implements prescriptive analysis to yield the maximum possible outcome with the best automation. The operations involving the data scientist are minimized to a great extent.

7.5.10.2 SwATEL

SwATEL stands for Swachh Adaptive Intelligence. The objective is to mimic the human brain and make decisions based on learnings from past experience or from in-house data. It enables communication and coordination between the machinery and equipment present inside the processing plants and bins. SwATEL promotes intelligence in the operating equipment to initiate a digital action. Every instruction is recorded on the open ledger blockchain, which may be kept private or public.

7.5.10.3 SwIoT

SwIoT stands for Swachh internet of things which refers to the customized application of IoT in waste management. Enabling of remote control, modification and adaptation of machineries in accordance with the requirements and instructions passed on by the controller are made possible with the help of IoT integration. A number of smart contracts are developed for Swachhcoin. All the transactions are initiated and recorded in a transparent, open ledger and in an immutable manner. Community-based consensus is adopted, wherein the generated revenue are transferred to the stakeholders in a transparent manner.

It can be inferred from the preceding applications that a set of processes are common for different applications of waste management. This section describes the possible research opening in each of the stages of waste management.

7.5.11 Blockchain for Managing E-Wastes

Waste management using blockchain techniques uses different strategies like digital tokens and digital ledgers. All transactions are traceable and verifiable as the digital ledger network is publicly available and arranged in chronological order. South Korea implemented blockchain technology for its eWaste management process [4]. People may adopt to dispose of wastes in the correct manner, if they are rewarded. Canada's PlasticBank model was implemented in Italy in another way. Miglianico, a town in Italy, uses the blockchain-based 'Pay as you Throw' (PAYT) model and RFID taggers in solid waste items. Also, the waste collectors wear wristbands that would scan wastes from residences. These preventive measures give 85% waste management efficiency to this town. Sometimes, the items to be recycled may not get updated, if a physical ledger is in use. These types of exploitations may be corrected if blockchain technology is used in the waste management process. All transactions are available in distributed ledger technology that provides tamperproof records.

7.5.12 Blockchain-Based Cyber-Physical System for Solid Waste Management

Blockchain and cyber-physical systems are used in the process of smart waste management [19]. The availability of the internet and smartphones makes this process easier. The system ensures verification and validation at each step and the involvement of all stakeholders, which in turn resulted in the success of the waste management process [19]. The municipal corporation, garbage collector and trash bin users are part of this system, using three components like public blockchain server, local server and mobile app for client interface. The status of bins is monitored continuously, and the block in the chain gets updated. A mobile app sends messages to the municipality worker when the threshold is reached in the bin. Blockchain is implemented with the use of ultrasound and humidity sensors to check the quantity filled in the bins, thereby notifying the sanitary workers.

7.6 Research Directions in the Application of Blockchain for Waste Management

7.6.1 Waste Generation

Waste generated by individual households and industries are being transferred to the bins manually. Development of a customized application that involves signup, location of the nearest bins, tracking of the waste level in the bins, registering complaints, reception of updates etc. according to the geographical region and culture is required.

7.6.2 Waste Segregation and Transfer to Collecting Agencies

The scope for automation in waste segregation involves identification of the waste contributor via a QR code mechanism, measurement of the quantity of the waste, evaluation of the quality of the waste and incentivizing the waste generator. Application of AI and deep learning techniques will improve the overall efficiency of the segregation process. There is a tremendous scope for research and development in selection and customization of classification algorithms for waste segregation. In case a bin needs to be emptied, integration with IoT can send alerts and notifications automatically to the officials for actions. The scope for optimization in transportation of wastes to the industry can be explored. Route optimization can be customized by application of sophisticated algorithms.

7.6.3 Transportation

There is a large research scope for use of machine learning algorithms for customizing routes and generating alerts to the transportation team for segregated waste collection. Also, machine learning algorithms can be used to predict the expected time of arrival of the next scheduled waste generation from the past data.

7.6.4 Waste Processing

Many research problems can be identified in the stage of waste recycling at the industry. Efficient smart contracts are to be developed to record all the transactions in the open and distributed ledger. Researches to customize the smart contracts according to the governing rules and demographic locations are required. Techniques for automation of waste recycling with minimal intervention of humans are to be evolved. Robots can be designed to handle the recycling process, thereby preventing humans from operating with toxic wastes.

7.7 CONCLUSION

The entire waste management industry can be revolutionized by application of blockchain technology. The use of blockchain technology enables the government to track and monitor waste from its generation to its disposal. Different steps like segregation, transportation, recycling, disposal and analysis of waste data in the waste management process can be simplified and made transparent to the public. The application of blockchain technology, with a decentralized architecture, in waste management resolves the complexity of handling large volumes of data and large numbers of stakeholders and ensures a sustainable solution for the development of smart cities. Adoption of blockchain for waste management coordinates producers, importers, retailers and recyclers [20]. Smart contracts also balance the organized and unorganized sectors with increased transparency by bringing all the stakeholders into the same blockchain platform. The technique includes the waste collection center as well as recycling units. The method increases the interactions between the stakeholders and ensures no inappropriate transactions happen, thereby establishing trust between the parties. This system also gives incentives or penalties for the appropriate and inappropriate actions by the parties, as defined in the contracts

REFERENCES

1. www.allerin.com/blog/revolutionizing-waste-management-with-blockchain-technology
2. https://medium.com/@johnsekhon/applying-blockchain-principles-to-waste-management-8e200a2da521
3. www.ibm.com/blogs/blockchain/2019/08/revolutionizing-the-waste-supply-chain-blockchain-for-social-good/
4. https://blockchain.news/analysis/how-blockchain-is-prompting-innovations-in-waste-management
5. Franca, A.S.L., Amato Neto, J., Gonclaves, R.F., and Almeida, C.M.V.B. 2019. Proposing the Use of Blockchain to Improve the Solid Waste Management in Small Municipalities. *Journal of Cleaner Production*, no. 244.
6. www.newsbtc.com/press-releases/how-blockchain-will-disrupt-the-waste-management-industry/
7. Zhang, David. 2019. Application of Blockchain Technology in Incentivizing Efficient Use of Rural Wastes: A Case Study on Yitong System. Proceedings of 10th International Conference on Applied Energy (ICAE2018), Hong Kong, Energy Procedia, no. 158, 6707–6714.
8. Ridda Laouar, Mohammed, Zaineb Touati Hanad, and Sean Eom. 2019. Towards Blockchain-Based Urban Planning: Application for Waste Collection Management. Proceedings of the 9th International Conference on Information Systems and Technologies. 1–6 https://doi.org/10.1145/3361570.3361619
9. https://hackernoon.com/smart-waste-management-and-blockchain-technology-887a8a185357
10. https://markets.businessinsider.com/news/stocks/hashcash-to-help-enterprises-with-blockchain-based-waste-management-platform-1028743216#
11. www.edie.net/news/5/Government-invests-in-blockchain-to-boost-waste-management-transparency/
12. www.globenewswire.com/news-release/2019/11/18/1948511/0/en/Can-Blockchain-Technology-Help-Clean-The-Oceans-And-Improve-Waste-Management-Practices.html
13. www.solidwastemag.com/product/decentralized-ai-blockchain-app-for-recyclables/
14. www.leewayhertz.com/blockchain-waste-management/
15. www.resourcesmag.org/archives/could-blockchain-technology-improve-water-waste-water-and-solid-waste-management-oil-and-gas-industry
16. https://theconversation.com/a-rubbish-idea-how-blockchains-could-tackle-the-worlds-waste-problem-94457
17. www.tudelft.nl/en/delft-outlook/articles/circularise-uses-blockchain-technology-to-trace-raw-materials/
18. Chidepatil, Aditya, Prabhleen Bindra, Devyani Kulkarni, Mustafa Qasi, Meghana Kshirsagar, and Krishnasamy Sankaran, 2019. From Trash to Cash: How Blockchain and Multi-Sensor-Driven Artificial Intelligence Can Transform Circular Ecnomy of Plastic Waste? *Administrative Sciences*, vol. 10, no. 23. doi:10.3390/admsci10020023.

19. Thada, Ayush, Uday Karan Kapur, Saif Gazali, Nikhil Sachdeva, and S. Shridevi, 2019. Custom Block Chain Based Cyber Physical System for Solid Waste Management. *Procedia Computer Science*, no. 165, 41–49.
20. Gupta, Neha and Punam Bedi. 2018. E-waste Management Using Blockchain based Smart Contracts. Proceedings of International Conference on Advances in Computing, Communications, and Informatics (ICACCI 2018), DOI: 10.1109/ICACCI.2018.8554912
21. www.nasdaq.com/articles/reducing-waste-introducing-garbage-collection-blockchain-technology-2018-05-08

8 Blockchain-Based Smart Supply Chain Management System

M. Vivek Anand and S. Vijayalakshmi

CONTENTS

8.1 INTRODUCTION

Supply chain management (SCM) is a process of delivering the goods from the manufacturer to the customer with various stakeholders. This is the process that transforms raw materials into the final product. Some of the current challenges and issues in SCM are causing problems for each SCM stakeholder at various levels. Maintenance of transparency is important from the process of manufacturing to delivery in SCM. SCM with central server monitoring is not trusted in a business in which any intruder can change the data in the server while it is connected to the internet. The customer won't believe in the quality without seeing the process of manufacturing (or) packaging goods with standards and procedures. Even though some of the standards are provided by various organizations such as ISO, the customer isn't able to believe in the standards. If any authority is giving rights to the poor-quality products, then whom can we trust? Manufacturers are expected to disclose the source of their product because the average customer is unable to know the social responsibility of the supplier.

The current SCM is not shared and does not have transparent data of the raw material, quality of the product, transportation, and delivery of goods. Identifying compliance violations is a difficult task for each stakeholder. If a product is delivered with a defect, the customer will not know where the defect originated. Because of different levels of the process in SCM, each stakeholder will blame others for the defect that occurred in the product. There is no monitoring system to monitor the activities of each stakeholder. If a central monitoring system is performing the monitoring process, trust in the centralized system is again questionable. In SCM, agreement between each stakeholder by seeing each process at the site will take a week or sometimes a month to come up with a consensus agreement. In the business world, it is not impossible for business people to spend more time in agreement and consensus for delivering the product. The disagreement between the stakeholders involved in each phase of the SCM will lead to the whole business falling in the market. Consensus among all stakeholders or entities in SCM has to be implemented in a shared, transparent environment. Maintenance of each product is essential after delivering the goods, and tracking of the product is essential for future maintenance because many customers forget to service their household products.

The current SCM is not well-equipped with the tracking the details of the product. A record of the particular product data should be maintained with a date of service, next service date, etc. These data should be in a sharable environment for all the users in the supply chain. Product recall cases have to be minimized by ensuring the safety and quality of the products. The environment is required to transparently show the process such as the selection of raw materials, procedures, and standards used for manufacturing, testing, and proving of the projects. A full report of the findings and recommendations should be specified in an open platform and automatic ordering of the product has to be enforced if it satisfies the specific condition. For the average customer, it is very difficult to survey the product quality, the procedures used for the production of the product and that the product has the right price and validity. Customers should know every detail in a shared environment and agreement between each stakeholder should be available transparently in a network. There should be a transaction environment where the transaction is trustworthy without requiring a third party. A smart contract has to be implemented in the supply chain where the agreement procedures are written as a program that should be shared with all the stakeholders in any block.

Figure 8.1 shows the supply chain without blockchain; the customer has to know what raw material is used for the product, what are all the procedures and standards used to produce the product, how it is transported, where it is stored, where it is delivered—all this information should be shared information on the network. Business people will experience a loss if they are not able to find where the problem occurred in SCM. Every stakeholder has to track the status of the product at any time. To avoid this problem, the suggestion from every scientist's research is to go to blockchain.

8.2 BLOCKCHAIN IN SUPPLY CHAIN COMMUNICATION

The supply chain is not only moving the product from the manufacturer to the end user, but it also goes backward if the customer wants to return the product. If the supply chain is properly implemented, the outcome will be increased sales, decreased fraud and overhead costs, reduced complexity in manufacturing, and also improved quality of the product.

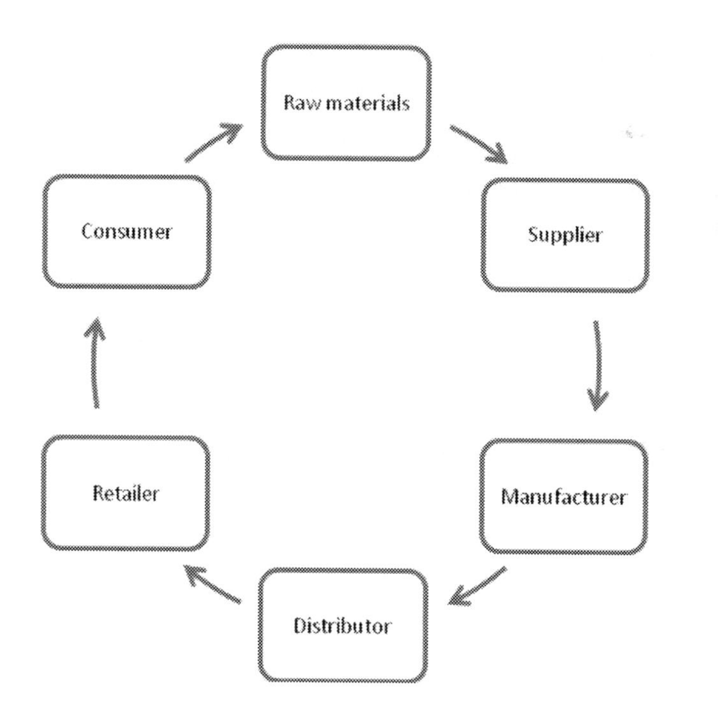

FIGURE 8.1 The current supply chain system.

8.2.1 Challenges in the Current Supply Chain

The following things are considered to be the challenges of the current supply chain:

8.2.1.1 Tracking

The consumer of the product usually doesn't know about the product—where does it is come from? The US suffered a major *E. coli* outbreak in October 2006. Spinach was the source of the *E. coli*, and this was found after 199 people were affected. Of the 199, there were 22 children below 5 years old. Thirty-one of the 199 developed kidney failure (hemolytic-uremic syndrome) and 3 people died. Finally, the source of spinach was found from the local market. The current supply chain is difficult to trace because there is no transparency in SCM.

8.2.1.2 Trust

Trust is questionable in SCM if there is no tracking mechanism. Trusting all of the participants is required when we use the supply chain even though it is not a trusted environment. In some cases, without considering the quality of the product and trusting the participants, someone can order the products for their business. There is no assurance that the trusted participants will provide a good quality product. In some cases, the order can be done unknowingly without checking the quality and knowledge about the product, such as if the raw materials for the new products are in demand or rare in the market. The quality of the products can be compromised by a bribe from any stakeholder in the supply chain. This will affect the trust among the participants in SCM. Our current SCM is not scalable such as the participants can leave or join at any time without maintaining the agreement process.

8.2.1.3 Cost

The procurement cost increases if we run a larger company and aren't able to visit the supplier to buy all the components. The procurement cost involves training costs, salaries, and bonuses for the employees. The transportation cost relies on time such as a decrease in time will cost more. A time compromise is required to reduce the cost. The inventory cost involves the cost of the place where the products are going to be stored and the guard for the inventories. The inventory cost will include the cost of lost, stolen, damaged, and expired products. The quality cost involves the money spent on quality. The quality cost will include the cost for a quality check, quality trainers, and the cost for maintaining the quality. The cost will increase because of globalization where each product is produced in different places and organizing those products in a single place will lead to the transportation cost.

8.2.2 The Steps toward Solving the Challenges of the Current Supply Chain

The current supply chain requires

1. Transparency among the participants
2. Avoidance of single third-party trust
3. Distributed data

4. Security
5. Scalability

8.2.2.1 Transparency

Transparency is required in SCM to find the process involved at each participant's site to maintain the integrity of the business. The customer has to know about the product, where it is coming from and what are the procedures and standards applied to the process of making the product. The transparency from the supplier of raw material to the end user has to be implemented to trust the process and procedures involved in making the product. Transparency has to be implemented in the movement of the product to track the product and find defective products in transport. Transparency in the network will make all the participants agree upon the process of the supply chain. Transparency will remove the activities of blaming other participants in the SCM.

8.2.2.2 Avoid Single Third-Party Trust

The current supply chain relies on a trusted third party for processing the data from the production to the customer. The storage of data is maintained in the centralized server that is vulnerable to data breaches. If the entire supply chain depends on the centralized server for the data or an agreement, the trust in the third party is questionable. Usually in SCM lot of IoT devices are involved in performing the entire process. In such a case, the IoT depends on a single centralized server that will make the process questionable. There is no assurance that the third party will not cheat. The IoT devices that are involved in SCM usually have three kinds of architectures such as client-server architecture, cloud architecture, and fog computing architecture. The vulnerability is there in all these architectures because of the central maintenance of the data.

8.2.2.3 Distributed Data

The data available in the supply chain should be shared and distributed across the network. This will ensure transparency and that every participant can view the process of what is happening in the SCM. The data should be geographically distributed across all the peers in the network to allow data access at every location. The storage of data in a single location will cause a failure when a threat enters a malfunction.

8.2.2.4 Security

The current supply chain does not assure security because of the centralized authority networks. If the centralized data system is affected by any malware, the entire system will be compromised. To ensure security, security measures have to be implemented. The cost of security also increases while securing the network. The updates for the security tools have to be done to protect the data in the supply chain. DDOS attacks are not allowing the peers to contact the server. If the peers do not receive any data from the server, no progress will be made in the supply chain. The centralized systems are vulnerable to a security breach. The security has to be assured with a new architecture for the supply chain.

8.2.2.5 Scalability

In the current supply chain, a node can add or leave the system at any time. The scalability is not maintained in the supply chain because of the various devices involved in communication. A single point failure can occur if any system relies on other nodes. The single point failure has to be avoided and scalability of the network has to be improved. The new architecture has to be implemented to avoid single-point failure.

8.2.3 Blockchain Introduction

Blockchain is most suitable for the supply chain because its properties overcome the challenges to the supply chain. Blockchain is a distributed, decentralized, peer-to-peer, shared, transparent ledger. Every node in the network is shared and the data are available to all the peers in the network. The special functionality of the blockchain is that it does not rely on a central server or a third party because its transactions are performed between the peers themselves. Every node is responsible for their transaction, and it is not dependent on a trusted third party. The security in a blockchain network is ensured by a tamperproof feature. The data shared in a single node will be available to all the peers in the network in the public blockchain. Tampering with the data in a blockchain ledger is highly impossible because, to make a change, the node wants to get approval from >50% of the network. Blockchain is viewed as a disruptive technology after the Internet came into existence in the19th century.

Blockchain was introduced by an anonymous person, Satoshi Nakamoto, in 2018 [1] in a white paper published as "The peer-to-peer electronic cash system". The first application of blockchain was the digital cryptocurrency called bitcoin. Bitcoin is the first cryptocurrency in the world that was designed using the concept of blockchain. In the bitcoin network, a transaction can be done without any third party such as a bank. If person A wants to send money to person B, person A broadcasts the message to the entire network of all the peers. Some of the peers called miners, will verify the transaction that is sent by person A. The miner verifies the transaction details such as sender, receiver, transaction amount, and required balance to send the transaction. The peers in the network called miners have the entire blockchain data in the network as a full node. The peer of a full node is capable of storing the entire blockchain ledger, and it can perform the mining process.

The miners of the network compete with each other to verify the transaction and add the transaction as a valid transaction to the permanent ledger called blockchain, then the transaction is completed.

Figure 8.2 depicts the working principle of the blockchain in the bitcoin network. The miner who verifies the transaction and adds the blocks of the transaction to the previous block of the blockchain will get the reward of bitcoins. Miners will do the mining genuinely to get the bitcoin as a reward (currently 6.25 bitcoins as a reward). To compete with each other, miners will create a puzzle and who solves the puzzle first is able to add the block of the transaction to the blockchain. Blockchain maintains its security through the cryptographic hashing technique. The hashing technique is an input given to the hashing algorithm that will generate the irreversible

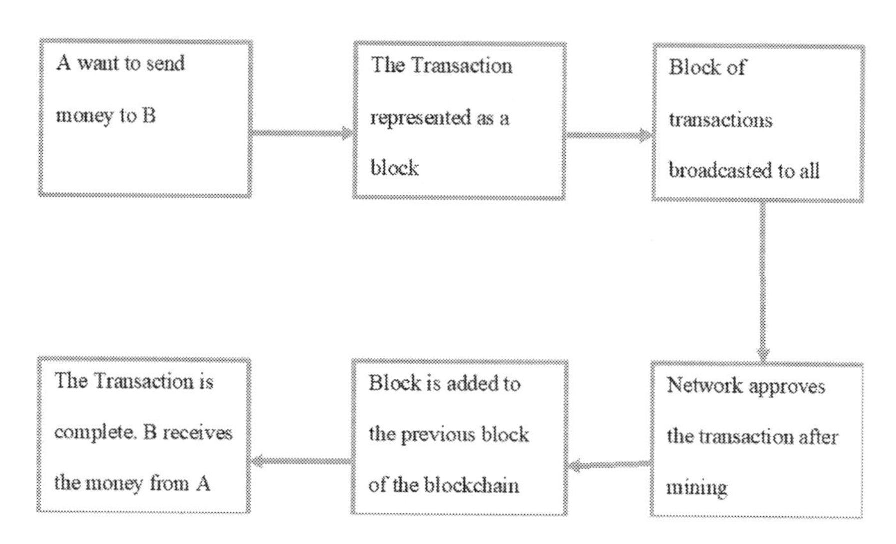

FIGURE 8.2 Blockchain working principle in the bitcoin network.

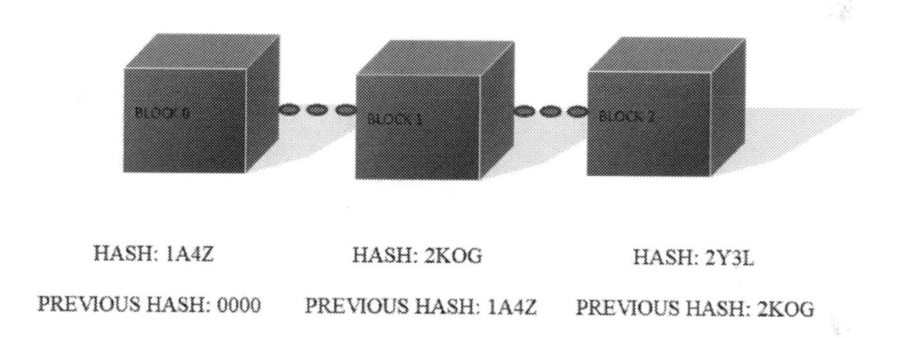

FIGURE 8.3 Chain of blocks with hash and previous hash.

hash code of alphanumerical values. SHA-256 provides the irreversible hash value with 256 character output. The same hash value will not be generated with different inputs; if that happens it is called a collision. The SHA-256 hashing algorithm is considered a secure hash algorithm without collision.

The hash value will be generated for each transaction in a block with the data such as sender, receiver, amounts to be sent, time, etc. More transactions can be added in a single block and for every transaction a hash value will be generated.

The hash of every transaction in a block is converted as a root hash to form a Merkle tree (Figure 8.5). The root hash value of the block will be added to the previous hash in the blockchain (Figure 8.3 and Figure 8.4).

Every block consists of the previous hash and the current generated hash for the block. Hashing links all the blocks by maintaining the previous hash in a current block. If any data of the transaction is changed that will affect the preceding hash value of the entire blockchain and it becomes invalid. Tampering with the data is highly

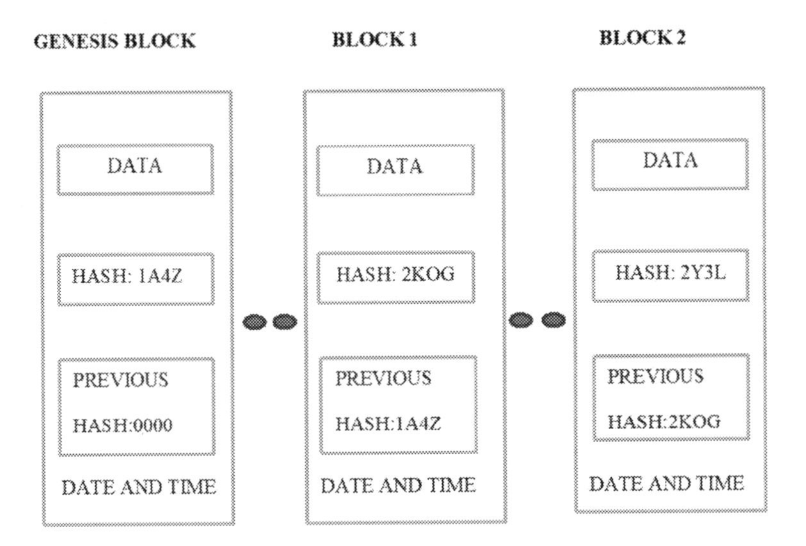

FIGURE 8.4 Chain of blocks with genesis block.

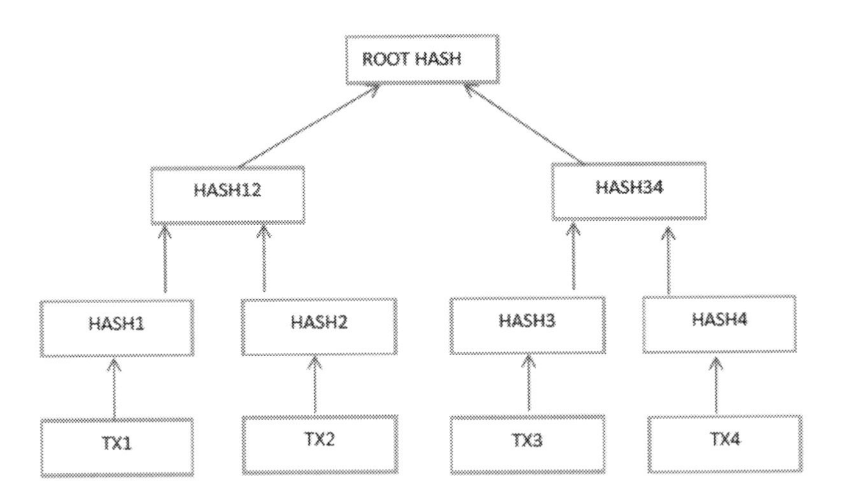

FIGURE 8.5 Merkle tree.

impossible due to the previous hash value associated with each block. Blockchain is cryptographically secure because of asymmetric cryptography in which the public key is used to send the data and the private key is used to receive the data.

8.2.4 Features of Blockchain to Adapt for the Supply Chain

The security of the blockchain is stable of its tamper-proof property, and the blockchain has been introduced in many industries and applications including SCM.

Blockchain technology will impact much of the supply chain by its useful properties. The expectation is high when blockchain is integrated into SCM because of its excellent functionality. The SCM system will have the features of transparency, robustness, and end-to-end traceability from integration of blockchain. Industries are finding ways to adapt blockchain to the supply chain. Blockchain with SCM will eliminate the inefficiencies of the current system.

8.2.4.1 Decentralization

Blockchain is a decentralized network where every node is responsible for safeguarding their data. Transactions are not performed through architecture such as banks, agencies, etc. Control of the network is divided among all the peers. The data is distributed throughout the network in all the peers. Tampering with the data is highly impossible because changes must be approved by most of the nodes in the network. IoT also requires decentralization to avoid the maintenance and updates of the centralized servers.

8.2.4.2 P2P Exchanges

Ina peer-to-peer network, every node has a connection with all the other nodes in the network. IoT also requires connectivity with each node and passing of information through gateways.

8.2.4.3 Payment System

The transaction is performed without third party assistance such as banks, agency, etc. No transaction fees or minimum transaction fees will be paid for the transaction. The transactions are carried out by the miners. Miners will collect all the transactions and verify the correctness of the transaction and finally some of the transactions are taken as a block that will be inserted with the previous block if the cryptographic hash is found. Miners will find the matching cryptographic hash by using computational resources. The miners will get bitcoins as a reward after adding the block of transactions in the exercises. Generally, more people are involved in mining, and they will share their amount after receiving the bitcoin as a reward. Miners will have the transaction carried by the users through the network as a broadcast message to all the miners. The miners who wish to take over their transaction will respond and do the transaction to get the reward.

8.2.4.4 Distributed System

The distributed system provides maintenance of a common ledger and the works that are assigned to the network will be distributed among all the nodes in the network.

8.2.4.5 Micro Transaction Collection

In the blockchain, every transaction will be noted in a common ledger to maintain traceability. IoT requires this feature of traceability to maintain the records properly in order to function.

8.2.4.6 Data transparency

Data stored in the blockchain is visible to all the nodes in a network. If a transaction is added to the blockchain ledger after finding the valid matching cryptographic hash, the data will be visible to all the nodes. IoT requires transparency to monitor the activities that are happening in different places. Based on the updated data, the activities may be triggered in the network.

8.2.4.7 Anonymity

Anonymity is a feature of blockchain in which the user can hide their identity and do a transaction in a network. The user of a blockchain network has a public key and a private key rather than no identity provided. IoT with blockchain will provide an environment that can hide the user identity and perform data sharing without identity. The public key is used as a transaction address. Every user can create a different transaction address for each transaction.

8.2.4.8 Cryptography

Cryptography is involved in blockchain as a public key and a private key. The public key is a transaction address where when the user wants to make a transaction, they can pick the transaction address from the wallet. The public key can be broadcast by the receiver in the network to receive the transaction from the sender. The public key is like a mail ID that should be available to the sender to initiate the transaction. If the transaction is finished then the user can give their public key to receive the transaction. Miners verify the transactions before adding them into the blockchain, and they will find the cryptographic hash to add the block into the previous block. If the hash value of the block is matched with the previous block, then it will be added to the previous block. If any miner who is trying to validate the wrong hash values the entire chain of the block becomes invalid.

Figure 8.6 shows the asymmetric key cryptography. A single key is used for both the sender and receiver in symmetric key cryptography. The disadvantage of the symmetric key is that each time the sender requires a different key for a transaction to a different receiver. In asymmetric key cryptography, different keys are used for sending and receiving the data such as the public key used to do the transaction by the sender and the private key used for receiving the transaction by the receiver.

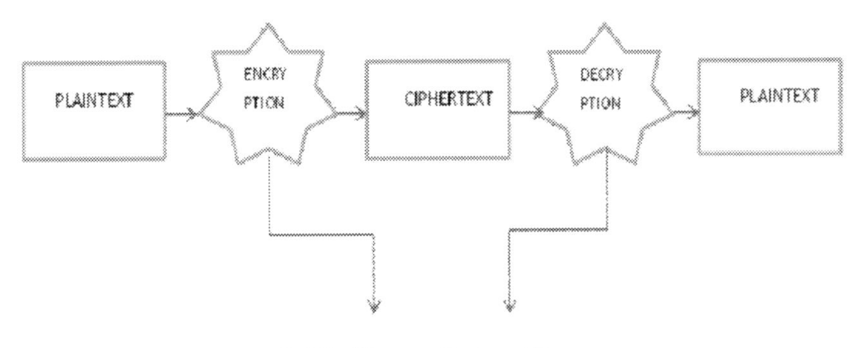

FIGURE 8.6 Asymmetric key encryption.

8.2.4.9 Consensus

The consensus is an agreement upon the valid transaction when the miners get the valid cryptographic hash. Every miner in the network will try to validate the cryptographic hash. If the hash is valid, miners will get a reward as bitcoins. To validate the hash, other miners will do the verification. The valid cryptographic hash is not acceptable if >50% of the miners do not agree on the transaction.

8.2.4.10 No Transaction Fees

In the blockchain, no transaction fees [2] will be deducted because after verifying the transaction and the transaction block is added into the network, the miner will get the bitcoin as a reward. The reward for the miners will be automatically credited to the corresponding address of the miner who solved the cryptographic puzzle. Although there were challenges that were raised while integrating blockchain and IoT, blockchain is inevitable nowadays in the IoT industry to make their applications secure. Finally, blockchain and the IoT will lead to a healthy environment in industry and developers, providing service to the people with ensured security.

Other than these advantages of blockchain, various research work is going on regarding integrating blockchain into various fields [3] and the size reduction and scalability of blockchain [4–6].

8.2.5 Blockchain Integration in the Supply Chain

Blockchain is called a disruptive technology because of its functionality and its applications. Internet was a disruptive technology in the 1990s; after the internet, many technologies developed and enormous applications came into existence. Blockchain is also considered to be disruptive where the security is maintained in the applications. Even though the first application of the blockchain in bitcoins faced some crucial problems of illegal transactions and money laundering, the core of the blockchain remains secure. Blockchain is applied in bitcoins without any third party such as banks. The transactions are done through the proper verification and consensus by most of the peers in the network. The miners who are persons in the peers will verify the transactions and add the transaction into the blockchain by solving puzzles. The miner will get the reward of bitcoins after the transaction is successfully added to the network. The bitcoin network is anonymous, and the person doing the transaction will only be identified by their public key as a transaction address. This feature of bitcoin led to illegal transactions in payment systems such as bitcoins. Bitcoin has many problems but the blockchain data cannot be tampered with. The security feature of the blockchain is assured by the hashing technique. The blockchain is now integrated into various fields of application. One of the applications is the supply chain where the blockchain is integrated with IoT. To integrate blockchain in SCM, some projects are using RFID and other sensors. The RFID uses electromagnetic fields to identify the objects. It is used to identify the product's starting location, where it is located at present, who has manufactured it, and who has handled it. The products are associated with data so that they can be easily tracked, and it allows the information to be instantly verified. Blockchain guarantees the traceability, authenticity, and reliability of the data, and smart contracts are possible by defining the programs in the business as a condition to be followed by all stakeholders of the business.

Blockchain can not only be used in the public sector, it can also be used in private sectors that may require some authority to take care of the blockchain data. The supply chain requires multiple participants with a private permission blockchain. In the supply chain, registrars provide unique identities to the participants appropriately. Organizers define the standards and insist that others maintain them. Certifiers provide certification to the user based on their role in the supply chain network. Manufacturers, retailers, and customers are the actors to maintain the system's trust [7]. The authentication and authorization have to be set for each participant to avoid discrepancies in the blockchain network. To have digital access, every product may have digital authenticity. The data must be collected for an analysis to be successfully run on the business process, where the data required includes the type of product, the status of the product, and the standards to be implemented for a product [8]. A tag is attached to a product to show the details about the product with a virtual identity in the blockchain [9]. The actor who wants to change the details of the product must have the permission to do so. The permission should be allowed to the actors by satisfying the smart contract rules written as a program with agreement and consensus. A digital signature is required for a product that is transferred between the actors or the participants to meet such a smart contract agreement requirement. The details will be updated on the blockchain only after agreed by the concerned actors with consensus. The data of the transactions will be automatically updated when a change happens in the data with approval from the actors with consensus [9].

The blockchain in the field of the supply chain provides transparency over the network and the details about the product can be maintained such as the type of the product, the quantity of the product, the quality of the product, and the location at which it is located, and details of the owner.

Figure 8.7 shows the advantages of blockchain when integrated into the supply chain. All participants of the supply chain are connected to the blockchain to receive immediate data from the blockchain. The recording of all the information with transparency removes the involvement of a central organization in the trusted environment. Even though a central third party is involved in much of our daily life, the trust in the central party is questionable. The third party can be vulnerable or the third party can do illegal activities. If the transaction is changed by the bank employee in a banking environment, the customer will complain about the data tampering to whom? Instead of relying on the trusted third party, the blockchain records information in ledgers as transactions. The material information will easily flow in the supply chain when it is inbuilt with blockchain. This environment will increase the sale of the product because of the transparent data about the product. The customer can know all the information about the product, where it was manufactured, who is the supplier, who is doing the transportation, whether the product is safe during transport, who is the retailer, and the details of the product. Production of products, not necessarily on material characteristics, also relies more heavily on knowledge of the product and information [10]. The customer's trust will be increased when the customer tracks the product to obtain detailed information on the product [11].

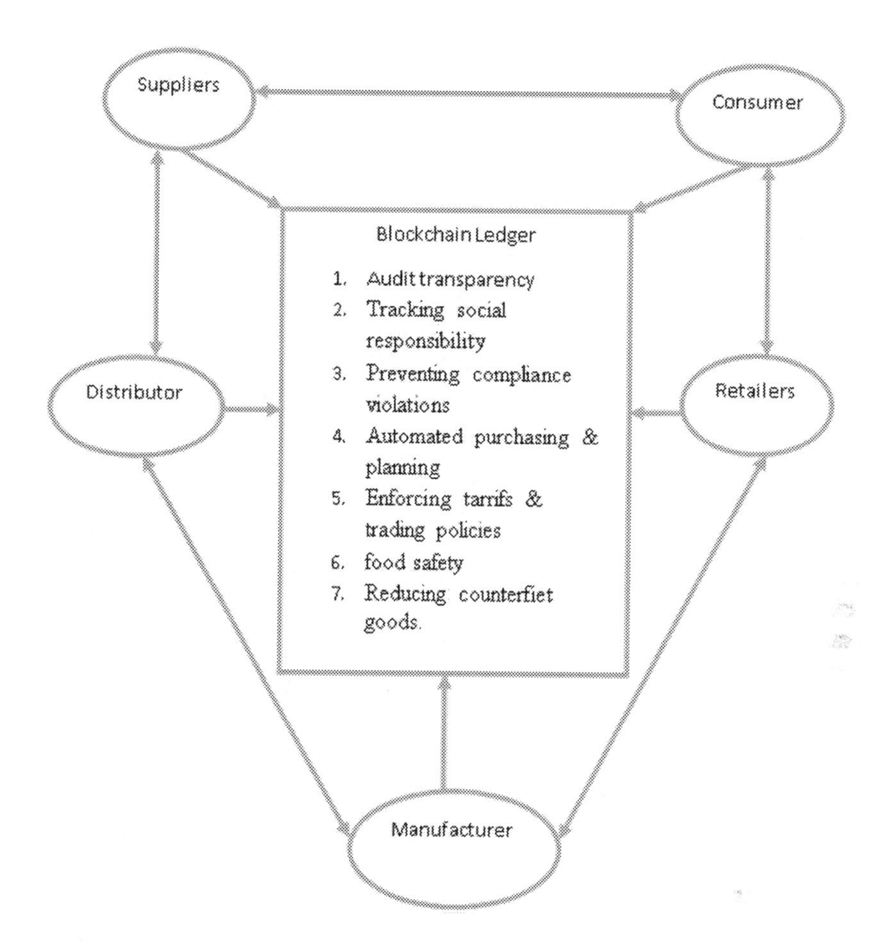

FIGURE 8.7 Blockchain integration in the supply chain.

Smart contracts are the rules that can be written as a program in the blockchain to agree with the business. Ethereum blockchain is used for smart contracts and the Solidity programs can be used to do the agreement. Smart contracts improve data sharing between the participants. A smart contract provides the certification and approval for the participants to access. No actors are permitted to change the rules without a consensus agreement. Smart contracts used for legal agreements and update the automated record of the goods with were they bought, how much was sold, and were they delivered to the user. Smart contracts are very useful to improve the business process of the supply chain [12]. Blockchain impacts every process of the supply chain, such as product management and financial transactions [13] in the supply chain. No intermediary is allowed in blockchain for financial transactions including payment systems, stock market exchanges, and any other money transfer services [14]. Blockchain is the best solution for financial transactions, which was proved by its first application (bitcoins).

8.2.6 Benefits of Supply Chain with Blockchain

The important benefit of the blockchain is transparency, which helps to reduce the time when the goods are stuck in any place in the supply chain. The possibility of misplacement is low because the product is tracked in real time with the blockchain ledger. Scalability is possible through blockchain in which any large database is accessible to a distributed environment from multiple geographical locations in the world. The security and standards will be maintained because of blockchain features such as cryptographic process and tamperproof nature, and it can be customized according to the data. Blockchain is not only used as a public network, but it can also be used as a private network with some permission given to the authorized person. Blockchain can be used to connect multiple ledgers with the appropriate consensus technique to maintain integrity among various participants. The integrity of the system is maintained in the blockchain; this helps the participants eliminate fraud. Some of the other benefits are listed following.

8.2.6.1 Increase Consumer and Partner Trust

The feature of transparency allows the customer to see the details about the product and the data about the product from manufacturing to delivery. This increases the trust among the customers, and it will impact the sale of the product. It also brings trust among the partners or the stakeholders in the supply chain. No one can change any data without a consensus.

8.2.6.2 Improving Inventory Management

Inventory management is used to organize and track all of the goods to properly maintain the product data. The blockchain data will provide details about the product, and it can be easily organized and stored. All the data is available in the blockchain to decide on inventory management.

8.2.6.3 Identify Issues Faster

The issues in the supply chain can be easily identified by any person, and they will be solved within the proper time. In the current supply chain, finding the issues is a big issue. The origin of the issue is not easily found in the traditional supply chain. The participants will blame each other for any issues in the current supply chain but in blockchain, everybody knows about everything so the issue can be easily identified and resolved.

8.2.6.4 Minimize Courier Costs

In the traditional supply chain, data or products are transferred through couriers without knowing the proper details. In blockchain-based supply chain, the quantity of the product required by the retailer will be easily identified by the supplier to courier the products. There is no possibility of miscommunication or misunderstanding in the blockchain. So, the supplier can send the products at a time as per the requirement, this will reduce the cost of the courier. The information is available to all the participants in the blockchain to decide on any activities. The cost of the product can be reduced because of the reduction in courier costs.

8.2.6.5 Reduce Delays from Paperwork

In blockchain, all the data can be shared as digital information, so there is no such paperwork and approval. All the processes can be digitally approved and the transaction will be carried out. The agreement between the participants can be written as a smart contract program. It will be executed if a certain condition is satisfied. This will be useful in doing the work smartly and quickly.

8.2.6.6 Reduce or Eliminate Fraud and Errors

In blockchain, there is no possibility of tampering with the data because of its cryptographic nature. If anyone in the network tries to change anything in the information, that will be broadcast to all the peers in the network. The changes will be made only after obtaining digital approval from most of the peers. So, nobody can cause malfunctions in the system and nobody can perform fraud activities. The error will be eliminated because of a consensus from the network. To do any changes in the public blockchain, >50% of the network peers have to approve the changes made by any peers. Practically, it is impossible to get approval from >50% of their peers in cases of fraud. If it is possible to compromise >50% of the network, then the network is not normally suited for blockchain. This kind of compromise of >50% of the network is called a 51% attack.

8.3 AUDIT TRANSPARENCY

The product is often audited by any of the participants or customers within the supply chain thanks to its transparent and shared network. The availability means that anytime we can do the audit of transactions or any data.

8.3.1 Supply Chain Transparency

Transparency of the blockchain provides readily available information to end users and firms. The trust among the customers increases thanks to this transparency. The organization can easily audit the methods involved in the entire supply chain process. In the current supply chain, the knowledge won't be passed from one participant to another and it's very hard to seek out the integrity among the participants. Miscommunication and misunderstanding can happen and this will impact the merchandise quality.

8.3.2 Information Transparency

Information transparency will influence the customer's purchasing behavior and cause the supplier to perform socially responsible practices. Information transparency will not create any anomaly for the customer while contracting or purchasing the product. The condition for every process can be written and that will be available to all the participants in the network. Nobody is allowed to make changes without approval.

8.3.3 Blockchain-Enabled Traceability Applications

Many companies and organizations have a lot of elements in their supply chains to make it convenient for their work. In multinational companies, it is impossible to trace each record because of big data storage. Transparency is required in the supply chain, otherwise, it will result in customer issues and that will decrease the market value of a product. We need a system for record-keeping and provenance tracking. This can be achieved with blockchain with SCM because the information is often accessed through the sensors that are embedded on the devices with RFID tags. The whole history of a product up to the product delivery will be maintained, such as where it's located, and the movement of the product from time to time can be traced through blockchain. The detection of fraud in any part of the availability chain is by accurate provenance tracking.

The features of blockchain are suitable for traceability applications. If the documentation of goods passes from one actor to another from the availability chain to another chain, the items are vulnerable to theft. To avoid this, blockchain technology involves the creation of a digital "token" that specifies the physical items. The tokens will tell us about the history of the items from their origin. No entity or group of entities can change the knowledge in the blockchain. The end users have good confidence in the product because their token provides them information about the product. Blockchain is a suitable technology to enable supply chain traceability with the help of a token with information about the product.

8.3.4 Blockchain-Enabled Supply Chain Traceability Applications

The first traceability application was enabled by Ethereum [15]. Yellowfin and skipjack fish were tracked throughout the whole supply chain (January to June 2016) from fishermen to distributors. The story of their tuna sandwiches could then be tracked by the end users via a smartphone, so that the end users could determine information about the suppliers, producers, and procedures. The digital "token" was used to verify a given fish's origin and it was tracked throughout the supply chain. Everledger [16] is another blockchain-enabled traceability application for the worldwide diamond industry. Here, 1,000,000 diamonds are registered on the blockchain to certify that the diamond was ethically sourced from conflict-free regions. Similarly, products such as wine are also getting certified by the blockchain.

8.4 TRACKING SOCIAL RESPONSIBILITY

Blockchain technology helps in environmental supply chain sustainability, and it can do so from several different viewpoints [17]. First, reliably monitoring substandard goods and recognizing more product transactions can help minimize rework and recall, which helps to reduce resource use and reduce greenhouse emissions. Traditional energy networks are centralized, whereas blockchain technology for energy systems powered by a peer-to-peer network will eliminate the need to transfer electricity over long distances and thereby save an enormous amount of the energy lost during long-distance transmission. It could also decrease the need for energy storage, which saves money. Several power networks, such as Echchain, ElectricChain, and Suncontract, promote blockchain technology to scale back the waste in the supply chain.

Second, blockchain may not be able to make sure that green goods are meant to be environmentally friendly. Details on the manufacturing of green goods are typically inaccessible and hard to verify. If a product's production process is checked to be green in terms of the number of greenhouse emissions, environmentally conscious consumers will also be more likely to buy green goods. In the sustainable Indonesian forest, for example, Ikea features a desk product consisting of a woodcut. To guarantee the desks are made of this particular wood, Ikea must follow the wood from the time of its cross-production to the ultimate product. This process is complicated, but with blockchain technology, it is also controlled. One such example is the endorsement of the Forestry Certification Scheme, which uses blockchain technology to track the provenance of about 740 million acres of certified forests all over the world [18]. A carbon tax is claimed to be another indicator of sustainability in the environmental supply chain. The carbon footprint of a product is difficult to determine in conventional structures. With blockchain technology, it becomes simpler to track the footprint of a particular company's goods [19]. It will help decide the amount of carbon tax that a business should be paying. If a product with an outsized carbon footprint is more expensive, purchasers may purchase a low-carbon footprint product. Such additional knowledge and customer or market pressure can cause companies to reevaluate and restructure their supply chain to reduce carbon emissions to meet buyers' demands. By providing the basics for supply chain mapping and implementing low-carbon product design, manufacturing, and transportation, blockchain technology will help minimize carbon emissions throughout the product journey. The Availability Chain Environmental Analysis Tool [20] (SCEnAT) proposes a system for the evaluation of the carbon emissions of all supply chain and product life cycle organizations involved. SCEnAT 4.0 can be a new platform that incorporates new technologies such as blockchain, IoT, AI, and machine learning to more efficiently handle big data and connect supply chain organizations to support industry 4.0 policies, emissions reduction, and green evaluations. Blockchain technology also has the power to rework the exchange in carbon assets. A green asset blockchain-based platform is being developed by IBM and Energy Blockchain Labs Inc. in China to help companies control and calculate their carbon footprint, meet the Carbon Emission Reduction (CER) quotas, and promote the production and exchange of carbon assets. Transparent, safe, and real-time blockchain information enables companies to collaborate more effectively and exchange their carbon assets within the green asset markets.

Third, recycling can be enhanced by blockchain. People and organizations may not be inspired to engage in initiatives for recycling. In return for the deposit of recyclables such as plastic containers, cans, or bottles, blockchain technology has been used to empower individuals in northern Europe through financial incentives in the form of cryptographic tokens. Meanwhile, the effect of varying recycling schemes is difficult to trace and compare. To determine the effect of different systems, blockchain allows data to be tracked. For example, a project based on blockchain technology called social plastic that aims to reduce plastic waste and turn plastic into money. Another blockchain technology that enables individuals to return plastic containers is RecycleToCoin. The probability of this kind of closed-loop supply chain initiative makes the blockchain ideal for evolving ideas such as the circular economy.

Fourth, by enhancing the effectiveness of emission trading schemes (ETS) [21], the blockchain benefits the emission trading method. Fraud is also avoided with the implementation of blockchain technology because of the fidelity and transparency of the blockchain. A reputation-based framework is then produced that addresses the inefficiency of the ETS and allows all participants to seek out a long-term solution to reducing pollution because the economic benefits of an honest reputation are encouraged by the participants. The use of blockchain technology as a mechanism for SCM and knowledge management will be challenging, particularly in a sustainable network. Typically, disruptive innovations face obstacles, whether in the short or long term.

8.5 PREVENTING COMPLIANCE VIOLATIONS

The prevention of compliance violation is required to run the business in a smoothy way. The few barriers of compliance violations are intraorganizational barriers, interorganizational barriers, system-related barriers, and external barriers.

8.5.1 Intraorganizational Barriers

Everybody within an organization has a consensus over the protocol and can maintain the standards. The violation of any rule is easily monitored by the higher authority with blockchain. It helps to make interior activities for organizations. Top management thinks about the successful implementation of the supply chain with blockchain. However, some organizations fail to possess a long-term commitment and have trouble adapting to new technologies. Lack of commitment by the management will decrease financial and market status. The new organizations have to adapt to the blockchain to get a good impact on their business but for that, the organization requires investing in new hardware and software that may be expensive for organizations and the partners of the network [22]. Lack of policies to clarify the usage of blockchain might be a challenge in an organization. Blockchain technology adoption in the current organizational structure can change the current organizational cultures [23]. Organizational culture covers the rules of the work culture and the appropriate behavior of the organization [24]. Supply chain with blockchain requires expertise to support adopting technology and new roles and responsibilities [23]. The limitation of a technical developer for the blockchain is a problem [22] because moving to blockchain requires good developers to build the blockchain as per the organizational requirements. The organization must have technical developers of blockchain when they want to add blockchain to their supply chain.

Blockchain technology is considered a disruptive technology that alters or replaces legacy systems [22]. The organizational hierarchy may change, and it may have an impact on the organizational cultures. If organizations wish to have blockchain for their supply chain they have to embed sustainability practices into their organizational vision and mission. To implement proactive plans, the organization requires the availability chain throughout the process of the supply chain. Lack of tools and methods hinders the successful implementation of blockchain in the supply chain process and sustainability practices [25].

8.5.2 Interorganizational Barriers

This identifies barriers between the supply chain partners. The supply chain has to show the relationships among partners to provide value for the stakeholders. Nowadays, maintaining relationships between partners might be challenging. Blockchain technology provides information sharing through a supply chain to all the peers in the network. Some partners don't want to reveal the data to other partners where the benefit of blockchain is questionable, but they can practice this with a private blockchain. Private blockchain is not a suitable solution for the supply chain. Lack of rules for information sharing among the users may lead to contradictory operational objectives and priorities [25]. Communication challenges would be worse for the supply chain where the partners are geographically dispersed with different cultures. Finally, combining conventional supply chain processes with sustainability practices isn't a simple process. Current technology like blockchain has to be implemented to get sustainable solutions for the supply chain. The designs, materials, and processes have to be improved to support sustainable practices, therefore the IoT and blockchain are two potential solutions to such a problem.

8.5.3 System-Related Barriers

The blockchain with supply chain needs the Internet of Things as a barrier for their communication. The new tools and techniques are a challenge for a few supply chain participants. The technology access limitation should be avoided during a supply chain entering into blockchain technology. Blockchain is a new technology to store data and it is in the early stages of development. Blockchain technology is thought of as an immature technology in terms of scalability and handling an outsized number of transactions. The bloat problem is that the number of blocks may be a storage dilemma for handling big data in real-time usage. This bloat problem was identified in bitcoins. For the supply chain, we need larger data requirements that transcend financial data and data processing. The storage management of the blockchain has to be improved with new approaches and several types of research are going on for minimizing the storage capacity of the blockchain.

Data manipulation requires serious attention in supply chain networks. Blockchain technology allows every participant of the supply chain network to verify the transactions, and consensus protocol is a possible way to go with the same frequency, but the security and privacy concerns also are challenges in blockchain technology [22]. Some of the security issues were addressed in the Bitcoin network including hacks and attacks in some research. Solutions were suggested to avoid blockchain security challenges, but the effectiveness of those solutions has not been evaluated. Most of the issues were reported in the bitcoin network such as the "dark web" reputation, which slows down blockchain technology adoption; However, the immutable property of the blockchain is provides a variety of applications to be integrated to the blockchain. The immutability means information can't be changed and removed in blockchain without consensus [8]. However, humans are still trying to write erroneously recorded data and this can be compromised when it is approved by >50% of their peers. The supply chain is a larger network in which getting approval from >50% of the peers is nearly impossible.

8.5.4 External Barriers

This category introduces challenges from external stakeholders, industries, institutions, and governments; those entities indirectly economically taking advantage of supply chain activities. Lack of maintenance on governmental and industry policy is leading to a big problem. To follow those policies and standards, we have to spend the resources and pay the cost. Governmental regulations and laws are still unclear about the usage of blockchain technology. Some governments provide an opportunity to use bitcoins for markets, but this will affect the broader usage of blockchain for business objectives [22]. The government should promote blockchain technology to make a sustainable solution in SCM. Organizations should look for investment in green products, sustainable processes, and new technology like blockchain and would be compensated by their customers. The improvements are needed in blockchain to adapt to the supply chain. So many researches are still going on for adapting blockchain in all the fields. The solution for each problem should focus on the environmental impact so that all the government agencies and non-governmental organizations (NGOs) will welcome the new world with the impact of blockchain.

The blockchain provides us positivity that, with some few changes, it can be applied in all the fields required without a central authority. Blockchain technology will provide advancement for supply chains but the adoption of blockchain technology should be faster which will make the transactional record more robust and reliable. Blockchain is suited to financial transactions where no physical goods move without any permission and everybody can be involved and see the status of the goods in various fields such as stocks, insurance, land-registry, taxation, medical records, financial instruments, and derivatives, etc.

8.6 AUTOMATED PURCHASING AND PLANNING

Blockchain allows the transfer of money without any third party such as banks anywhere within the world as transactions happen from peer to peer. It is stable and the transaction will be done within 10 minutes when the miners validate the transaction and put the transaction into a block and add the block to the blockchain. The transaction is complete if the block is added to the blockchain. Transfers to Bitcoin have lower transaction fees. For its transaction with the manufacturers, Australian vehicle manufacturer Tomcar uses bitcoin. Tomcar's suppliers have agreements to use the standard of bitcoin as the transaction currency. The advantage of using bitcoin as the transaction currency is that there is no third-party involvement and the transactions are tamperproof. Although the government has the dilemma of accepting bitcoin in their country, most of the users worldwide trust bitcoins and use it as transaction money. Blockchain can be used for automated purchasing and planning with smart contracts. The smart contracts, the written programs of the business agreement, will execute the relevant process to be done on the network when the condition is satisfied. The customer can buy the product whenever the price is satisfied with the customer and also the retailer. The suppliers can supply the components automatically when the condition of the manufacturer is satisfied. Ireland's Moyee Coffee provides transparency across the whole supply chain to avoid the complexity of handling multiple parties. Because of counterfeit goods, the automobile industry is suffering. For car manufacturers, this is a major problem. Counterfeit spare parts are not reliable as the quality standards degrade and the cars can crash. The consumers will be disappointed and their trust in the brand will be revoked. Blockchain technology proves to be useful with regard to counterfeit goods because it offers an ability to uniquely identify replacement parts. Digital spare part recognition provides the system with accountability and establishes a sharable environment within the network between different parties.

Smart manufacturing is provided by blockchain, and inbound logistics will help the automotive supply chain to be successful. The monitoring of individual components of an inbound supply chain is currently vulnerable to mistakes. For the smooth functioning of the supply chain, a collaboration between third-party logistics, multitier suppliers, and transport companies is critical, but the blockchain makes available accurate and real-time information between different parties. The status can be checked at any time by supply chain participants. There is no common data-sharing model for the current outbound supply chain, which makes it difficult to share information. Blockchain's transparency ensures quicker transactions and lower settlement periods.

8.7 FOOD SAFETY

8.7.1 The Challenges of Food Safety

Food safety, which leads to food recalls leading to a threat to profit, is significant. Food Safety magazine counted 456 food safety recalls last year, with each recall estimated to cost $10 million in addition to the effect on society and industry. Large amounts of food are wasted day by day, which shows the consumer's distrust of the food product. The companies cannot quickly identify the remedy for food wastages. Incident tracing is required to avoid food wastages because of an inadequate process for food safety. The response for tracing food safety will not provide immediate results because of the long supply chain. It may take days or weeks and companies have struggled to trace. Paper-based food data is not sufficient for food recalls in a growing network. The digital documentation also should have the availability to trace the chain. In the event of any weaknesses in the control of the supply chain, this will create vulnerabilities, manufacturing deficiencies, and control processes that make the food system vulnerable. The researchers indicated that a large number of food safety warnings are needed to prevent a lack of transparency in the supply chain. The lack of accountability has allowed food fraud and corruption to happen. With new technology such as blockchain, food traceability practices can be controlled. The blockchain provides food safety by its transparent and traceable features that will avoid food recalls. The availability of the blockchain provides a way to trace at any time.

8.7.2 Blockchain for the Food System

In the digital model of a food system, access tools and maintenance of data in digital storage make the data easy to retrieve. Some of the features not involved in the current digital food system for storing the food data can be provided by a blockchain. In a decentralized network that enables us to foster trust and accountability, blockchain technology stores the same digital form records. With immutable data storage, Blockchain will enhance the food system and ensure food protection. The integration of blockchain in the food supply chain is shown in Figure 8.8.
It has the following features.

1. Transparency
2. End-to-end traceability
3. Food trust

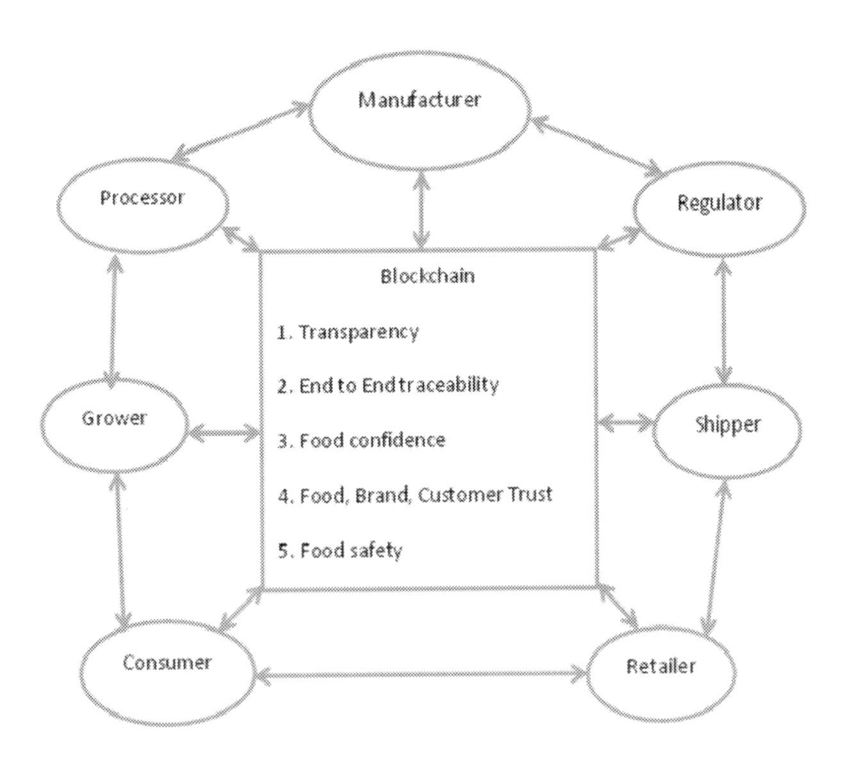

FIGURE 8.8 Blockchain integration in food supply chain.

4. Food safety
5. Customer trust
6. Brand trust

Transparency: Top-to-bottom visibility enables brands to manage damaged products without disrupting the whole supply chain. The consumers and supply chain partners can easily know about their brands, the quality, and the origin of their foods, which results in trust in the products.

End-To-End Traceability: The food safety is assured with immediate clearance about the food products when defects occur. The end-to-end traceability provides details about who is affected and who should react to it.

Food Trust: Food fraud can occur if the network does not have transparency. The organizations know about which foods are grown or produced with the process, but the customer does not know about the ingredients used for the food product. The blockchain that generates a safe, shared and permitted transaction record ensures food trust. During each stage of the food supply chain, this enables visibility. Through transparency, blockchain provides food customers with trust, making food safer and smarter from farm to end user.

Food Safety: Blockchain stores digitized records that are transparent and tamperproof, which earns trust and boosts food safety. Food trust is founded on approval by all, such as producers, manufacturers, distributors, and retailers, within the food system. Food safety is assured with blockchain monitoring all the participants.

Customer Trust: The customer has numerous options regarding where to buy his or her food. The brand is chosen by the customer for their food product in the competitive food industry. The customer must know the origin of the food, where it was grown and how it was processed,. Consumers concentrate predominantly on where they are raised. Brand loyalty does not occur without food protection and consistency. If it is not achieved, the individuals will turn to another brand. The consumer wants more details about their food product.

Brand Trust: The food supply chain will likely be the primary one to undergo a major national blockchain. Recalls are a serious problem for the brand to retain trust among the customers. Several major foodborne bacteria outbreaks recently occurred within the United States. This encourages the companies implement blockchain to increase the visibility and traceability of the products. Consumers are putting pressure on businesses to supply more provenance and authenticity. Most consumers can pay a premium for ethically made goods.

Walmart Company has been working with IBM on a food supply chain blockchain-based traceability solution since 2016. The system makes it possible for the distributor to track food supplies from farm to market. The corporation is also looking for how blockchain can be extended to monitoring and controlling the spread of foodborne illnesses. This helps to minimize costly recalls. Other businesses are exploring how blockchain can certify the origin and paths of products sold and supply data on the authenticity. As an example of this, OriginTrail in partnership with TagItSmart has recently tested the IoT with blockchain to stop wine fraud.

Blockchain and SCM is a game-changer for food safety. In the upcoming years, global grocers will adopt blockchain for food safety and traceability to improve their production and quality. The food supply chain has complexity and it is becoming normal when it is integrating with blockchain. The blockchain helps food manufacturers, retailers, and distributors to ensure the provenance of the goods. With traceability, food safety problems can be solved. Only via blockchain is finding the source of the contamination and identifying the root cause achievable. In the current system, finding the root cause of the problem is difficult. One-out-of-ten individuals fall sick and about 420,000 die from tainted food, according to the World health Organization (WHO). Consumers are aware of tainted food and expect honesty concerning the food they eat. Ninety-four percent of consumers expect the data about the food products to buy.

Blockchain can overcome the uncertainty in the supply chain by offering neutrality in the network. There are no third parties involved in the deals and all operate based on an agreement between the participants. The system will be maintained as a written software of intelligent contracts with a set of rules to keep the system running effectively. The originality of the food suppliers and the consistency of their deliveries can be ensured by monitoring the system for tampering attempts, such as where the food item travels in the supply chain. When fraud occurs, the supplier will be notified and the alert will be sent to the manufacturer before the food item reaches its destination. When a defective food product is found in the shops, identification is easily achieved. Instead of testing the entire batch of goods, this will provide the room to delete only those offending objects. Consumers have accountability and openness so that the products they need to purchase can be reviewed. This enables high-quality food to be identified by consumers. Because of a lack of accountability in the system, the seafood supply chain has been associated with negative headlines. Manual record-keeping makes it more open to mistakes. Owing to the prevalence of unregulated activities and inappropriate food storage conditions, the seafood supply chain is inefficient. Because of poor maintenance of the devices, the quality and protection of the food are not guaranteed to the end user.

Blockchain technology is an appropriate solution to the problems of seafood authentication because it can track seafood from production to delivery. With Hyperledger, several ventures will adopt blockchain technology to solve the problems of the seafood supply chain. One of Hyperledger Sawtooth's ventures is to fully change the traceability and transparency of the supply chain. In Hyperledger Sawtooth, the proof of the elapsed time consensus algorithm is used to allow actors in the system to have a consensus in an area. In Sawtooth, many sensors are used that allow the seafood to be tracked from place to place in the supply chain and can transmit each location's time frame.

The coffee supply chain is complicated and needs reform. Production of coffee is fragmented in the different locations of the world. Volatile prices and the effect of changing climates are the factors that influence the production of coffee. This should be taken into consideration to maintain the supply chain. In the coffee supply chains, farmers and laborers work mostly work in remote areas where cases of abuse have been reported. The coffee supply chain involves the implementation of a blockchain, adding accountability and quality to the system. Bext360 is a Denver startup using a "beatmatching" blockchain-enabled machine that analyses the farmed coffee beans and assigns them numerical values for their traceability. This is reviewed for a performance measure; about 50 kg of coffee can be processed by the machine in a minute.

In the coffee supply chain, blockchain offers greater productivity, which brings more reasonable deals for producers. When their goods are sold, accountability guarantees immediate compensation to the producers. The end customer can still track the details and track their coffee's source. In blockchain, the records are permanent and immutable. So, it can be used by companies to track each product to its source. Blockchain is used by the major retailer Walmart to trace pork sales in China. The system provides data inside the supply chain, such as where each bit of meat comes from, packaging, and storage steps. In the case of a product recall, blockchain has the mechanism to find which batches are impacted and who purchased them,.

8.8 REDUCING COUNTERFEIT GOODS

Reducing counterfeit goods is required to improve business and make it profitable. The certificates for each item that are given by the central authorities are not sufficient for trust. The agency or an organization can give the certification for a component even if the component does not have sufficient quality. This leads to poor quality products and that leads to dissatisfaction among the customers so that the business drops in the market.

The provenance of valuable goods also relies on paper certificates that can stray or be manipulated. Is the certification of a diamond true or false? It isn't always easy to find out whether a diamond has been stolen. The same situation happens for pricey wine, watches, handbags, etc. The Everledger [26] start-up solution is regarded as an alternative that reveals 40 data points that uniquely define a diamond in the records. A prospective buyer may decide whether or not the seller is the actual owner of the diamond in the blockchain with publicly accessible records on the ledger. For some high-value products, such as gold, Everledger plans to extend this fraud detection system into a provenance network.

Blockchain could improve patient safety in the medical sector by using IoT devices with sensors to establish supply chain transparency from manufacturers through wholesale and pharmacies to the individual patients. Barcodes or RFID technology are used to do the transaction easily and the chip can be inserted in the patient's body to show the changes in the patient's body. The tablet or the necessary medical items can be easily passed over the supply chain immediately. It is possible to use a private registered blockchain to monitor supply chain failure. If a drug is not prescribed by the doctor, it can't move through the supply chain. This will prevent product recall and avoid the patient taking the wrong medicines. Blockchain makes it tougher to tamper with goods or to channel in goods of illicit origin.

Cross-border exchange includes manual procedures, physical documentation, various intermediaries, and several ports of entry and exit checks and verifications. There is a possibility of a product recall if it's not good. The amount of spending on transportation is high and the proper product's data is required priorily before proceeded to the supply chain. Transactions are very sluggish, expensive, and the status of the product is poorly noticeable. The blockchain removes all the cross-border problems due to its transparency. The status of the product is visible all the time. Blockchain makes it much easier for enterprises to confirm sales orders, invoices, and payments without spending more money and risk. The blockchain can use bitcoins for the transactions, which will help in performing international transaction. The account can be created with bitcoin wallets that provides the public key for the transaction. The provider receives an order with links to the blockchain and can acquire the sum of bitcoin for the same instantly. Bitcoin is also transferable to our normal currency with the agencies running an exchange for bitcoins. The promptly accessible applications are machine driven that uses smart contracts with their protocols.

Blockchain allows manufacturers, third-party retailers, participants, customers, and regulators to verify any product's legitimacy. The products pass across the price chain, making it more difficult for ersatz products with adulterated labels to avoid detection. Authentication happens in real-time and ensures that sales and revenues go to legitimate brands and owners. Regulators can digitally provide the certification and trace the provenance and chain of custody for any product that is being sold. The applications linked to the blockchain platform make it easy for customer support professionals to verify whether a claim is genuine or not. A faster response can be given to the claim, and the customer can have trust in the process of the supply chain. The accurate authentication allows resource access for legitimate users, high-value purposes, reduced waste, improved responsiveness, and increased customer satisfaction.

Blockchain platforms give manufacturers an efficient, scalable platform that provides an easy way to identify counterfeit parts and goods. Blockchain with IoT has a wide range of applications including the supply chain. Here stakeholders across the chain can check whether a component or a product is legitimate, where it was sourced, and how it was transported and stored with the help of various IoT devices with embedded sensors. Considerable adoption of blockchain in the supply chain will yield success. Blockchain platforms become more mature for their use cases and technologies, which are evolving day by day to make enterprise solutions become more readily available. The blockchain is more and more useful in counterfeiting the product to improve the business and sales.

8.9 CONCLUSION

Blockchain is safe and secure for the application that wishes to share the information in the standard ledger, avoid third party assistance, maintain consensus among all the peers, remain tamperproof, and maintain safety with cryptography. Blockchain is suitable for SCM by considering all the aspects of blockchain within the above statements to take care of the activities in SCM like (i) securing the information, (ii) audit transparency, (iii) tracking the product and social responsibility, (iv) preventing compliance violations, (v) automated purchasing with smart contracts, (vi) agreement and consensus on a business process, and (vii) reducing product recall cases. Blockchain will be a good solution to address the challenges and issues in SCM. Blockchain is the future of the supply chain to regulate the activities and processes in SCM. Blockchain will have a positive impact on the trust of the customers and the stakeholders. Finally, blockchain is revolutionizing the world as a descriptive technology by showing its performance in the supply chain as a showcase for others.

REFERENCES

[1] Nakamoto, S. 2008. "Bitcoin: A Peer-to-Peer Electronic Cash System." http://bitcoin.org/bitcoin.pdf Blockchain.info. 2013.

[2] *A Review on the Use of Blockchain for the Internet of Things*. May29, 2018. https://ieeexplore.ieee.org/iel7/6287639/6514899/08370027.pdf

[3] Vivek Anand, M., and S. Vijayalakshmi. 2020. "Image Validation with Virtualization in Blockchain Based Internet of Things." *Journal of Computational and Theoretical Nanoscience* 17, May.

[4] Bruce, J. D. "The Mini-Blockchain Scheme." https://www.weusecoins.com/assets/pdf/library/The%20Mini Blockchain%20Scheme.pdf (accessed on 10 April 2018).

[5] Blockchain.info. 2013. "Blockchain Size Data." https://blockchain.info/charts/blocks-size

[6] Homomorphic Mini-blockchain Scheme B.F. Francabrunoffranca88@gmail.com 24 April 2015.

[7] Steiner, J., and J. Baker. 2015. "Blockchain: The Solution for Transparency in Product Supply Chains." https://www.provenance.org/whitepaper.

[8] Tian, F. 2017. "A Supply Chain Traceability System for Food Safety Based on HACCP, Blockchain & Internet of Things." 2017 International Conference on Service Systems and Service Management (ICSSSM).

[9] Abeyratne, S. A., and R. P. Monfared. 2016. "Blockchain Ready Manufacturing Supply Chain Using Distributed Ledger." *International Journal of Research in Engineering and Technology* 5 (9): 1–10.

[10] Pazaitis, A., P. De Filippi, and V. Kostakis. 2017. "Blockchain and Value Systems in the Sharing Economy: The Illustrative Case of Backfeed." *Technological Forecasting and Social Change* 125: 105–115.

[11] Tian, F. 2016. "An Agri-food Supply Chain Traceability System for China Based on RFID & Blockchain Technology." 13th International Conference on Service Systems and Service Management (ICSSSM).

[12] Christidis, K., and M. Devetsikiotis, 2016. "Blockchains and Smart Contracts for the Internet of Things." *IEEE Access* 4: 2292–2303.

[13] Hofmann, E., U. M. Strewe, and N. Bosia. 2018. "Discussion – How Does the Full Potential of Blockchain Technology in Supply Chain Finance Look Like?" In *Supply Chain Finance and Blockchain Technology*, 77–87. Cham: Springer.

[14] Tapscott, D., and A. Tapscott, 2017. "How Blockchain Will Change Organizations." *MIT Sloan Management Review* 58 (2): 10.

[15] "Provenance Has a Big Year Ahead Delivering Supply Chain Transparency with Bitcoin and Ethereum." *International Business Times*. http://www.ibtimes.co.uk/provenance-has-big-year-aheaddelivering-supply-chain-transparency-bitcoin-ethereum-1537237 (accessed on 13 September 2017).

[16] Roberts, J. J. "The Diamond Industry Is Obsessed with the Blockchain." *Fortune*. http://fortune.com/2017/09/12/diamond-blockchain-everledger/ (accessed on 13 September 2017).
[17] Futurethinkers. 2017. "7 Ways the Blockchain Can Save the Environment and Stop Climate Change." http://futurethinkers.org/blockchain-environment-climate-change/.
[18] Rosencrance, L. 2017. "Blockchain Technology Will Help the World Go Green." *Bitcoin Magazine*.
[19] de Sousa Jabbour, A. B. L., C. J. C. Jabbour, C. Foropon, and M. GodinhoFilho. 2018b. "When Titans Meet – Can Industry 4.0 Revolutionizes Environmentally-Sustainable Manufacturing Wave? The Role of Critical Success Factors." *Technological Forecasting and Social Change* 132: 18–25.
[20] Koh, S. L., A. Genovese, A. A. Acquaye, P. Barratt, N. Rana, J. Kuylenstierna, and D. Gibbs. 2013. "Decarbonizing Product Supply Chains: Design and Development of an Integrated Evidence-Based Decision Support System – the Supply Chain Environmental Analysis Tool (SCEnAT)." *International Journal of Production Research* 51 (7): 2092–2109.
[21] Khaqqi, K. N., J. J. Sikorski, K. Hadinoto, and M. Kraft. 2018. "Incorporating Seller/Buyer Reputation-Based System in Blockchain-Enabled Emission Trading Application." *Applied Energy* 209: 8–19. doi:10.1016/j.apenergy.2017.10.070
[22] Mougayar, W. 2016. *The Business Blockchain: Promise, Practice, and Application of the Next Internet Technology*. Hoboken: JohnWiley & Sons.
[23] Mendling, J., I. Weber, W. van der Aalst, J. V. Brocke, C. Cabanillas, F. Daniel, and S. Dustdar. 2017. "Blockchains for Business Process Management-Challenges and Opportunities." *ACM Transactions on Management Information Systems (TMIS)* 9 (1): 1–16. arXivPreprint ArXiv:1704.03610.
[24] Gorane, S., and R. Kant. 2015. "Modelling the SCM Implementation Barriers." Journal of Modelling in Management 10 (2): 158–178.
[25] Mangla, S. K., K. Govindan, and S. Luthra. 2017. "Prioritizing the Barriers to Achieve Sustainable Consumption and Production Trendsin Supply Chains Using Fuzzy Analytical Hierarchy Process." *Journal of Cleaner Production* 151: 509–525.
[26] Lomas, N. (2015). "Everledger Is Using Blockchain to Combat Fraud, Starting with Diamonds." *TechCrunch*.

Abbreviations

1. **SCM**—Supply Chain Management
2. **ISO**—International Organization for Standardization
3. **DDOS**—Distributed Denial of Service
4. **SHA**—Secure Hashing Algorithm
5. **IoT**—Internet of Things
6. **RFID**—Radiofrequency Identification
7. **SCEnAT**—Supply Chain Environmental Analysis Tool
8. **CER**—Carbon Emission Reduction
9. **ETS**—Emission Trading Schemes
10. **TAM**—Technology Acceptance Model
11. **NGO**—Non-Governmental Organization
12. **WHO**—World Health Organization

9 Enhancement of Interoperability in Health Care Information Systems with the Pursuit of Blockchain Implementations

N.S. Gowri Ganesh, R. Roopa Chandrika and A. Mummoorthy

CONTENTS

9.1 INTRODUCTION TO HEALTH CARE

Healthcare provides medical care to people in an organized manner. It comprises hospitals, medical devices and medical equipment, clinical trials, outsourcing, telemedicine, medical tourism and health insurance agencies. Doctors, therapists, nurses, paramedics and other frontline clinicians are not the only participants who contribute to healthcare as technology professionals also have a place in supporting the people in the community.

Healthcare providers will lend digital services to people by associating with 5G technology that will support the extensive digitization of medical records to discover disease patterns for future analysis. Blockchain technology, internet of things, internet of medical things, robotics and industrial sensors connected to edge computing devices will also transform the way healthcare professionals provide and manage care. Healthcare, as with any other digital service operation, also requires systematic innovation efforts to maintain competitive, cost-efficient and up-to-date digital services.

9.1.1 Components of the Health Care System

In "Introduction to the U.S Health Care System", 2007, the authors have defined the healthcare system into five major components:

- The health care facilities
- The work force that staffs the facilities (doctors, medicos, etc.)
- Health care therapeutics
- Research and educational organizations that help to improve the healthcare system with necessary feedback and knowledge
- Financial management

9.1.2 Components in Digital Healthcare

- Medical staff and experts
- Patients
- Data and value
- Competitors in Medical field

9.2 WHAT IS THE VALUE IN HEALTH CARE?

Value in health care is based on the outcome of the patients. It is centered around the results and not on the input to the health care system and its services. Health care services are rendered by numerous organizations that offer either multiple services or a single service. The services follow a cycle known as the care cycle. It is the treatment procedure from the day the patient is admitted in the hospital until they recover and return to normal activities. The cycle involves interventions of numerous specialties for the effective outcome or the sustainable recovery of the patient. The activities of each specialty organization are interrelated and are responsible for the patient's health. Different patients have varied medical conditions and the treatment regime may follow particular patients' results and treatment procedure. Medical reports and their outcomes are shared among multiple departments for better decision-making in improving the patient's health. Each department is usually seen as a separate unit. Porter (2010) focuses on the cost incurred for health care by the patients. If the medical billing is taken care by one of the departments in health care system and it is functioning separately, then the pharmaceutical expenditure tries to disguise the overall value of the care and misleads the decision-making process in reducing the cost of pharmaceutical spending. For example, if the drugs needed to treat the patient are considered as a separate cost or unit in health care system.

There is a tendency that health care is provided in over-resourced facilities. There are circumstances in expensive medical centers in which the resources are under-utilized. There are skilled physicians and other medical experts and staff who are not making use of their good skills and training. Because of the prevailing situation, there are excess administrative costs, delays in diagnosis and treatment and costly delay in the time required for the treatment. Most of the physicians, administrators and other medical staff are unaware of the total costs that the patient would have to bear for his/her treatment. But the medical experts' fees caring for patients are truly much less than the total cost of care for a particular medical condition.

9.3 DIGITAL REVOLUTION IN HEALTH CARE

Rapid technological innovations accelerate the realization of scalable solutions in global healthcare. Reports say that we are in the middle of the third revolution. This revolution does not directly tell about health care but involves the subjects as shown in Figure 9.1: patients, knowledge and information technology.

9.3.1 Digital Medical Services and Information Growth

Telemedicine applications provide an interface between the doctors and patients and allow an exchange of information. The information gathered may help the medical experts suggest the right follow up in the treatment regime.

Patients and clients seek to gather information from many digital sources, digital community groups and patients with similar ailments. They are anxious to know the treatment procedure and the recovery rate. People are provided access to telemedicine to improve health care.

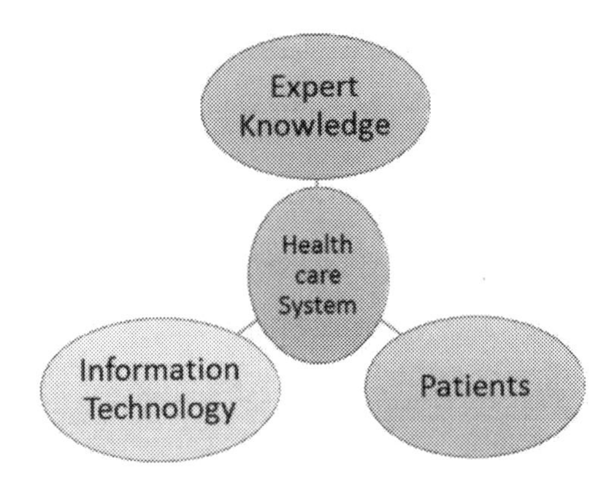

FIGURE 9.1 Three forces involved in the health care revolution.

The rapid change has encouraged development of a good relationship between patients and doctors. The revolution is dynamic, pervasive and fast such that human genome analysis is on the World Wide Web. But initially, the emerging technologies are finding it difficult to penetrate into traditional systems. The traditional methods are also not manageable and are threatened if the system fails. Considering the other side, where patient chronic diseases are continuing in the aging population and health care costs are also increasing, a solution to choose is a qualitative technological health care. The majority of adult and aged populations are living with multiple chronic disorders. About 60% of the global population would be aged above 60 and the health care costs will be five times more in the future. The different nature of disease in patients makes it hard for the doctors to diagnose and treat them. There are cases in which patients without any history have to be treated and medical experts have to fight with the insidious infections. They fall victim to human errors in diagnosing the infection or become infuriated. Because of this, managing traditional methods would not be a better solution.

Patients do not have knowledge of their health conditions. But real-time data will reflect the measure of their health and will be conveyed to the physician. In this case, the manual records maintained are inefficient and any missing crucial information would make the treatment futile. The digital innovations will allow the future doctors to tailor the treatment for each individual based on their genetics, life style, height, weight, race and other factors to improve the quality of life.

The health care revolution will provide an opportunity to increase the economic growth of a country providing employment. The revolution has to cater the health care or medical services in an easy, affordable and understandable manner. There are daunting challenges in developing countries, but there can be a flip in the traditional health care model from serving the sick to prevention.

9.4 CHALLENGES IN TRANSITION FROM MANUAL HEALTH CARE SYSTEM TO DIGITAL RECORD SYSTEM

According to WHO (World Health Organization), there is a shortage of medical staff members. Research shows an optimistic evidence of improvement in the quality of health care with the advent of electronic health records. The number of patients with chronic diseases and the treatment provided are simultaneously rising with the life expectancy. Both a hardware and software revolution are emerging in the health care industry. In traditional systems, patients were not involved in any treatment decision process. The decision-making was out of patients' control and was also one of the significant steps to adapt to newer technologies.

In 1960, at the University of Vermont, USA (Wright et al., 2014), Weed proposed two versions of problem-oriented medical information system that were adopted worldwide in later decades. In 2014, the American Medical Association reported dissatisfaction with the electronic health record (EHR) systems. The most important disadvantages were high cost, privacy, disrupted workflows and inefficiency. The reasons that EHRs were not widely adopted worldwide were because of the dearth of uniform standards, physician resistance, high costs and less knowledge in the system. Zandieh et al. (2008) noted the remarks on the challenges in the process of transition from legacy systems to digital systems as given in Table 9.1.

Any record is the observation of an event. Gary et al. stated, "If an event is not recorded or documented, then the event did not happen at all". The document preserves the information and is a historical artifact transforming the event into actionable object. The medical encounters are documented and are an essential process in healthcare management. The modification in the document creation process by EHR systems might lead to consequences in the usability of health information. The documents may be viewed in different contexts among health care organizations during the information exchange. In developing countries, there is no streamlined process

TABLE 9.1
Paper-Based Respondents' Challenges in Implementing an EHR.

Perception	Remarks
IT issues	
Acquiring IT workstations and peripherals	Doctors and paramedics have no knowledge of the requirements of the digital equipment until the medical staff are trained on the system.
Medical staff have to be designated to learn the system	
Comfort level of physicians and supporting crew with IT	The time spent to learn the system will have a great impact on patient volume and budget
Decreased productivity	Medical management expects a few weeks of inefficiency that in turn leads to 2 to 6 months of system inefficiency
Safety and quality	

Source: Zandieh et al. (2008)

to integrate the legacy existing systems. The digitized system is customized based on the circumstances of the organization. So there are also disparities in the technology stack of an EHR system. The software system and the standards implemented in a few countries with a smaller population might not be suitable in countries with a huge population because the system might not scale to that extent. The standards might be relaxed for certain valid reasons, making it difficult to integrate with other digitized systems. The solution given by researchers for such problems is testing and preimplementation prototype checking.

Another real threat is security breaches in health data. Measures have to be taken to strengthen information access internally and externally. Maintaining the legacy system without digitizing is not a solution rather to handle the risks with caution and the need to not compromise the patients' health and their records. Even though the data exists, there are only a few data scientists or researchers who can make meaningful data from the data store. Another major challenge is the availability of the data store in different locations and integrating them. A successful health care infrastructure can be built based on the quality of the data collected and reported. Knowledge and training in systematic processes, policies and frameworks will remove the barriers or resistance in adopting new technology.

9.4.1 Challenges in Implementing the Paradigm Shift

While the technological innovations are gearing up, certain areas will become crucial in coping with the future demands in data security, health care quality and health care-based applications. Fragidis and Chatzoglou (2012) discussed the major challenges of implementing digital health records like standardization, funding, interoperability and communication. EHRs have been implemented in different countries with different approaches, namely top-down, bottom-up and middle-out and the lessons and consequences of adapting the system were learnt. The United Kingdom followed the top-down approach in implementing EHR, spending 6.2 million pounds. The local national health organizations in the country were told to collaborate and adapt to the new system. After 8 years in practice, it was noted that there were repetitive changes in specifications, delay, compatibility issues, increasing cost and other technical issues.

9.5 Requirements of the Digital Health Care Infrastructure

The resources that exist in healthcare are patient health records (PHRs), EHRs, clinical results, IOT devices for continuous monitoring of vital signs, telemedicine records, biological lab diagnosis and insurance companies. The resources provide information in activities like health monitoring surveillance, cost monitoring, data security, health management and hospital management. For a better digital healthcare system, issues like security and privacy, interoperability and governance must be addressed. The development of the Internet and World Wide Web have allowed the growth of information agencies that exchange data in the form of web applications among public health care centers, pharmaceutical laboratories, medical institutions and other medical-related organizations.

Other than the digital resources available, there are developing applications for customized hospital management, billing and scheduling, robotic-assisted systems, patient decision support system, clinical support system, acquisition of vital signs and disease-monitoring system, PHR maintenance and data aggregation, retrieval, analysis and reporting. The storage and computation facilities have increased extensively in multiple fields. There are public and private organizations emerging and working for the progress of health care. The organizations are taking initiatives to allow information exchange among the stakeholders across the world. Information exchange is required for large scale analyses from the historical data of patients, their diseases and outcomes. The clinical evidence of patients have to be combined in the health care industry for optimal health care, but this is not yet prevalent because of obstacles in data sharing and data consistency. The digital health infrastructure provides a herculean task to maintain and manage information.

9.5.1 Motivation to Interoperability: Data Incomparability and Inconsistency

Patient EHRs and related repositories for best therapies are shared to medical practitioners, researchers and biostatisticians on agreed conditions by the supervisory personnel. But in recent days, sharing of information is facing obstacles like security, confidentiality, privacy and intellectual property rights. These factors make the vision of better service and patient outcome impractical in the real world. The information also is not consistent and comparable among the patient information provider agencies, medical practitioners and researchers. One reason is the standards used for representing patient data [4]; it stops scalable analyses and the harmonization of the data. The EHR providers are contributing to and adopting systems that follow standards that lead to a better place in the healthcare information technology. Interoperability will result in a better way for health care information exchange.

9.6 INTEROPERABILITY IN HEALTHCARE

Healthcare Information and Management Systems Society Inc. (HIMSS)(*Interoperability in Healthcare*, 2020) defines interoperability as "The ability to communicate, share data and use the information exchanged by various information management systems and software applications". It is a global advisor involved in the transfiguration of the information systems used in the health industry with the aid of information technology. It is accepted by the global experts (Who's to Blame for Healthcare's Interoperability Struggles?, 2017) that the interoperability needs to be promoted extensively so that no patients suffer because of the unavailability of information.

The actual expectation in terms of interoperability is that the legal digital information in any form in the software on any hardware devices needs to work together. HIMSS defines interoperability in three levels:

1. Foundation level systems exchange information, protecting the data without the ability to interpret the data;
2. Structural level defines the data formats, keeping the syntactic information; and
3. Semantic level at which healthcare systems exchange data and interpret it using standard codes.

9.6.1 The Benefits of Interoperability

The interoperability in the field of health care satisfies various criteria as given in Table 9.2.

Need for healthcare standards ("Overview of Healthcare Interoperability and Healthcare Standards," 2017) for implementing interoperability:

The main hurdles to implementing interoperability are:

1. Different formats of the data used in healthcare software

   Each software system because they were developed in a silo prefers defining their own formats. For example, the patient's disease can be described in text form but elsewhere it may also be described in images based on the disease and category.
2. Software systems developed with older formats and unique formats

   The software systems built a while back still cater to the hospital's needs but were developed with formats that may not be compatible with the commonly defined formats of the healthcare department.
3. Support from real practitioners for software developers to create the proper workflow

   Software developers can build very good software in technical terms, but the usability of the software is obtained with the involvement of the actual practitioners who work in the healthcare field to create proper workflows.

TABLE 9.2
Benefits of Interoperability.

S.No.	Benefits	Description
1.	Benefit to the patient	Patients will be able to get the best possible diagnosis and treatment irrespective of geographical boundaries
2.	Health history can be viewed	Because of interoperability, the patient's history can be viewed, which will avoid wrong and harmful injections resulting in further deterioration of the patient's health
3.	Promotes clinical research	As there is a huge amount of data in the database of healthcare software, it may aid researchers in the promotion of research

9.6.2 Contributions toward Interoperability

The National Health Information Infrastructure (NHII) (Aspden et al., 2004) in the United States believes that patient safety can be assured by various factors, of which an important one is the implementation of common data standards for interoperability, which is possible with the involvement of both industry and government. These organizations are expected to develop the following types of standards:

1. Data interchange standards
2. Adoption of terminologies
3. Clinical knowledge representation

9.6.2.1 Data Interchange Standards

There are a number of standard organizations, including American National Standards Institute (ANSI), Health level seven (HL7) and the Institute of Electrical and Electronics Engineers (IEEE), who develop data interchange standards and work on various categories such as patient information, EHRs, clinical laboratory, data interchange, vocabulary, object modeling, security, accounting, medical devices, XML and templates.

9.6.2.2 Adoption of Terminologies

The Health Insurance Portability and Accountability Act (HIPAA) compels the use of common clinical terminologies. To ensure interoperability, data must be stored in a terminology that is recognized by the computer that receives the data; local terms are not understood by these computers. Adoption of such terminologies are developed through public-private partnerships. The new clinical terms should be added into the main reference terminology (Systemized Nomenclature of Human and Veterinary Medicine Clinical Terms [SNOMED CT]) of the National Committee on Vital and Health Statistics (NCVHS) core terminology group, with alignment to next level classifications. WHO maintains the International Classification of Diseases (ICD) and the International Classification of Functioning, Disability and Health (ICF).

9.6.2.3 Clinical Knowledge Representation

The advancement in clinical knowledge leads to modifications in the best practices of the design of the care processes resulting in a change in the clinical data requirements. For example, in 1981, it was identified that early identification of certain eye conditions resulted in diabetes mellitus. Therefore, the American Diabetes Association released guidelines to recommend annual eye examinations. The MedBiquitous consortium, in 2003, released the XML framework to have a mapping between this evidence-based to care process.

The standards that are adapted in India can be tabulated (sourced from National resource centre for EHR standards—tentative list) as follows in Table 9.3.

TABLE 9.3
Standards Adapted in India.

S.No.	Type	Standard
1.	Patient Identity	UIDAI Aadhar
2.	Software architecture requirements	ISO 18308:2011 Health Informatics—Requirements for an Electronic Health Record Architecture
3.	Functional requirements	ISO/HL7 10781:2015 Health Informatics—HL7 Electronic Health Records-System Functional Model Release 2 (EHR FM)
4.	Terminology	SNOMED Clinical Terms (SNOMED CT)
5.	Data exchange standards	ANSI/HL7V2.8.2–2015HL7 Standard Version 2.8.2—An Application Protocol for Electronic Data Exchange
6.	Discharge/treatment summary	Medical Council of India (MCI) under regulation3.1 of Ethics
7.	E-prescription	Pharmacy Practice Regulations, 2015 Notification No.14–148/2012-PCI as specified by Pharmacy Council of India
8.	Data privacy and security	ISO/TS14441:2013 Health Informatics — Security & Privacy Requirements of EHR Systems for Use
9.	Data Integrity	Secure hash algorithm (SHA) used must be SHA-256 or higher

9.7 ELECTRONIC HEALTH RECORD

The main ingredient in the automation of healthcare is the data and this inevitably forms the primary thing to be standardized for the purpose of interoperability in healthcare.

The data stored in the healthcare information management system are EHRs. In some countries, the terms electronic medical record (EMR) and patient health record (PHR) are used interchangeably.

To maintain these records, electronic health management systems are required. Each country adopts these systems stage by stage according to their defined adoption models. The adoption models are used by each country according to the way they want to implement the system. To name a few are Capability Maturity Model Integration (CMMI), Enterprise architecture approach, Australian national health interoperability model and HIMSS EMR Adoption Model.

EHR is defined (Iakovidis, 1998) as “digitally stored health care information about an individual’s lifetime with the purpose of supporting continuity of care, education and research, and ensuring confidentiality at all times”.

The article (Eichelberg et al., 2005) discussed various standards of electronic health records for interoperability and concluded that the true interoperability could be achieved with semantic interoperability.

OpenEHR (*What Is OpenEHR?*, n.d.) is an initiative created by a virtual community with the focus of building interoperability among EHR.

The advantages of using EHRs (Jabbar et al., 2020) in healthcare as in Figure 9.2:

1. Smooth access to information enhances the physician's patient care and efficiency.
2. Information is available in a structured form so that it reduces the administration costs and avoids clinical errors.
3. A database of the patient's health information aids data mining and analytics facilities.
4. Convenient access to health records improves the patient's control of their health and wellness.
5. A database of the patient's records leads to proper decision-making resulting in value-based reimbursement eligibilities.

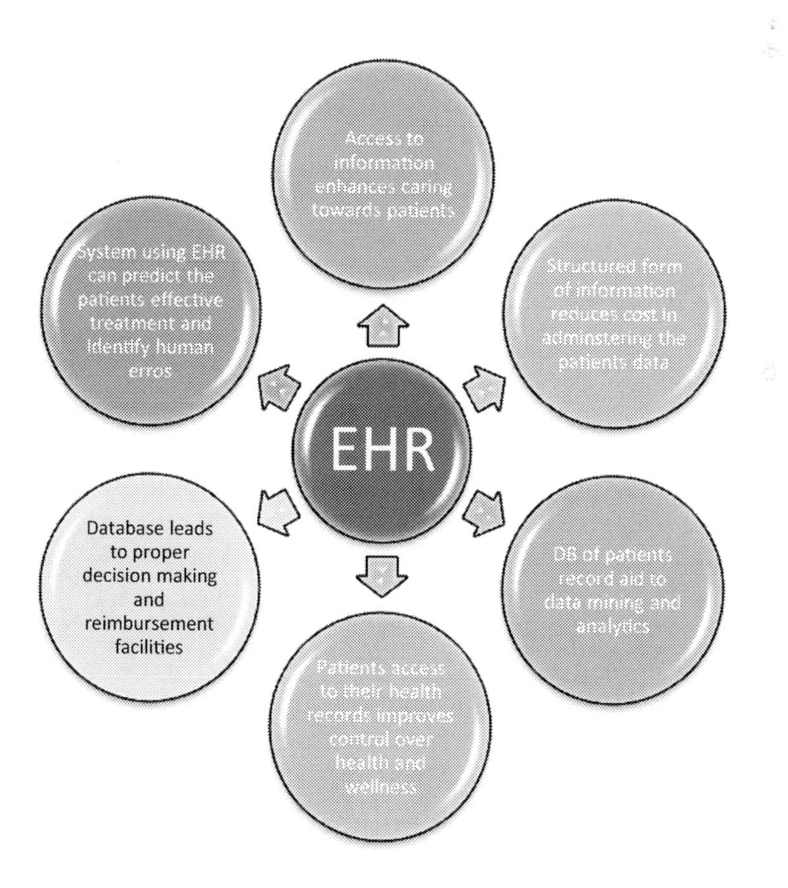

FIGURE 9.2 Advantages of using EHR.

6. A mechanism within the system that uses EHR can track the patient's records and predict effective treatment and also identify if there are any human errors during treatment.
7. The stakeholders of the interoperability are the clinics, physicians, patients and pharmaceutical companies. The data created by anyone of these stakeholders are the beneficiaries of the interoperability.

9.8 INTRODUCTION TO BLOCKCHAIN AND HEALTH CARE

A blockchain is a cryptographically secured chain of blocks containing defined data structures connected according to the timeline of their addition. The concept of blockchain originated from the concept paper for the cryptocurrency bitcoin defined by Satoshi Nakamoto (n.d.) in 2008.

The blockchain is maintained in a peer-to-peer network. Any anonymous members of this network can participate by actively proposing and constructing new blocks in the chain as shown in Figure 9.3. Adding such a new block is called mining and those who participate in adding the new block are called miners. The miners ensure consensus—that all the participants reach an agreement to add a new block and each one of them has an identical view of the previous block. This mining process provides irreversibility to new transactions making them secure and tamperproof. The algorithm implemented to reach such agreement is called a consensus algorithm. In bitcoin, it is referred to as a proof of work algorithm. Each added block must contain a random quantity that is equivalent to that of the cryptographic hash. The property of a cryptographic hash function is that it cannot be inverted and finding such a nonce could be done with a brute force search. Such a process is called work and if a miner is able to satisfy finding a nonce, then it is termed the proof of work. When a miner is able to create a block, they will be compensated with a transaction fee. Sometimes there is a possibility that more than one miner will be able to find the nonce and create a new block. This is referred to as a fork. But in blockchain only one chain is allowed. The proof of work consensus algorithm tackles this situation by allowing only one block considering only the longest chain that contains more blocks among the forked chains.

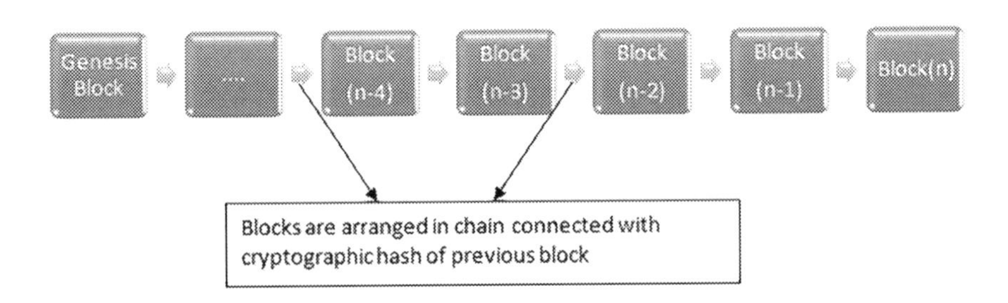

FIGURE 9.3 Structure of the blockchain.

9.8.1 Types of Blockchain

Public blockchain: As discussed previously, if any new transaction is proposed, this will be broadcast to all the participants of the network. The transactions or events happening during the mining process are stored in a data structure termed a decentralized ledger or blockchain ledger. The ledger is created with the series of blocks containing an ordered sequence of entries. Public blockchain is also called a permissionless blockchain as it is designed to allow public participation. Examples: Bitcoin, Ethereum.

Permissioned blockchain: Unlike public blockchain, permissioned blockchain restricts the participants in the network either as a miner or observer because of scalability and privacy reasons. This is also called a consortium blockchain or federated blockchain.

Private blockchain: A private blockchain is a type of permissioned blockchain in which an entity who is treated as the owner of the blockchain has the control and ability to add the blockchain. Big businesses who want to implement blockchain in their applications have a preference for permissioned blockchain as it gives them complete control. Multichain is an example of private blockchain.

Consortium or federated Blockchain: This is also called a partially private blockchain. It is operated by the consortium of organizations participating in the blockchain. The organization is allowed to take control over the selection of nodes participating in the consensus protocol instead of having complete control of the blocks. Example: Hyperledger Fabric.

Hybrid blockchain: This is the combination of permissionless and permissioned blockchain. It gives the freehold to verify the immutable record on the permissionless blocks and can make the transactions private. Example: XinFin

9.8.2 Attacks Encountered in the Blockchain

Double spending (Zhang & Lee, 2019) is a threat in bitcoin. Once the transaction is initiated by the attacker intimating the merchant to transfer money for goods from the merchant and the goods appear in the main block, it will be captured by the attacker, then the attacker secretly mines his own branch and adds the block to his branch before the original transaction has added its branch such that the attackers branch is the longest one. Because of the longest chain rule, all the miners agree to this branch created by the attacker instead of the branch created by the original honest miners. As a result, the attacker gets both the goods and the money. Such double spending threat for a merchant is a great loss to his business.

Sybil attack is also a potential threat in the peer-to-peer system which generates many fake identities to break the trust and the redundancy mechanism of a system.

9.8.3 Consensus Protocols

Proof of work (PoW): Satoshi Nakamoto used this protocol in bitcoin. It solved both the double spending and Sybil attacks, which involved the peers in the network competing with each other by solving puzzles which is called mining. The peer who is able to solve the puzzle will be announced as the winner to the peers and is allowed to create a block. Though it uses a brute-force search procedure, the answer can be verified with the hashing process, which has O(1) complexity.

Proof of Stake (PoS): PoS uses stake to replace the mathematical computation in PoW. The participants will be chosen to create the block based on the more stake a participant has. The attackers need to have the majority of coins in circulation to perform an attack. Peercoin and Shadowcash uses this protocol.

Delegated Proof of Stake (DPoS): The participants (Ago, 2017) of the peer-to-peer network delegate the responsibility of producing the blocks to a group of nodes. There are no mathematical computations. The selected block producers are monitored by the majority of the stakeholders.

9.9 DECENTRALIZED APPLICATIONS (DAPPS)

A blockchain application (Dapp) (*What Are Dapps?*, 2019) is an opensource software program hosted in a P2P decentralized blockchain environment in which validators of the blockchain are incentivized and use PoW (in Bitcoin) or PoS (in Ethereum) that can run in their own blockchain or use already existing blockchain or use only protocols of any blockchain. Based on how they run, it is termed as Type I, Type II and Type III Dapps.

9.9.1 SMART CONTRACTS

Nick Szabo in his paper (Szabo, 1997) introduced the term smart contract, in which he explained that the many kinds of contractual classes such as violation of the agreement made between a buyer and the seller of a property can be made expensive or prohibited by embedding a small piece of code in hardware or software. He defined smart contracts (Szabo, 1996) as "a set of promises, specified in digital form, including protocols within which the parties perform on these promises".

Bitcoin uses smart contracts written in scripting language embedding simple mathematical, crypto operations by validating transactions only if particular conditions are satisfied.

Ethereum (*Ethereum*, n.d.) uses Ethereum virtual machine (EVM) to execute the smart code instructions written in the programming languages like Solidity (Maintainers: Alex Beregszaszi, chriseth, n.d.) and Serpent (*Ethereum/Serpent*, 2014/2020).

Ethereum provides the platform for the developers to develop smart contracts that can be applied for many real-time use cases such as automatic execution of an agreement made in a property transaction. In this case, the agreement can define the rules to on various real time cases like what needs to happen in case the buyer is not able to complete his payment or the seller is not able to provide the defined property. These smart contracts act as the blueprint for Dapp.

Other than Ethereum, there are two more popular platforms for smart contract: EOS (*EOSIO—Blockchain Software Architecture*, n.d.) and Tron (*TRON Foundation: Capture the Future Slipping Away*, n.d.).

9.9.2 Ecosystem of Blockchain

Blockchain can be generalized into layers (Bashir, 2018) as depicted in Figure 9.4. At the bottom is Layer 0, which is the communication layer using networks such as meshnet connecting various trusted hardware. The network may be a public or private network. Layer 1 will be blockchains, sidechains and consensus algorithms. Layer 2 has channels for private peer-to-peer medium representing the state of the blockchain and payment for the decentralized application, like smart contract that establishes a set of rules. Also it can provide the platform for decentralized application execution and storage like Interplanetary File System (IPFS) and Swarm. Layer 3 is the decentralized applications. Layer 4 is the user interface like BAT browser, MetaMask.

9.9.3 Applications of Blockchain for Interoperability in Health Care

Blockchain has several properties that could benefit the healthcare domain to ensure interoperability among the healthcare management systems. They are listed in Table 9.4.

The healthcare system involving blockchain can be organized as follows:

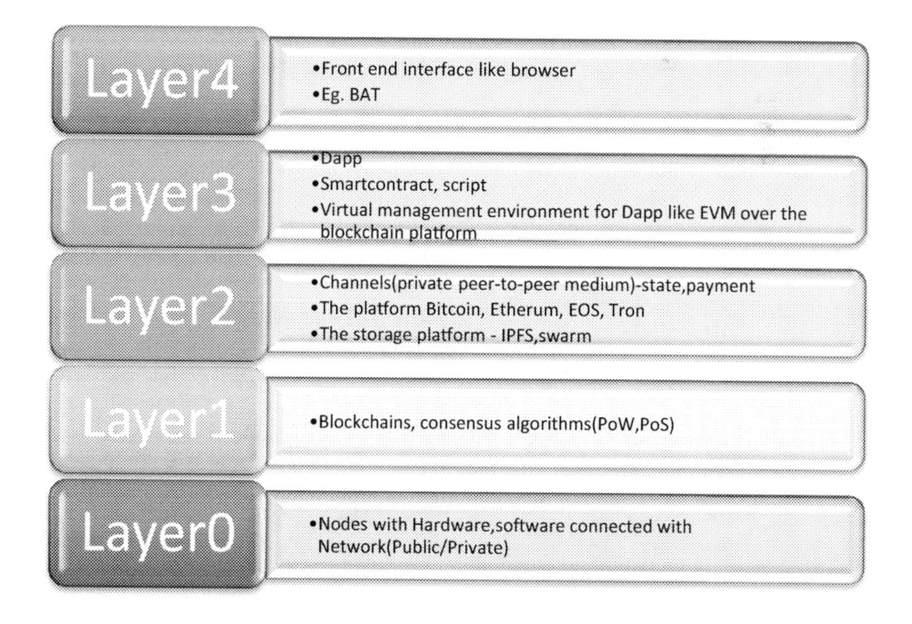

FIGURE 9.4 Blockchain ecosystem.

TABLE 9.4
Blockchain and Its Benefits in Healthcare Industry.

S.No.	Property of blockchain	Benefit in healthcare
1.	Decentralized management	The patient, physicians in various hospitals, health advisors and insurance advisors are stakeholders in curing the health issues and placing insurance claims through the decentralized collection of nodes. Blockchain block or ledger maintains a copy of the patient data and other relevant information and each node gets the local copy
2.	Cryptography (key management and other techniques)	Patient identity is kept private and anonymous
3.	Immutability	Patient data integrity and authenticity. EHR/PHR can be securely transferred between the stakeholders
4.	Data distribution and aggregation	Participation of healthcare experts, physicians and patients. The data from multiple hospitals can be aggregated and sent to the receiver healthcare expert for use to make necessary decisions.
5.	Digital access rules	Blockchain smart contract ensures that the rules enforced by any healthcare stakeholders are appropriately executed.
6.	Transparency	The electronic health record can be accessed by the experts/researchers without revealing the identity of the patients
7.	Non repudiation	Patient and related healthcare data can be validated and proof of origin can be retrieved
8.	Traceability	The treatment given to the patients can be easily tracked with this feature of blockchain. Any wrong treatment given to the patients can be identified quickly for appropriate measures to be taken. Also, this feature plays a very important role in tracking the pharmaceutical supply chain.

1. Patient data management for interoperability
   a. Patient identity

Patient identity can include the patient unique number along with their personal information and treatment history. The patient may require the option to hide their identity with some unique alphanumeric address or show the proof of identify to others. Data anonymization is this concept and when applied on the patient's identity it gives complete privacy for the patients.

b. Electronic health record (EHR)/Patient health record (PHR) management

EHR/PHR are the information about the treatment currently undertaken. The metadata of the EHR will contain the doctor id, the advice/opinion suggested for the treatment to be undertaken and all the relevant information that is easily recognized by the health management systems. There is a slight difference between the EHR and PHR. In the PHR, the ownership of the data is completely under the control of the patient. To view or update the health record requires patient consent in the case of the PHR.

c. Health information exchange (HIE)

HIE refers to the healthcare information exchange between the stakeholders of the healthcare ecosystem like patients/caretakers to that of the physicians/hospitals. Three HIE models are (i) direct: the patient knows the designated member to whom data is to be sent. The information is encrypted before sending. Both the parties can be experts in 2 different hospitals for consultations; (ii) query-based: a central repository contains electronic records, and a patient caretaker can query the database and request information about the patient from other experts; iii) patient-centered exchange: laboratory results, discharge summary, relevant medication and diet can be viewed by the patients directly. Patients are harmed because of mistakes (Makary & Daniel, 2016) made by individuals (health practitioner, health experts) or by hospitals. An error can bring an end to the patient or cause long-term incapacitation. A medical error is defined as an unintended act of omission or commission or one that resulted because of a wrong plan. This error can be avoided by providing quick attention to the affected patient by sharing the information with the right skilled person and at right time.

d. Taking the preceding relevant information into consideration, Figure 9.5 clearly depicts the flow of EHR data in the direct exchange model. The EHR data is uploaded after the appropriate permission from the patient and encrypted with the appropriate SHA-256 algorithm for security purposes. The data are now sent to the blockchain. The block containing the data along with the sender and receiver information will be added into the blockchain as a new block so that the receiver can mine for the required information and add to his/their ledger for further processing.

Variations in patient data management for healthcare with blockchain was discussed in manuscripts as follows:

(i) A patient-centric health information framework (Zhuang et al., 2020) was prescribed in this paper, which considered a blockchain adapter that can extract the EHR information and store it in the secured database and retrieve it with the same apparatus. The EHR reports are in JSON format and stored in smart contract.

(ii) This work (Esmaeilzadeh & Mirzaei, 2019) examined the HIE data exchange model by adding factors regarding privacy concerns of the owners of the data and the sensitivity of the health information. The

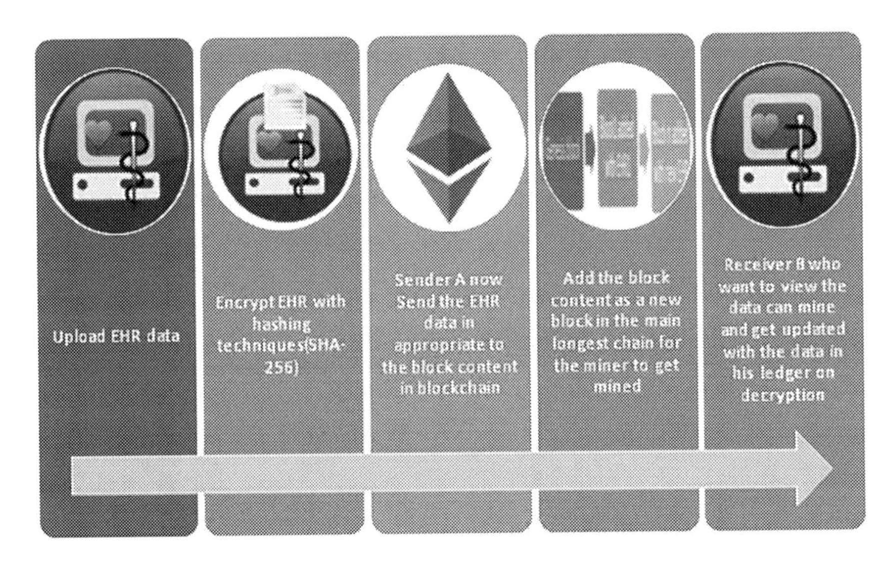

FIGURE 9.5 Direct exchange of EHR in blockchain.

HIE models were examined by adding a strong privacy policy and a weak privacy policy with the blockchain environment. They concluded that the blockchain can contribute to HIE in EHR management.

(iii) The EHR is sent across different countries and tested satisfactorily using blockchain with security and privacy of the data. This work (Lee et al., 2020) experiments on precision medicine that requires careful collection and management of clinical data. It also calls for the cross country standard blockchain architecture for the exchange of the EHR data as the future scope of the work

(iv) A framework (Patel, 2019) to enable access for the patients to delegate electronic access of medical imaging and PHR to various healthcare experts was designed using blockchain, which depicts the good illustration for the healthcare interoperability.

(v) It is inappropriate to consider the health care systems without IOT/ubiquitous/body sensor devices. This work (Uddin et al., 2018) considers the data acquisition from a body sensor network to collect the health status of a patient for continuous monitoring. In this a patient centric agent (PCA) and healthcare provider agent (HPA) are used before and after the blockchain environment. The main purpose of the PCA is to manage the data that is coming in huge volumes from various sensors, to manage the selection of miners and provide security by generating a key to hide the patient identity. The HPA will restore the data by the key management mechanism and analyze the data.

(vi) OmniPHR (Roehrs et al., 2017) is the work for the interoperability of the PHR in the healthcare system. It maintains the healthcare data in the form of blocks and the data exchange follows the publish subscribe protocol. To maintain the interoperability of the data, it uses the individual or combined techniques of the interoperable standards: Dublin core metadata model (DC), natural language processing (NLP), ontologies and software agents. The devices need to have the capability of reading PHR data in the metadata form and can join the network as providers or consumers. This architecture is devised based on the recommendations by the OpenEHR intiative.

(vii) ActionEHR (Dubovitskaya et al., 2020) is the system devised for data sharing and integration in the permissioned blockchain. It uses the hybrid data approach; the metadata on the shared data is stored in the block of the chain, whereas the encrypted shared data is stored in the HIPAA-complaint cloud storage.

(viii) MedRec (Azaria et al., 2016) is one of the earlier works in this direction, which has the EMR manager at both the provider and patient node using Ethereum permissioned blockchain environment. The EMR manager is the management module that interacts with the other modules. It supports HIPAA regulations and open standards of FHIR, HL7 for health data exchange.

2. Digital access rules in healthcare using blockchain for interoperability

   a. Digital access rules are exhibited by the smart contract in blockchain. For example, if a consumer wanted to purchase an item at a particular price and with warranty then the transaction is executed according to the rule set. Smart contract is a software program that verifies and executes certain conditions at the time of the occurrence of the predetermined events. Miners create the smart contract by posting a transaction in the blockchain. The transactions are validated and authenticated by consensus among the peer participants. The transactions are coded with algorithms before being added into the blockchain. Once added, the contract cannot be modified and execute according to the instructions (Figure 9.6).

There are many research contributions that exhibit the smart contract in the blockchain, such as:

(i) SHealth is a smart health management system designed to work in private blockchain. The participants are health providers in a licensed node who provide healthcare services and health partners that provide health insurance services. The patient visits the hospital, and the doctor advises having a blood test to determine if there is a bacterial or viral infection. A smart contract setting the rules to prescribe the patient with the necessary medication based on the outcome of the blood test results is created and termed a smart prescription.

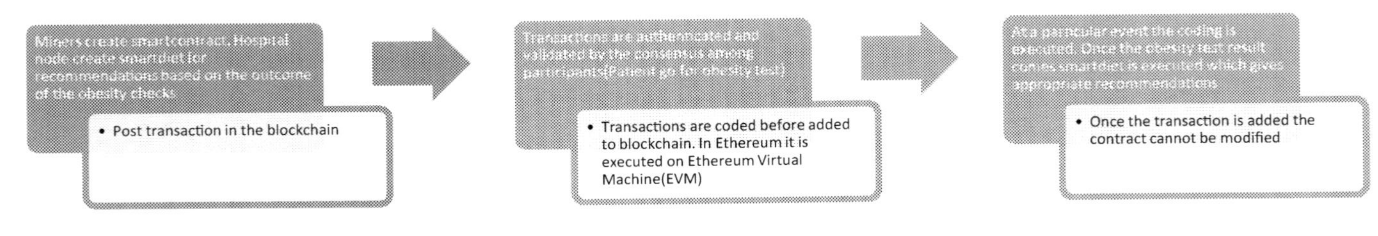

FIGURE 9.6 Smart contract in healthcare.

(ii) This work (Khatoon, 2020) exemplifies many scenarios for smart contracts in the healthcare system. A smart contract could be designed to give viewing permission to a third- party person by health care professional and patient after a set of condition is met. The reimbursement procedure for the patient in the insurance section of the hospital management system can be codified with the help of a smart contract. The medical device manufacturer looking for the clinical trials can be recommended by the hospital after the appropriate acceptance by the concerned patient is included in the health management system with smart contracts. Similarly, the cost estimation for a particular treatment is recommended to have a scope for smart contract in the health care management system.

(iii) Pharmacogenomics (Gürsoy et al., 2020) data deals with the results of the particular gene–drug interactions. This helps researchers and doctors to observe the nature of the reaction by the patients when allowed to occur. To satisfy an index-based multiple mapping is designed in smart contract to store the data.

(iv) Smart contract for attestation authority (Hasan & Salah, 2018) is created in the Ethereum blockchain for the delivery of the digital assets between two parties. The code allows for attestation provided it satisfies all the terms and conditions between all the participating entities.

(v) This manuscript (Wang et al., 2019) describes the basic terminologies, process flow, vulnerabilities and application of smart contract in various domains like finance, management, IOT and energy.

(vi) Drug counterfeiting (Sahoo et al., 2019) is a big problem that not only results into financial loss to the genuine drug manufacturer but also affects the health condition of the patients. The main problem lies in the loopholes of not able to track the goods and finance in the supply chain management system. A traceability system is required to track the goods transported from the production site to other places such as the warehouse, dealers and vendors. By applying origin chain smart contract, this paper (Lu & Xu, 2017) is able to solve the traceability of goods and sevices in the supply chain management system. This paper used pharmcrypt (Saxena et al., 2020) tool using blockchain involving smart contracts.

(vii) Many developing countries still do not have accessible health centers located in their vicinity. The solution is telemedicine (Abugabah et al., 2020), which allows interaction with the people who seek support from health experts. The framework for telemedicine support with blockchain is developed with the features that guarantee secure transactions between patients and hospitals, store medical records in IPFS and use smart contracts.

9.9.4 Challenges Related to Blockchain Implementations in Health Care

Blockchain as a technology started as cryptocurrency and is now applied in most software applications because of its inherent features of decentralized management, security, immutability and transparency. Despite these many advantages, there are challenges to implementing the blockchain technology (Figure 9.7). They are:

1. The current research (Esmaeilzadeh & Mirzaei, 2019) is focusing on implementing the blockchain technology toward interoperability with the already existing standards. It is expressed in various contributions that a standard modified appropriately for the gaps identified for this technology, such as blockchain-based HIE, is sought. Most of the implementation had a mix of decentralized and centralized architecture to achieve HIE standards.

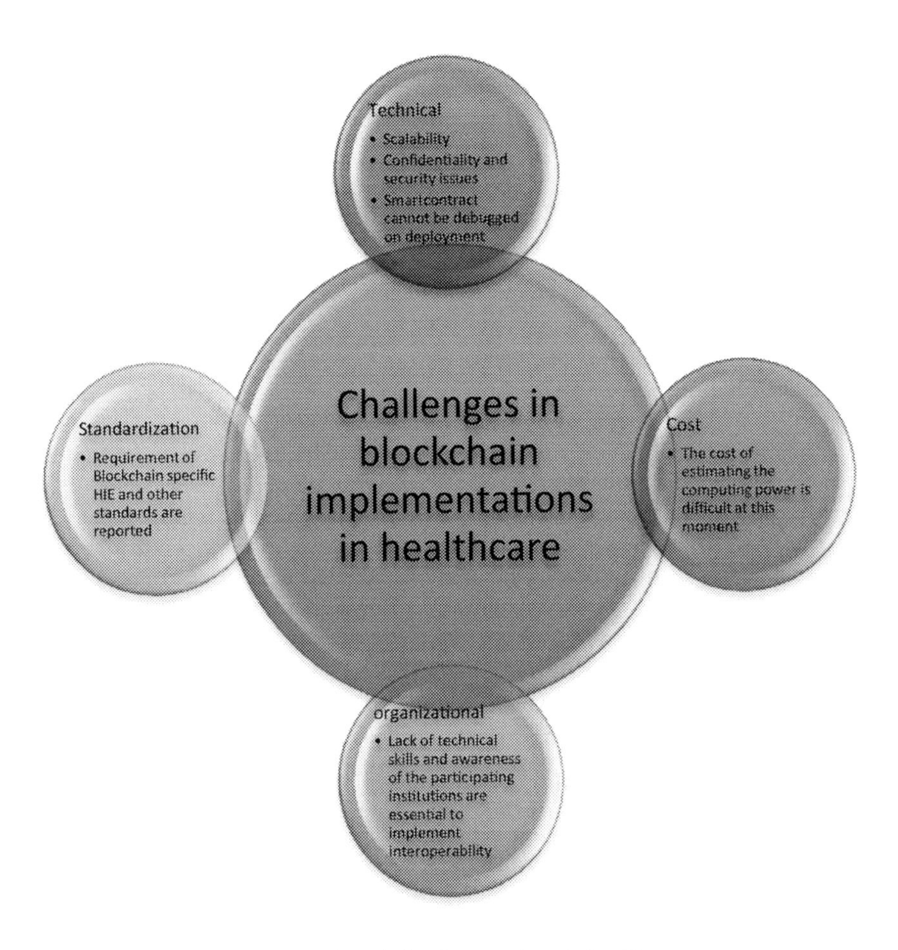

FIGURE 9.7 Challenges in blockchain implementations in healthcare.

2. Similarly (Houtan et al., 2020) blockchain-based EHR/EMR/PHR are expected to realize the fully decentralized system.
3. Though Dapps like Solidity is described as equivalent to java, it cannot be debugged after deployment. So necessary testing needs to be done only before the deployment.
4. Healthcare applications (Gordon & Catalini, 2018) may require a storage facility for transferring the medical imaging when sharing with other stakeholders. The storage size to store medical imaging is still in development although IPFS and Swarm are in the scope of a blockchain accessible file system.
5. Although blockchain technology prevents fraud of the data in the block, guaranteeing the stakeholders (Durneva et al., 2020) identity (such as patients, doctors identity) is still a question that need to be improved in this direction.
6. The operating costs (Rathore et al., 2020, p. 8) of blockchain need to be estimated and studied in the earlier adaption, which is not done as all the work currently undertaken is only to demonstrate it as the proof of concept. The computing power is directly proportional to the volume of the transactions, it is necessary to have guidelines for estimating the cost of operation on implementation of the blockchain technology powered healthcare system.
7. The high-end technologies always face the challenge in operation because of the lack of awareness (Esmaeilzadeh & Mirzaei, 2019) of the HIE empowered blockchain implementations among the patients and health service providers.

9.10 CONCLUSION

Healthcare, as with any other service operation, requires systematic innovation efforts to remain competitive, cost efficient and up-to-date. It involves subjects namely patients, knowledge and information technology. The digital resources that are exist in healthcare are PHRs, EHRs, clinical results, IOT devices etc. The related medical data and its repository are shared for best therapies among medical practitioners, researchers and biostatisticians on agreed conditions by the supervised personnel. The documents may be viewed in different context among health care organizations during information exchange. Interoperability means that the legal digital information in any form in the software on any hardware devices need to work together. The main hurdles to implementing interoperability are different formats and transition from the old format to unique formats. Blockchain has several properties that could benefit the healthcare domain to ensure the interoperability among the healthcare management systems. They are decentralized management, cryptography, immutability, digital access rules, data identity, traceability, transparency, non-repudiation and digital identity. Despite these many advantages, there are challenges like implementation standards, operating costs, lack of awareness in high-end technologies and preserving patient identity using blockchain as the implementing technology.

REFERENCES

Abugabah, A., Nizamuddin, N., & Alzubi, A. A. (2020). Decentralized Telemedicine Framework for a Smart Healthcare Ecosystem. *IEEE Access*, 8, 166575–166588. https://doi.org/10.1109/ACCESS.2020.3021823

Ago, D. (2017, May 29). DPOS Consensus Algorithm—The Missing White Paper. Steemit, #dpos • 3 Y. https://steemit.com/dpos/@dantheman/dpos-consensus-algorithm-this-missing-white-paper

Aspden, P., Institute of Medicine (U.S.), & Committee on Data Standards for Patient Safety. (2004). *Patient Safety: Achieving a New Standard for Care*. National Academies Press. http://public.eblib.com/choice/publicfullrecord.aspx?p=3376726

Azaria, A., Ekblaw, A., Vieira, T., & Lippman, A. (2016). MedRec: Using Blockchain for Medical Data Access and Permission Management. 2016 2nd International Conference on Open and Big Data (OBD), 25–30. https://doi.org/10.1109/OBD.2016.11

Bashir, I. (2018). *Mastering Blockchain: Distributed Ledger Technology, Decentralization, and Smart Contracts Explained*, 2nd Edition. 2nd Revised edition. Birmingham: Packt Publishing.

Dubovitskaya, A., Baig, F., Xu, Z., Shukla, R., Zambani, P. S., Swaminathan, A., Jahangir, M. M., Chowdhry, K., Lachhani, R., Idnani, N., Schumacher, M., Aberer, K., Stoller, S. D., Ryu, S., & Wang, F. (2020). ACTION-EHR: Patient-Centric Blockchain-Based Electronic Health Record Data Management for Cancer Care. *Journal of Medical Internet Research*, 22(8), e13598. https://doi.org/10.2196/13598

Durneva, P., Cousins, K., & Chen, M. (2020). The Current State of Research, Challenges, and Future Research Directions of Blockchain Technology in Patient Care: Systematic Review. *Journal of Medical Internet Research*, 22(7). https://doi.org/10.2196/18619

Eichelberg, M., Aden, T., Riesmeier, J., Dogac, A., & Laleci, G. B. (2005). A Survey and Analysis of Electronic Healthcare Record Standards. *ACM Computing Surveys*, 37(4), 277–315. https://doi.org/10.1145/1118890.1118891

EOSIO—Blockchain Software Architecture. (n.d.). EOSIO. Retrieved October 19, 2020, from http://eos.io/

Esmaeilzadeh, P., & Mirzaei, T. (2019). The Potential of Blockchain Technology for Health Information Exchange: Experimental Study From Patients' Perspectives. *Journal of Medical Internet Research*, 21(6), e14184. https://doi.org/10.2196/14184

Ethereum. (n.d.). Ethereum.Org. Retrieved October 19, 2020, from https://ethereum.org

Ethereum/Serpent. (2020). [C++]. Ethereum. https://github.com/ethereum/serpent (Original work published 2014)

Fragidis, L. L., & Chatzoglou, P. D. (2012, July). Challenges in Implementing Nationwide Electronic Health Records: Lessons Learned and How Should Be Implemented in Greece. In *Proceedings of the 10th International Conference on Information Communication Technologies in Health* (pp. 221–229).

Gopal, G., Suter-Crazzolara, C., Toldo, L., & Eberhardt, W. (2019). Digital Transformation in Healthcare—Architectures of Present and Future Information Technologies. *Clinical Chemistry and Laboratory Medicine (CCLM)*, 57(3), 328–335. doi: https://doi.org/10.1515/cclm-2018-0658

Gordon, W. J., & Catalini, C. (2018). Blockchain Technology for Healthcare: Facilitating the Transition to Patient-Driven Interoperability. *Computational and Structural Biotechnology Journal*, 16, 224–230. https://doi.org/10.1016/j.csbj.2018.06.003

Grossmann, C., Powers, B., & McGinnis, J. M., editors. (2011, September). *Digital Infrastructure for the Learning Health System: The Foundation for Continuous Improvement in Health and Health Care: Workshop Series Summary*. Washington, DC: National Academies Press (US); Institute of Medicine (US); Growing the Digital Health Infrastructure.

Gürsoy, G., Brannon, C. M., & Gerstein, M. (2020). Using Ethereum blockchain to Store and Query Pharmacogenomics Data via Smart Contracts. *BMC Medical Genomics*, 13. https://doi.org/10.1186/s12920-020-00732-x

Hasan, H. R., & Salah, K. (2018). Proof of Delivery of Digital Assets Using Blockchain and Smart Contracts. *IEEE Access*, 6, 65439–65448. https://doi.org/10.1109/ACCESS.2018.2876971

Houtan, B., Hafid, A. S., & Makrakis, D. (2020). A Survey on Blockchain-Based Self-Sovereign Patient Identity in Healthcare. *IEEE Access*, 8, 90478–90494. https://doi.org/10.1109/ACCESS.2020.2994090

Iakovidis, I. (1998). Towards Personal Health Record: Current Situation, Obstacles and Trends in Implementation of Electronic Healthcare Record in Europe Disclaimer: The View Developed in this Paper is that of the Author and does not Necessarily Reflect the Position of the European Commission. *International Journal of Medical Informatics*, 52(1), 105–115. https://doi.org/10.1016/S1386-5056(98)00129-4

Interoperability in Healthcare. (2020, July 30). www.himss.org/resources/interoperability-healthcare

Jabbar, R., Fetais, N., Krichen, M., & Barkaoui, K. (2020). Blockchain Technology for Healthcare: Enhancing Shared Electronic Health Record Interoperability and Integrity. 2020 IEEE International Conference on Informatics, IoT, and Enabling Technologies (ICIoT), 310–317. https://doi.org/10.1109/ICIoT48696.2020.9089570

Jonas, Steven, Raymond L. Goldsteen, & Karen Goldsteen. (2007). *An Introduction to the US Health Care System*. New York: Springer Publishing Company.

Khatoon, A. (2020). A Blockchain-Based Smart Contract System for Healthcare Management. *Electronics*, 9(1), 94. https://doi.org/10.3390/electronics9010094

Lee, H.-A., Kung, H.-H., Udayasankaran, J. G., Kijsanayotin, B., Marcelo, A. B., Chao, L. R., & Hsu, C.-Y. (2020). An Architecture and Management Platform for Blockchain-Based Personal Health Record Exchange: Development and Usability Study. *Journal of Medical Internet Research*, 22(6), e16748. https://doi.org/10.2196/16748

Lu, Q., & Xu, X. (2017). Adaptable Blockchain-Based Systems: A Case Study for Product Traceability. *IEEE Software*, 34(6), 21–27. https://doi.org/10.1109/MS.2017.4121227

Maintainers: AlexBeregszaszi, chriseth. (n.d.). Ethereum/solidity. Retrieved October 19, 2020, from https://github.com/ethereum/solidity

Makary, M. A., & Daniel, M. (2016). Medical Error—The Third Leading Cause of Death in the US. *BMJ*, 353. https://doi.org/10.1136/bmj.i2139

Nakamoto, S. (n.d.). *Bitcoin: A Peer-to-Peer Electronic Cash System*. 9.

Overview of healthcare interoperability and healthcare standards. (2017). *International Journal of Latest Trends in Engineering and Technology*, 8(3). https://doi.org/10.21172/1.83.019

Patel, V. (2019). A Framework for Secure and Decentralized Sharing of Medical Imaging Data via Blockchain Consensus. *Health Informatics Journal*, 25(4), 1398–1411. https://doi.org/10.1177/1460458218769699

Plsek, Paul E., & Greenhalgh, T. (2001). The Challenge of Complexity in Health Care. *BMJ*, 323(7313), 625–628.

Porter, M. E. (2010). What Is Value in Health Care. *The New England Journal of Medicine*, 363(26), 2477–2481.

Rathore, H., Mohamed, A., & Guizani, M. (2020). Chapter 8—Blockchain Applications for Healthcare. In A. Mohamed (Ed.), *Energy Efficiency of Medical Devices and Healthcare Applications* (pp. 153–166). Academic Press. https://doi.org/10.1016/B978-0-12-819045-6.00008-X

Roehrs, A., da Costa, C. A., & da Rosa Righi, R. (2017). OmniPHR: A Distributed Architecture Model to Integrate Personal Health Records. *Journal of Biomedical Informatics*, 71, 70–81. https://doi.org/10.1016/j.jbi.2017.05.012

Sahoo, M., Singhar, S. S., Nayak, B., & Mohanta, B. K. (2019). A Blockchain Based Framework Secured by ECDSA to Curb Drug Counterfeiting. 2019 10th International Conference on Computing, Communication and Networking Technologies (ICCCNT), 1–6. https://doi.org/10.1109/ICCCNT45670.2019.8944772

Saxena, N., Thomas, I., Gope, P., Burnap, P., & Kumar, N. (2020). PharmaCrypt: Blockchain for Critical Pharmaceutical Industry to Counterfeit Drugs. *Computer*, 53(7), 29–44. https://doi.org/10.1109/MC.2020.2989238

Szabo, N. (1996). Smart Contracts: Building Blocks for Digital Markets. www.fon.hum.uva.nl/rob/Courses/InformationInSpeech/CDROM/Literature/LOTwinterschool2006/szabo.best.vwh.net/smart_contracts_2.html

Szabo, N. (1997). The Idea of Smart Contracts. www.fon.hum.uva.nl/rob/Courses/InformationInSpeech/CDROM/Literature/LOTwinterschool2006/szabo.best.vwh.net/idea.html

TRON Foundation: Capture the future slipping away. (n.d.). Retrieved October 19, 2020, from https://debug.tron.network

Uddin, M. A., Stranieri, A., Gondal, I., & Balasubramanian, V. (2018). Continuous Patient Monitoring with a Patient Centric Agent: A Block Architecture. *IEEE Access*, 6, 32700–32726. https://doi.org/10.1109/ACCESS.2018.2846779

Wang, S., Ouyang, L., Yuan, Y., Ni, X., Han, X., & Wang, F. (2019). Blockchain-Enabled Smart Contracts: Architecture, Applications, and Future Trends. *IEEE Transactions on Systems, Man, and Cybernetics: Systems*, 1–12. https://doi.org/10.1109/TSMC.2019.2895123

What Are Dapps? The New Decentralized Future. (2019, May 1). Blockgeeks. https://blockgeeks.com/guides/dapps/

What Is openEHR? (n.d.). Retrieved October 17, 2020, from www.openehr.org/about/what_is_openehr

Who's to Blame for Healthcare's Interoperability Struggles? An Expert Weighs in—and Urges Stakeholders to Move Forward. (2017, February 28). Healthcare Innovation. www.hcinnovationgroup.com/policy-value-based-care/blog/13028191/whos-to-blame-for-healthcares-interoperability-struggles-an-expert-weighs-inand-urges-stakeholders-to-move-forward

Wright, A., Sittig, D. F., McGowan, J., Ash, J. S., & Weed, L. L. (2014). Bringing SCIENCE TO MEdicine: An Interview with Larry Weed, Inventor of the Problem-Oriented Medical Record. *Journal of the American Medical Informatics Association*, 21(6), 964–968.

Zandieh, S. O., Yoon-Flannery, K., Kuperman, G. J., Langsam, D. J., Hyman, D., & Kaushal, R. (2008). Challenges to EHR Implementation in Electronic-Versus Paper-Based Office Practices. *Journal of General Internal Medicine*, 23(6), 755–761.

Zhang, S., & Lee, J.-H. (2019). Double-Spending with a Sybil Attack in the Bitcoin Decentralized Network. *IEEE Transactions on Industrial Informatics*, 15(10), 5715–5722. https://doi.org/10.1109/TII.2019.2921566

Zhuang, Y., Sheets, L. R., Chen, Y.-W., Shae, Z.-Y., Tsai, J. J. P., & Shyu, C.-R. (2020). A Patient-Centric Health Information Exchange Framework Using Blockchain Technology. *IEEE Journal of Biomedical and Health Informatics*, 24(8), 2169–2176. https://doi.org/10.1109/JBHI.2020.2993072

10 Securing the Blockchain Network from Cyberattacks

S. Vijayalakshmi, Kiran Singh, Vaishali Gupta and Savita Dahiya

CONTENTS

10.1 INTRODUCTION

Blockchain is built upon the distributed mechanism where the database comprises a record list that grows incessantly with each record called a block. Each record needs to be protected against tampering as well as modification efforts. Nowadays it is gaining more popularity because of the promised alleviation of cybersecurity. This technology has a transparent design. A message encryption mechanism is an added security by which the blockchain is capable of safeguarding the associated nodes beyond cross platforms. The decentralization mechanism here enables data safety even with social media platforms. In this way, the secured data flow is confirmed. This is a revolutionary disruptive technology in which every transaction involving financial matters is executed in a very much reliable manner. Currently blockchain is being applied in various areas. Cryptocurrency is now the catchphrase in academics as well as corporate. In this regard, bitcoin had a major victory in the capital market with more than 110 million dollars during 2016 [1]. Bitcoin technology has been supported with a specifically made storage architecture so that the dealings will be compiled in the network with no external intervention. Blockchain serves as the central tool for bitcoin. As the transactions are compiled as a chain of blocks, it tends to grow incrementally. Its main features are a decentralized system that is persistent as well as anonymous and auditable. It is economically as well as performance wise efficient. In order to know how blockchain can be secured against cyberattacks, one needs to understand the concept of the blockchain network as well as the cyberattacks related to it.

10.1.1 Overview of the Blockchain Network

Previously existing digital technologies, even though they provided many good features, most of them were lacking in regard to security. Blockchain provides value exchange with security-enabled digitized trust by eliminating third party intervention.

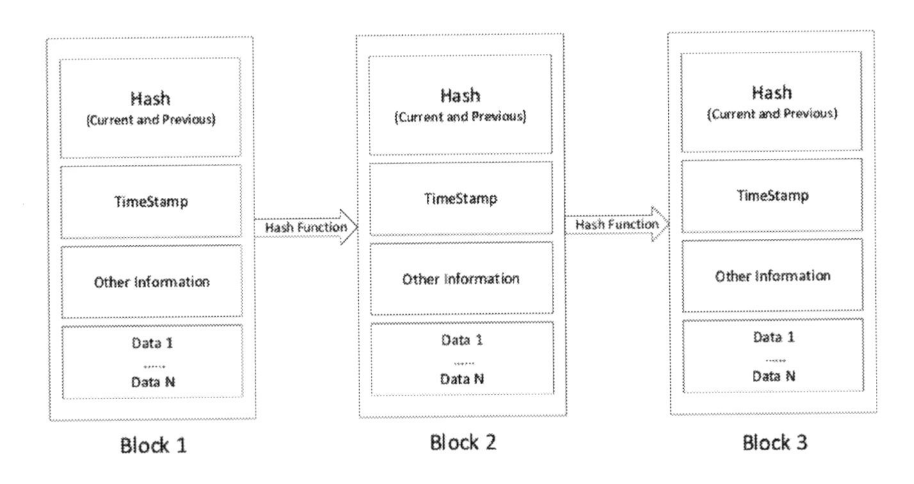

FIGURE 10.1 Blockchain structure.

The transactions are wrapped into blocks in a timely manner like the pages written in a ledger book [2]. Every newly formed block is a dependent of the prior existing block thereby forming a chain of blocks from the initial block as shown in Figure 10.1. Whenever any updating is to be done on a specific block, it needs the recomputation of all the proceeding blocks. Usually it uses the dispersed ledger having an agreement system with encryption based on public or private keys. It is a specially designed database with a unique type of storage architecture. As discussed prior, the blocks are stored with data. Whenever fresh data arrives, a newly created block is ready to hold this. As the blocks are getting filled with the data, they are being attached to the chain existing along with the previous blocks so that the blockchain is now in sequential order.

Any kind of information is inserted in the nodes of the chain, and the common practice is to store the transaction details. For example in bitcoin, the specific individual user has no authority over the blockchain, but all users as a collection hold the control. The important feature is that here the data entry is not reversible as in bitcoins one can see that the transactions have a stable recording that is visible to everyone in the network. Blockchain differs from normal databases in its storage structure. i.e., databases follow a tabular approach of storing the data whereas blockchain follows the chain of blocks holding the data. Every block that is a part of the chain has been provided with an accurate time stamp that tells the correct time of that block being added to the chain.

The blockchain security technology is the framework that stores sensitive information like transaction details also known as a "chain" or "block" in a connected network. We can call this process a "digital ledger". The three key elements of the blockchain security framework are: distributed ledger technology, immutable, and contracts. Here distributed ledger means all authorized participants can have access to the ledger. Immutable means the record is immutable, which means no person can make a change to a transaction after recorded in the ledger. Smart contract means a set of rules are created, stored, and executed. These are some conditions that are used for corporates transfer [3].

The blockchain network can be defined as a technical infrastructure that provides a ledger and a set of rules (chain code) to applications. Here the chain codes used to create different transactions are distributed to peer nodes where they have immutable records. The end users can use applications of the blockchain network administrator. The blockchain network is created by multiple organizations that come together as a consortium, here consortium means a set of non-order organization in the blockchain network, to form a network and some set of policies are also designed for their permission and these policies are agreed by nonorganizations in this network. The designed policies can be changed or modified according to the requirements [4].

Blockchain network can be classified into different categories like private blockchain, public blockchain, permissioned blockchain and consortium blockchain.

Public blockchain network (PBN): In PBN, any organization can join or participate in the network, like bitcoin. But there are some drawbacks in public blockchain networks; they are less secure, require computational power and have little privacy.

Private blockchain network: this network is like PBM; it is also a decentralized peer-to-peer (p2p) network, the only difference is that only one organization controls the entire network. This one organization can decide who can participate or join the network, also execute a policy, and maintain the entire shared ledger. It can be run behind a corporate firewall and be hosted on the premises.

Permissioned blockchain network: A business or any organization that uses a private network is known as a permissioned network. A public blockchain network can also be a permissioned blockchain network [5]. This type of network has restrictions, which means this network provides the permission as to who can join or participate in the network. Participants and organizations need an invitation or a permission letter to join a network.

Consortium blockchain network: more than one organization can join the network and can share the responsibilities of maintaining the network. In comparison with the permissioned blockchain network where only one organization can give the permission to access the data; in the consortium network, more than one organization can determine who can access the data or network. The consortium blockchain network is best suited for business organization where all participants need permission and have responsibilities for the blockchain network [6].

10.1.2 The CIA Triad in the Blockchain Network

CIA is an information security model used by organizations for the development of security policies. In CIA, C means confidentiality, I means Integrity and A means Availability. The responsibility of confidentiality is to ensure that only authorized persons can access and modify the sensitive information and to keep the information safe from the wrong hands [7]. In confidentiality, data or information is protected from the wrong people. Some security mechanisms that can be followed in confidentiality are username, encryption, passwords, and access control list. The responsibility of integrity is to ensure that information or data should be original or in the correct order. Also ensure that the receiver of the information receives the information in the same format that the sender sent it. During the life cycle of the information, modifications and changes are not allowed accidentally or maliciously. In the integrity process (trustworthy and accurate), only an authorized person has the permission to make a change in the original information. The security mechanisms followed by integrity are hashing and data consistency [8–9]. The responsibility of availability is to ensure that data and information should be available at that time for those who really need the information. Availability is implemented using hardware and software maintenance, adequate communication bandwidth and software patching. Blockchain security fulfills all the requirements of availability and integrity but faces a lot of challenges in achieving confidentiality.

10.1.2.1 Confidentiality in Blockchain

In blockchain security technology, digital technologies face some security challenges like privacy exposure, identity theft, and confidentiality disclosure to a third party. Blockchain computing technology is a digital ecosystem and has more focus on security. Confidentiality in blockchain means sensitive information is hidden from unauthorized persons. But getting a better confidentiality rank is a very challenging task because of the permissionless nature of blockchain, like bitcoin. Blockchain requires confidentiality in business. When we talk about the role of confidentiality in business, it becomes a critical task; here confidentiality is used to make a strong relation between the customer and stakeholders. In this situation, permissioned blockchain can work better because it gives the permission to access the data only to the preselected customers. When two businesses interact with each other, it does not mean how much information can be shared but also considers who can access the information under what condition [10]. Hyperledger fabric is a blockchain framework, in this IBM suggested some key points needs to be followed:

- In every transaction, it needs to be mentioned in the contract how much information can be seen by a participant: none, a small part, or all of the information.
- Understanding the nature of the network, whether it is static or flexible. So, confidentiality parameters can be changed according to the needs.
- The regulator should confirm the limit of the data.

Some library files are used by Hyperledger Fabric to achieve confidentiality: attribute-based access control (ABAC) and Hyperledger Fabric encryption literacy.

10.1.2.2 Integrity in Blockchain

Blockchain on integrity: in cybersecurity, encryption can provide confidentiality against internal attacks, but it cannot protect information from external sources like configuration errors, espionage, and software bugs. Here blockchain can provide a solid approach in the form of the hashing algorithm and Merkle tree model. Tampering with the data can be avoided with the help of integrity. Here blockchain uses hashing to keep the ledger tamperproof. One-way is a characteristic of the hashing function that means that data cannot come back from the hash function result. For integrity, the Blockchain hashing function can be used as follows:

1. Hashing public function key with SHA-256 algorithm is used to compute the Bitcoin address.
2. Hashing public function key with SHA-256 algorithm can be used to create Ethereum account identifier.

10.1.2.3 Availability in Blockchain

Business applications can be accessed from public and private networks. Blockchain software technology runs on the cloud to keep the value until it is broken and distributed. Blockchain technology is used as a decentralized application and it should be kept available all the time at the front end and back end of the system. As mentioned previously, availability means that the information should be available at the time when the authorized user wants to access the information. But distributed denial-of-service (DDoS) attacks can make the information unavailable or inaccessible. But the decentralized characteristic of blockchain security technology can make it harder for DDoS attacks to do these things. In this technology, even when one node is not working well, it is possible for the rest of the nodes to access the information in the network. All the nodes have a copy that is always up to date, and all these are decentralized from their ledger.

10.2 ERADICATING CYBERATTACKS IN BLOCKCHAIN

Blockchain is popular because it is secure and so it is essential to eradicate cyberattacks. So following cyberattacks related to blockchain are discussed first followed with how to eradicate them.

10.2.1 Cyberattacks in Blockchain

Blockchain is a distributed ledger that stores various information very securely by using cryptography. Blockchain is a data structure that stores the data in a secure manner where hacking is not possible. Organizations believe that blockchain is one of the most secure technologies in which information hacking is not easy because the stored information on blockchain uses cryptographic technologies for secure transactions. The blockchain architecture is like a spreadsheet in which row and column are available and store transaction details. The first block in blockchain is very special and is known as a genesis block. Every block has the previous block except the genesis block and is empty which also means has no data. Each block has some properties like index, time stamp, previous hash, hash, and data. Blockchain provides a unique approach to verify transactions and to store transaction details. The goal of cybersecurity is to provide a secure environment where hackers and third parties cannot enter. Like other security technologies, blockchain can also be attacked by different cyberattacks and it is denoted in Figure 10.2. The different cyberattacks include malware, phishing, man-in-the-middle attack, denial of service, SQL injection and DNS tunneling [11–13].

Denial-of-service attack (DDoS attack): The DDoS attack is a powerful weapon that has the ability to crash the targeted website or make the website unavailable to the users. By making connections and transferring fake data packets, the targeted website is flooded with messages. The first DDoS attack was executed in 1974 by a 13-year-old boy, David Dennis, who wrote a "external or ext." command program that forced the nearby research lab computer to power off. A DDoS attack on blockchain security involves bogus traffic and submitting more transactions. In blockchain, DDoS attacks overloads the blockchain with bits of data, which can force the blockchain security to the server to use processing power. In this situation, the server can lose connection at the time of crypto wallet, connected application and crypto exchange [14].

Endpoint security attack: In this platform, a human being can interact with the system. The enterprise controls the end points but in the blockchain environment, the users use their personal system to use the blockchain services. A private key is required to access the blockchain services. So, this private key can be hacked by attackers by attacking the email id and personal system. Mt. Gox, based in Japan, was the largest bitcoin exchange in 2014; it was attacked by hackers trying to gain access to the auditor computer system of the company. By getting access, the hackers changed the bitcoin value and transferred more than 2,000 bitcoins from different customers' accounts [15–16].

Sybil attack: this is an online security attack in which one person can take the entire network down by using multiple accounts, or nodes. Sybil attack is the same person who uses different social accounts. Here the "Sybil" word is derived from a woman named "Sybil Dorsett" who was treated for multiple personality disorder. In this online security attack, the attacker pretends to be multiple nodes in the blockchain network at the same time. A large sybil attack means that the attackers control most of the network computing and hash rate. In this situation, the attackers can control the transactions; they can change the sequence of the transaction or can also reverse it [17].

Eclipse attack: In the Sybil attack, the entire network is attacked by attackers but in eclipse attacks, a specific node is hacked by attackers. In the eclipse attack, a would-be bad actor hacks the particular node and eclipses these nodes from the entire blockchain network. Attackers send transaction details by showing payment proof details to the attacked node and after that they can send it on the entire network by using the same link or token [18–20].

Code vulnerability: Code vulnerability is related to software security. A flaw or some error in the program code can be a risk to security. The hacker takes advantage of this flaw to steal your data. By attaching some end point, hackers can erase everything by using this flaw. Hackers can create smart contracts that can run on the top layer of blockchain. A big hack due to overflow vulnerability was discovered in 2010 against the bitcoin network.

Routing attack: A routing attack comprises two attacks: a partitioning attack and a delay attack. In the partitioning attack, fake internet service providers partition the network in more than two groups by hacking some important key points. In the "delay attack", fake internet service providers can delay the propagation program that makes the network suspectable [21].

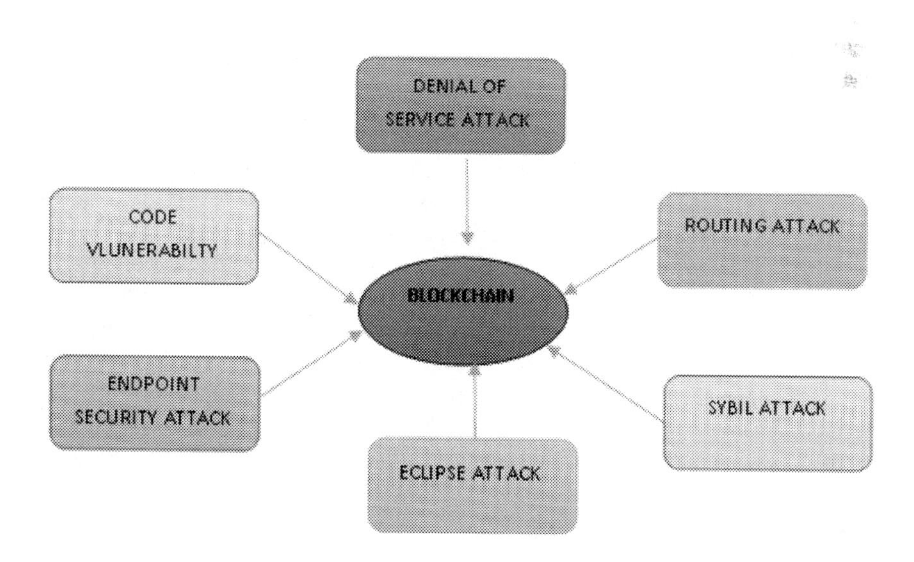

FIGURE 10.2 Attacks in blockchain.

10.2.2 Blockchain for Cybersecurity

No information security system is a hundred percent secure. We have virtual banks to transfer money, online platforms to order things like household, personal, educational, and many more goods. All these virtual platforms require accounts and passwords. Our personal data is available online and spread everywhere. From the virtual environment we have an amazing experience of shopping, money transfer, and entertainment. But all these come with a cost namely "privacy". We use dozens of accounts for banks, social media platforms and shopping websites and all these are protected by weak passwords like birthday date, bank account number, names, and phone numbers. Our personal information is not secure on online platforms; attackers can easily steal our data and misuse that information. So, a cybersecurity protocol is required to secure our online information. The largest threats in companies, industry, and in professions is cybercrime. Blockchain is a distributed ledger technique that focused on creating trust in an unsecured system. In blockchain, everything is transparent to the user of the specific domain. Members can see, record, and pass transactional information on their personal blockchain environment. Unique features of blockchain provide the benefits of cybersecurity. Blockchain security puts an impenetrable wall between attackers and information. This ledger technique creates a password-free environment. Blockchain security uses a biometric secure mechanism that includes retinal and fingerprint scan for uncrackable entry into personal data. Here we discuss some cases that make the beneficial use of blockchain technology to strengthen the cybersecurity [22].

Secure messaging: Everyday a number of people join social media platforms and communicate with each other via messaging. Huge amount of data are created during the message interaction. These social media users protect their data with weak and unreliable usernames and passwords.

Many social media applications like Facebook, WhatsApp, Instagram, and Twitter are joined by millions of people every day. These platforms use the blockchain service as a superior option for securing their users' data. It can be used to create a secure protocol. The application program interface framework uses blockchain to enable messaging communication between two parties. Numerous attacks were made against Twitter and Facebook in which millions of accounts fell into the attackers' hands. If blockchain security technology is implemented well in these messaging platforms, it can protect the users' data from future attacks very well.

Security: Blockchain technology is also involved in IOT security. Currently, attackers use edge devices like routers or thermostats to gain easy access over the entire system. In artificial intelligence-based home automation systems, it is very easy for the attacker to get access to the entire system just by using a "smart" switch edge device. In this situation, blockchain can secure the entire home automation system by using a decentralizing approach. In this approach, devices can secure themselves without depending on the central admin, because the attackers attack the central system so that they can automatically control the other devices very easily. So this blockchain decentralizing approach ensures that attackers cannot hack the devices.

Decentralizing medium storage: For every organization, business data theft is a primary concern. Most organizations use a centralized system to store the information. But an attacker can easily hack this stored information just by using a single vulnerability point. In this attack, a hacker can get confidential and sensitive information like financial records very easily and can misuse this information. Here blockchain security can protect the sensitive information by using decentralized data storage. This method would make it impossible for attackers to steal the data. Apollo Data Cloud is the best example; it is a blockchain technology that is used by the Apollo currency organization to protect data from hackers [23].

Secure data transmission: In the future, blockchain can be used to secure the data from unauthorized users at the time of transmission. An attacker can alter or delete information at the time of transmission, resulting in a huge communication gap. The encryption feature of blockchain can secure the data from malicious users at the time of transmission. This approach would increase the confidence and data integrity through the blockchain security mechanism.

Securing distributed denial of services: A DDoS attack occurs when users of a specific target resource like a network resource or server are denied service by the target resource. In this type of attack, the resource system can shut or slow down. The domain name system is centralized and is a particular target for attackers who can enter between the website and IP address. In this attack, hackers can make the website inaccessible or can redirect users to other websites. Here blockchain can be used to reduce such attacks by using the decentralized approach (domain name system entries). This approach can remove the single vulnerable point.

Blockchain in cybersecurity is used in different industries: cryptocurrencies, traditional banking, health care, defense and the military, and internet of things [24].

10.3 CYBER RISK MANAGEMENT

Before discussing cyber risk management, we first need to discuss why cyber risk management is necessary and what are the conditions that make the arrangement of cyber risk management systems in blockchain. Blockchain security works like other applications in which a small coding error can be met with large cyberattacks. Because for attackers, only a small key point is necessary to get access over an entire system. In 2016, the Decentralized Autonomous Organization DAO, which is a virtual system built on Ethereum public blockchain technology, was attacked by hackers who altered the source code of the system. The result of the attack on this organization was the loss of Ethereum tokens. Blockchain technology uses encryption algorithms for security purpose. Encryption techniques are reliable and vetted but they are also vulnerable to attacks. The vulnerability can make any blockchain platform and end user vulnerable. Private key management, wallet control, and phishing malware are a few end user vulnerabilities that need any organization to look into while applying blockchain technology [25].

Risk management is a process of identifying risk and assessing the risk impacts. It is a similar concept as a company that has assets to protect. Let us take the example of insurance. Various insurance plans are designed to protest the person from losses.

Risk management can be compared to physical devices like doors and locks that are used to protect homes from physical risks. On the other side, internet technologies departments depend on strategies and technologies. These departments use the user educational knowledge to protect the information from attackers. The need for cybersecurity risk management depends on the growth in the volume of cyberattacks. Cybersecurity risk management involves the process of identifying a vulnerability and applying an administrative approach and action plan to protect the organization from attackers [26].

For every organization, regardless of size, it can small or it can be a big industry, development of a cybersecurity risk management plan is necessary. Some point that need to be considered at the time of planning a risk management by every organization include: design an organization culture, distribute the responsibilities to every employee of the organization, give training to the employees, provide cybersecurity risk information to everyone in the organization, design the cybersecurity framework, prioritize risks, and plan for incidents [27].

10.3.1 Approach for Cyber Risk Management in Blockchain

Currently different applications like industry, academia, and many more are attracted to blockchain for the secure environment it provides. The blockchain system that depends on the "proof-of-work protocol" has gained popularity because of its capabilities like decentralized ledger and provision of an environment for data-driven organization. Even then they are vulnerable to double-spending attacks. So, an approach was designed for the blockchain service market cyber risk management. This approach is known as a "Game Theoretic Approach". This approach can jointly handle risks and security. In this approach, cyber insurance is taken as a tool for making cyberattacks ineffective. This approach uses the "blockchain service market", which is integrated with infrastructure, user, blockchain provider, and cyber-insurer. The "Stackelberg game model" is a two-stage model: upper stage and lower stage. In the upper stage, the aim is to get the maximum profit. Here the blockchain provider and cyber insurer work together to make a strategy for getting the maximum profit. The user is involved in the lower stage; here the users manage their demands with respect to the cost and security of the blockchain service provider. In this approach, balancing between a proactive strategy like computing power investment and cyber insurance investment is considered by the blockchain provider. For the user of the service, the impact of social externality and service security is considered in the blockchain service valuation. And last, for the cyber insurer, the risk adjustment mechanism is incorporated in the premium adaptation [28].

A blockchain is a computerized record of exchanges. The name comes from its structure, in which individual records, called blocks, are connected together in a single list, called a chain. Blockchains are used to record exchanges made with digital currencies, for example, Bitcoin [29], and have numerous different applications.

Cybersecurity: "Cyber security is principally about people, cycles, and advancements participating together to join the full scope of danger decrease, weakness decrease, prevention, worldwide duty, occurrence reaction, adaptability, and recuperation strategies and exercises, including PC network activities, data confirmation, law authorization, and so forth" (javatpoint.com).

Network safety is the protection of Internet-associated frameworks, including hardware, programming, and data from computerized attacks. It includes two aspects: one is digital and the other is security. Digital is identified with the innovation, which contains frameworks, organizations and projects or information. Whereas security is identified with the insurance, which includes frameworks security, network security, and application and data security.

It is the assortment of advancements, cycles and practices intended to ensure networks, gadgets, projects, and information from assault, robbery, harm, change, or unapproved access. It might likewise be alluded to as data innovation security.

We can likewise characterize network safety as the arrangement of standards and practices intended to protect our processing assets and online data against dangers. In an advanced industry, because of the heavy reliance on PCs that store and send a bounty of private and fundamental data about the individuals, network protection is a basic function and required for the protection of numerous organizations.

10.4 USE-CASES OF BLOCKCHAIN FOR CYBERSECURITY

Blockchain [29] was created to have one of the most foolproof types of execution in the computerized network domain. The advancement has been credited for its data incorruptibility. Numerous frequently used areas can profit from blockchain, and because it has various uses, blockchain can be incorporated into various jobs. Likely the best uses would employ its integrity affirmation to build network protection solutions for some different advancements. Following are some cases of future significant use of blockchain to strengthen digital protection:

Securing Private Messaging: With the web turning the world into a general town, a growing number of individuals are joining on the web media. The amount of online media stages is also on the rise. More social applications are being dispatched with every dawn as conversational exchange becomes conspicuous. Colossal proportions of metadata are assembled during these interactions. Most online media stage customers guarantee the organizations and their data with delicate, sensitive passwords.

Most educational associations are warming up to blockchain to ensure their customer data as a superior option than the beginning to end encryption that they are now starting to use. Blockchain is used for standard security. To enable cross-messenger correspondence limits, blockchain can be used to shape a bound together API framework.

In the past, different attacks have been executed against social platforms like Twitter and Facebook. These attacks achieved data breaks with countless records being infiltrated and customer information showing up in some unwanted hands. Blockchain use in these systems may thwart such future computerized attacks.

IoT Security: Hackers use edge devices, for example, indoor regulators and switches, to gain admittance to general frameworks. With the current interest in artificial intelligence (AI), it turns out to be extremely simple for programmers to get to all frameworks like home robotization through edge devices like "brilliant" switches. In practically all cases, countless IoT [29] devices have Scrappy security highlights. For this situation, blockchain can be utilized to secure such frameworks or devices by decentralizing their organization. This methodology will allow the device to settle on security choices on their own. Not relying upon a focal administrator or authority makes the edge devices safer by identifying and following up on dubious orders from unstable organizations.

Ordinarily, programmers infiltrate the focal organization of a device and naturally oversee the devices and frameworks. By decentralizing such devices' authority frameworks, blockchain guarantees such assaults are more difficult to execute.

Securing against DNS and DDoS: A DDoS attack occurs when clients of an objective asset, for example, an organization asset, worker, or site, are denied admittance or administration to the objective asset. These attacks shut down or hinder the asset frameworks. Then again, an unblemished domain name system (DNS) is highly consolidated, making it an ideal objective for programmers who penetrate the association between the IP address and the name of a site. This attack makes a site difficult to reach, cashable and even divert to other trick sites. Luckily, blockchain can be utilized to reduce such types of attacks by decentralizing the DNS passages.

Decentralizing Medium Storage: Important information hacks and burglary are turning into a leading and obvious reason to worry for associations. Most organizations actually utilize the consolidated type of capacity medium. To access all the information stored in these frameworks, a programmer basically abuses a solitary weak point. By utilizing blockchain, delicate information might be secured by guaranteeing a decentralized type of information stockpiling. This alleviation strategy would make it harder and even impossible for programmers to enter the information stockpiling frameworks. Numerous capacity administration organizations are surveying ways that blockchain can shield information from programmers. Apollo Currency Team is a good illustration of an association.

The Provenance of Computer Software: Blockchain can be utilized to guarantee the integrity of programming downloads to forestall unfamiliar interruption. Similar to how MD5 hashes are used, blockchain can be applied to check exercises, for example, firmware updates, installers, and patches to forestall the installation of noxious programming in PCs. In the MD5 situation, the new programming traits are contrasted with the hashes accessible on the seller's site. This technique isn't totally secure as the hashes accessible on the supplier's site may be undermined. Be that as it may, on account of blockchain innovation, the hashes are forever recorded in the blockchain. The data recorded in the innovation isn't variable or inconsistent; therefore, blockchain might be more productive in checking the integrity of programming by contrasting its hashes against the ones on the blockchain.

Verification of Cyber-Physical Infrastructures: Data altering, framework misconfiguration along with segment failure affect the integrity of data produced from digital physical frameworks. Nonetheless, the abilities of the blockchain innovation in data trustworthiness and confirmation might be used to verify the status of any digital physical frameworks.

Protecting Data Transmissions: Blockchain can be used in the future to prevent unauthorized access to data while it is in transit. By utilizing the complete encryption feature of the technology, data transmission can be secured to prevent malicious actors, be it an individual or an organization, from accessing it. This approach would lead to a general increase in the confidence and integrity of data transmitted through blockchain. Hackers with malevolent aims tap into information in the midst of travel to either change it or totally erase it. This leaves an enormous hole in wasteful correspondence channels, for example, messages.

Diminish Human Safety Adversity Caused by Cyberattacks: Thanks to creative innovative technological advancements, we have recently observed the release of automated military hardware and public transportation. These robotic vehicles and weapons are conceivable thanks to the Internet, which encourages the exchange of information from the sensors to the controller data sets. Nonetheless, hackers have been at work to break into and access the organizations, for example, Car Area Network (CAN). When taken advantage of, these organizations offer unlimited oversight permission to the authoritative car capacities to the programmers. Such events would directly affect the well-being of people. Yet, through information confirmation directed on blockchain for any information that goes in and through such frameworks, numerous difficulties would be prevented.

10.5 DECENTRALIZED DATA STORAGE IN BLOCKCHAIN

Decentralized [30] capacity is a good arrangement that blockchain organizations are investigating and executing. It is an arrangement of having the option to store your reports and documents without depending on enormous, concentrated storehouses of information that don't decide significant qualities, for example, protection and accessibility of your information and data can keep a full record of our lives, putting away in a real sense everything from individual photographs and recordings from our cell phones to work archives. At first sight, this solution makes our lives easier, but some unexpected threats can be found under a covering of comfort and extensive customer care.

Centralized data storage has its own advantages—fast speed and availability, quick throughput and low latency—but includes some major disadvantages. Huge distributed storage organizations, for example, Google and Amazon, that dominate the business are frequently associated with helping out specialists and giving them access to private information. It could be usefully refined on the basis that customers' documents are not scrambled, stored in a single location and are defenceless against any controls.

Governments can likewise confine admittance to certain substances for political reasons, as was done by Turkish authorities in 2017 when Wikipedia was prohibited in the nation. China went much further as the world's most famous online media, cloud storage and video platforms have been prohibited in the state and supplanted by local analogs.

In contrast to centralized cloud storage, decentralized storage services brag of being safer and private. They don't store client information on a solitary, centralized server. Instead, they divide files into numerous pieces and send them to different servers or hubs and in this manner diminish the chance of outside command over client information. Regardless of these improvements, decentralized capacity also has a few limitations.

Since blockchain began emerging, there have been enthusiasts who guarantee that it makes everything from banking to medical care and from voting to fundraising better.

10.5.1 Decentralized Cloud Storage Operation Principles

Cloud storage systems [30] store data on remote servers accessed from the internet and called "clouds". These servers are maintained by cloud server providers. In contrast to conventional cloud server, decentralized distributed storage doesn't keep the customers' information on one specific unified server. All things considered, it utilizes various hubs situated around the world, which are autonomous of one another. The nodes are not hosted by a single entity and are not controlled by service providers; anyone can run a node.

It all started almost 20 years ago with the Bit Torrent protocol, which was designed for peer-to-peer file sharing. Bit Torrent users download various video, music, and text files to their local storage and then can share them with other users.

The Interplanetary File System (IPFS) convention is another progression in the advancement of decentralized stockpiling. It appeared in 2015 and later became the foundation for some of the currently developing blockchain-based decentralized storage solutions, for example, File coin. Similar to HTTP, IPFS is a hypermedia convention for the web intended to move information among clients and servers on the web, yet it deals with numerous hubs rather than a focal server. At the point when somebody transfers a document to the IPFS network [30], the record is separated into parts called blocks. Every one of them gets an individual hash. The squares can later be found and recovered into one record by their hash or name utilizing content-based addressing, which contrasts from location-based addressing in HTTP.

10.5.2 Blockchain-Based Solutions in Cloud Storage: Off-Chain and On-Chain

BitTorrent and IPFS are both a long way from perfect and have various difficulties. With the rise of blockchain innovation, utilizing it to improve information stockpiling has been engaging different engineers around the world. Blockchain-based decentralized [2] cloud arrangements have gained from their archetypes and focused on improved security, protection, and the clients' power over their information. One of their distinctive highlights is encryption. At the point when you transfer a document to the organization, it naturally encodes the record. From that point forward, you can gain admittance to your record with an encryption key; without the key, nobody can access and peruse your document.

What blockchain-based arrangements share practically speaking with BitTorrent and IPFS is concealment. In basic terms, it is a cycle of breaking a solitary record into various pieces so these pieces could be put away on various hubs. No single hub holds your whole record, all things being equal; they just keep a section of it. Those sections are copied, which prompts repetition in the information; regardless of whether a specific hub separates with a part of your document, a similar piece can be found on different hubs.

In a general sense, there are two various methodologies in blockchain information stockpiling arrangements: off-chain and on-chain.

The on-chain methodology suggests that all customer data is taken care of inside each block on the blockchain. The verifiable advantage of this technique is that even in the event of an assault, the data can be restored and resynchronized. The overhauled security incorporates some huge traps to maintain full hubs: Every hub ought to contain, from a genuine perspective, all transferred data, which is a verifiably more expensive option. It is acknowledged that the blockchain isn't adequately adaptable to store the customers' entire records. Any running node should keep a duplicate of all transferred clients' information, and all nodes should continually synchronize with each other. In the event that every client transfers only a couple megabytes of information, the organization will get overburdened. Additionally, it will cost a fortune in organization expenses. This issue is known as blockchain swelling. That is the reason practically all information stockpiling arrangements available are off-chain. They are attempting to take care of the adaptability issue by not storing the clients' information in the blockchain, restricting themselves to simply putting away metadata on-chain and utilizing blockchain for encouraging the data in initial stages. The obvious sensitive motivation behind off-chain game plans is more delicate security. If the system gets attacked, speculatively, there can be a circumstance in which metadata will be essentially the primary concern left, whereas data will be lost completely. Albeit, off-chain courses of action are more financially savvy and have different use cases. The off-chain arrangements use miners who give their hard disk to store other clients' records for a prize, and the blockchain is utilized to encourage the capacity market among miners and clients. Persuading clients to store another person's information on their disks and to run hubs may be testing yet is basic for scaling the initial system of off-chain arrangements, and blockchain helps decentralized clouds with that. One of the most broadly spread alternatives is to utilize the platform's local crypto coins as an impetus. This persuades clients to lease their extra disk space, along these lines permitting this trustless initial system to develop.

In the File coin organization, blockchain is likewise used to interface clients who need to store their information with the individuals who can give extra room—they are additionally called "miners". A customer presents an offer on the blockchain, and when a coordinating request from a miner is discovered, parties sign an arrangement request. Miners are then compensated with coins. Decentralized distributed storage has its focal points, and blockchain adds some more in contrast to conventional concentrated cloud services like Amazon or Google drive; blockchain-based decentralized distributed storage has various convincing advantages.

Security: Blockchain-based decentralized distributed storage makes holding and transmission of information more secure. Records are scrambled with private keys, which make it unthinkable for anybody without the key to get to the document.

Documents are additionally partitioned into pieces to be kept on various hubs so that there is no single point of failure. In the event that a unified server disconnects, you'll presumably lose access to your information. On the off chance that a specific hub fails, you will secure your documents.

Changelessness: Because there is no focal power, nobody can remove your document, limit access or make alterations to it for control. The document's hash is kept in the record.

Lower cost: While incorporated distributed storage providers like Amazon S3, Google One and Dropbox offer 1 GB of space for $0.023, $0.02 and $0.005 every month, respectively, their rivals utilizing blockchain have costs as low as $0.002.

Awards for storage: various decentralized cloud ventures utilize blockchain and local digital currencies to boost clients. The individuals who have extra storage room—unused hard drives, plates, server farms—can lease it for a prize. Blockchain-distributed storage spaces interface clients ready to share their extra space with those who need it, making it a mutually beneficial arrangement.

10.6 EDGE DEVICES

Edge devices [31] are pieces of hardware that control data flow at the boundary between two networks. Cloud computing and the internet of things (IoT) have increased the role of edge devices because of the need for more intelligence, calculating power, and advanced services at the network edge.

Most cyberattacks aim to threaten the confidentiality, integrity, and availability of data and to disrupt an organization's business processes. To minimize the confidentiality or integrity of data, the attacker must get access to it. A breach of confidentiality results in only reading the data or exposing it to those who should not have access. To spoil the integrity of the data requires an attacker to have the ability to change the data in some way. Compromising the availability of the data, on the other hand, does not require the attacker to get access to the data. Disrupting operations also does not necessarily require deep access to a network. For external attackers to achieve these goals, they have to find a way into a network remotely. Edge devices are a common target for attackers wishing to get inside because of their position and functions. These devices give interconnectivity between various organizations by communicating, checking, sifting, interpreting, or storing the information that passes from one organization onto the next. They primarily serve as an entry or exit point for networks, making them inherently attackable by outside entities. This analysis will not cover IoT devices—such as home appliances, wearable technologies, and other smart gadgets—that are often talked about in tandem with edge computing and edge hardware. Four main categories of devices will be covered in this report: (i) network edge devices: routers, switches, wide area network (WAN) devices, VPN concentrators; (ii) network security devices: firewalls; (iii) network monitoring devices: network-based intrusion detection systems (NIDS); and (iv) customer premise devices: integrated access devices.

10.6.1 Attacks Targeting Edge Devices

Edge devices might be run in a variety of different services at once. These devices must withstand a constant onslaught of inbound, unsolicited traffic, much of which mimics the legitimate requests that originate with the intended users of these services. A survey of honeypot data obtained over an extended period reveals that most of the attacks involved brute force attempts to pass default or common username and password credentials. These attacks focus on a variety of services, including the remote access Virtual Network Computing (VNC) [32] or Remote Desktop Protocol (RDP) protocols, remote terminals over Telnet or SSH, internet telephone adapters, or database servers. Many of these automated attacks appear to use widely available default credentials from a broad range of network-connected devices, including routers, network-attached storage (NAS) devices, cameras, Wi-Fi access points, DSL and link modems, and IoT devices or IoT control hubs. Attackers can employ these methods to install malicious code onto the device or change a configuration in such a way as to benefit the attackers, for example, changing the DNS servers to point to an IP address under the attackers' control to subtly manipulate the destination of network traffic. In addition to attacks against popular services, CTA members have also seen a swath of attacks leveraging publicly disclosed vulnerabilities on a range of enterprise- or consumer-grade networking products. Exploits against, and the attempted use of, default administrative credentials for routers and other networking equipment from Huawei, Cisco, Zyxel, Dasan Networks, Synology, D-Link, TP-Link, TrendNet, MikroTik, Linksys, QNAP, and many others are now part of the common vernacular of scripted attacks and brute force attempts observed on a daily basis. Because some of these devices now have high-end processing capabilities, they can be targets not only for penetration but also for malware designed to carry out an array of malicious activities, including illicit cryptocurrency mining, storing stolen files, or leveraging the infrastructure to stage future attacks. Attackers often leverage these infected devices to mount attacks against (and deploy copies of themselves to) similarly vulnerable devices elsewhere on the internet.

10.6.2 Blockchain to Secure Edge Devices

Blockchain innovation can be used to protect systems and devices from attacks. Furthermore, blockchain security means that there is no longer a centralized authority controlling the network and verifying the data going through it. The edge blockchain that is maintained by all edge servers [32] is a decentralized chain that stores the transactions among IoT devices and between IoT devices and edge servers, which require unification, consistency, and transparency. The edge servers have computational abilities and extra room for correlative administrations. Each edge server has an administration regulator and capacity pool. The administration regulator functions as an outsider to oversee exchanges from IoT devices. As the assets of IoT devices, for example, extra room, calculation force, and memory are compelled, an assortment of exchanges should be put away in edge servers. At the point when the exchanges are created by the IoT devices [32], they are communicated to the web upheld by the edge servers. For security and protection confirmation, the exchanges created by IoT devices are obscure and mixed and associated with computerized marks of the IoT devices. The edge servers filling in as blockchain chiefs will occasionally coordinate exchanges into a block with the agreement convention (for example, Confirmation of Storage) and broadcast the block to other edge servers for checking. The edge servers with the most commitment in the organization are compensated throughout some undefined time frame, which is a motivation to urge edge servers to offer enough help for keeping up the blockchain.

10.7 PREVENTING DDOS ATTACKS USING BLOCKCHAIN

A Denial of Service (DoS) attack is a situation in which the intended users cannot access a machine or network because the attack floods the network with so much traffic or data that no one else can use it. The attack can be performed by overloading or shutting down a network by sending so many invalid, malformed or overwhelming connection requests in such a short period of time that it overwhelms all resource consumption such as memory, processing cycles, storage space, bandwidth etc.

The Distributed Denial of Service (DDoS) attack can be considered as a specific kind of DoS attack in which the attack originates from multiple computers simultaneously to target a single victim. The targeted victim in the DDoS attack could be the whole network or a particular host. The impact of this kind of attack is more lethal because it utilizes multiple computers at the same time and thus harms the digital availability.

DDoS attacks that are severe and complex are increasing day-by-day in the digital world and affecting enterprises from all sectors, all sizes, and all locations. The DDoS attacks have become a severe cyber threat because of the high availability of the Internet and use of new networking technologies such as wireless, IoT, etc. The most well-known recent DDoS attacks are Amazon Web Services (February 2020), Github (February 2018), Google (2017), and Dyn (October 2016).

There are many technologies available to prevent DDoS attacks and reduce the risk of DDoS attacks. One such technology to prevent the DDoS attacks is blockchain technology. The blockchain could be a promising new technology solution for the DDoS attack problem because of its salient features such as stability, decentralization, and secure architecture.

Blockchain is an innovative technology based on the concept of public key cryptography and digital signatures that fulfills the most important security requirements, that of integrity, reliability, authenticity, immutability, etc. The distributed ledger system of the blockchain makes the processing infrastructure decentralized, which helps in avoid failures at a single point.

Because blockchains are decentralized distributed ledgers that work independently without any third party in peer-to-peer networks with cryptographic authority, blockchain technology can be considered a DDoS attack solutions. There are several ways to prevent DDoS attacks using blockchain:

- Blockchain technology can support creation of a secure peer-to-peer network in which all nodes would interact and communicate in a secure and reliable manner to avoid threats and DDoS attack commands from malicious sources.

- With the help of the decentralized feature of Blockchain, the type of DDoS attack could be addressed on the DNS servers. A blockchain-based decentralized DND naming system could be built in which only legitimate parties or people (domain owner) are allowed to update the domain record using a private key. Because IP addresses would be stored on the blockchain, they would be copied ubiquitously across all the nodes, which would guarantee no single point of failure and hence no possibility of DDoS attacks.
- One more ideal use of Blockchain technology could be the use of blockchain in reengineering the entire structure of the Internet, that is, from a centralized client/server model to a decentralized model without specific servers. There would be no servers; hence no possibility of DDoS attacks. The decentralized model can be built by using the collective processing power and storage capacity of the millions of connected devices in a peer-to-peer network secured by blockchain.

10.8 BLOCKCHAIN FOR PROHIBITING HACKERS ON IOT DEVICES

With the boom of smart devices and high-speed networks, the IoT has acquired significant acceptance in the form of smart homes, smart hospitals, smart cars, smart cities, and smart everything. According to Cisco Inc., there will be 50 billion connected IoT devices by 2020 [33]. These smart devices are more vulnerable to attacks because of their limited computational power, storage capability and network capacity. Therefore, hacking of most of the IoT devices would be easy. Because IoT devices are placed and connected in unsecured IoT environments, security of these IoT devices is a main concern in order to protect the data that is collected and passed through these devices from hackers.

In order to achieve a secure IoT environment, blockchain is a potential solution.

The blockchain technology can be considered as a core technology for providing feasible security solutions to IoT device security issues. The blockchain technology plays an important role in management, control, and security of IoT devices.

A blockchain is a decentralized, distributed, shared, and immutable database ledger that stores records and transactions across a peer-to-peer network. It is a sequence of blocks of data and uses cryptography and digital signature concepts to provide data authentication and integrity. Blockchain can be built either as private network wherein joining is limited to only specific participants or as a public network wherein anyone can join. A private network in blockchain technology provides greater control of privacy and access.

10.8.1 Security Requirements for IoT Devices

Data Privacy: In the IoT environment, devices, services, and the network are integrated in a diversified manner. This makes IoT devices and their data more prone to privacy threats through security compromise of the devices in a network.

Data Confidentiality: Because information in the IoT environment travels through several IoT devices, there is a need for an efficient encryption technique in order to ensure the confidentiality of the information across the network.

Data Integrity: In the centralized model, attackers can gain unauthorized access to IoT devices and violate data integrity by making changes in the data that is stored on IoT devices for malicious purposes.

Data Authentication and Authorization: Authentication and authorization are required to provide secure communication between IoT devices. Devices must be authenticated to enable privileged access to services. To ensure authorized access to devices or information, authorization techniques must be provided.

Availability of Services: The conventional denial-of-service attacks on IoT devices may adversely affect the provision of services through device malfunctioning, resulting in deterioration in the quality of service to the IoT device users.

Energy Efficiency: As highlighted earlier, IoT devices have limited computing power and storage capacity. Because of these characteristics of IoT devices, when attacks happen, the energy consumption of these devices may go up because these attacks create several false or obsolete service requests that overload the network and exhaust these devices.

10.8.2 Blockchain Solutions for IoT Devices Security

The salient characteristics of blockchain can be helpful for ensuring the security of IoT devices. These are highlighted as follows:

Address Space: Blockchain has a 160-bit address space that allows generation and allocation of addresses in offline mode for approximately 1.46×10^{48} IoT devices [34].

Data Authentication and Integrity: Another salient characteristic of the blockchain network is that it is cryptographically proofed and digitally signed by the authenticated user. This ensures authentication and integrity of transmitted data by the devices that are connected to the peer-to-peer network. All transactions related to the IoT devices are stored on the decentralized distributed ledger of the blockchain, allowing their secure tracking.

Data Privacy: Blockchain smart contracts could be useful for ensuring data privacy. These contracts can set the rules for accessing the network and can lay down the timing and other conditions for accessing and controlling the IoT devices by specific individuals or groups. In addition, smart contracts are also responsible for providing the right to update, modify, or reset the IoT device, initiate or terminate a service, ownership rights and provision of the IoT device.

Removal of third-party risks: The decentralized characteristics of blockchain technology enables the IoT devices to communicate with each other without any third-party risk.

Secure Communications: The peer-to-peer blockchain network allows IoT devices to have their unique key GUID and asymmetric key pairs that would totally eliminate the key management and distribution in order to achieve secure communication among the IoT devices.

Fault tolerance: Decentralized IoT devices in the peer-to-peer blockchain network are less likely to fail because each and every device has the same copy of the record. Hence, failure of any single device would not affect the entire network. Therefore, blockchain prevents the IoT environment from having a single point of failure.

10.9 IMPROVING DATA VERACITY USING BLOCKCHAIN TECHNOLOGY

Big data is a collection of huge and diversified data sets having structured, semistructured, and unstructured data obtained through multiple sources in varying sizes ranging from terabytes to zettabytes. The defining factors of big data are the 4Vs, namely, volume, velocity, variety, and veracity. Volume means the quantity of data whereas velocity states the accelerating speed at which data is generated. As big data is generated on a real-time basis on a very large scale and comes from diverse sources, variety refers to the type of data that is structured, semistructured, and unstructured in nature. The fourth dimension, veracity, refers to the level of trust or credibility of the data. With an increase in volume, velocity, and variety of data, the veracity decreases and leads to low creditability of data or less trustworthy data. The uncertain nature of big data leads to untrusted results that further lead to poor decision-making. Uncertainty with regard to the veracity of data comes from different sources such as measurement errors from sensors data, lack of credentials from social media data, etc. Therefore, big data veracity improvement is essential to control business risks associated with decision-making.

As data veracity is a huge problematic area in big data analytics, blockchain can provide useful solutions for improving data veracity and security. Blockchain technology offers a new mode of trusted communication between two parties by eliminating third-party interference in a network. In a shared ledger blockchain system, all transactions are secure, authenticated, and verifiable because of agreement among all the parties in the network.

Another way in which Blockchain can help improve data veracity is through the use of token-based crowdsourcing and source identification [35]. The process of making data or information publicly available for use and verification is called crowdsourcing. Crowdsourcing is used to engage a "crowd" or group of unknown users for a common objective. In token-based crowdsourcing, tokens are offered to crowd as a prize for data input and validation. To check the validity of input and genuineness of a user, identity authentication must be used in token-based crowdsourcing. For identity management, a decentralized ID based blockchain may be used. The decentralized feature of blockchain where no central authority is required and each user in the network can record on the ledger allows for consensus in the network without the requirement of a third party enhance data veracity in the context of big data. The more people that are in the network, the more difficult majority collusion becomes for subverting the veracity of the information on the blockchain.

10.10 OPEN RESEARCH CHALLENGES AND FUTURE DIRECTIONS

Blockchain has a great potential but it also faces some challenges that can limit the broad use of blockchain. Some major challenges are [36]:

10.10.1 Scalability

Lots of transactions are done every day so blockchain looks bulky. Every node in blockchain stores all the transaction information to validate and to check the current transaction source status. The restriction on block size or time interval is used to make a new block. Bitcoin blockchain can process only seven transactions at one time so processing of millions of transactions is not possible in real time. Block size is also very small so transactions can be delayed due to the block size. The scalability problem can be categorized into two different types: storage optimization and redesigning.

Storage optimization of blockchain: A single node cannot operate full ledger. A cryptocurrency scheme is given by Bruce [37] in which old transaction information is erased by the blockchain network and an account tree is a database that stores the balance of the non-empty node addresses. VerSum is another approach that allows lightweight users to exist; correct computations are ensured by comparing results from different servers [38].

Redesigning blockchain: The idea behind the designing of the next generation of bitcoin is to separate the conventional block into two different parts: key block and micro block. Here the key block is responsible for leader election and the micro block is used to store transactional details. Time can be divided into epochs and in each epoch miners create a key block. After key block generation, the node is known as a leader that is responsible creating micro blocks [39].

Privacy leakage: Blockchain uses public and private keys to preserve privacy. Users operate with public and private keys without using their real identity. Blockchain cannot guarantee the of privacy of transactions because all transactional values are available publicly for every public key. A bitcoin user's transaction information can be used to reveal other user information. According to a study, user pseudonyms can be linked to the internet protocol address at the time of user involvement in network address translation. From a set of connected nodes, the user can be uniquely identified [40–42]. Mixing and anonymous are two methods that improved the anonymity of blockchain. In the mixing method, user details are written in under a false manner [43–45]. But the address can be linked to the user's personal identity because most of the time users perform the transaction with their real identity. Anonymity is possible in the mixing service in which funds can be transferred from different input addresses to different output address [46–47].

Selfish mining: Blockchain technology is susceptible to different attacks of colluding miners. The network can be vulnerable even if a small amount of hashing is used to cheat. Selfish miners' mined blocks are kept confidential, and a private branch is revealed publicly only when predefined requirements and conditions are satisfied. Stubborn mining is an approach in which miners amplify its security by combining attacks with eclipse attacks. Trail-stubbornness is another stubborn strategy approach in which miners mine the block when only the private chain is left. ZeroBlock strategy is designed and built on the simple scheme that a block is generated and accepted by the network in some time interval. Within Zero Block strategy, it is not possible for selfish miners to achieve the targeted rewards [48–49].

10.10.2 Future Directions in Blockchain Security

Blockchain has potential in almost all areas such as industry, education, health, and academia. Here we discuss possible future directions of blockchain in four areas: testing, blockchain application, tendency to centralization, and big data [50–51].

Blockchain testing: Currently, different types of blockchain appear in the IT sector and more than 500 cryptocurrencies are listed to date. But most developers provide incorrect information about blockchain to attract investors. If anyone wants to integrate blockchain with some specific application, they first need to verify the necessary requirements. For example, if someone wants to integrate blockchain in a business application they first need to check which kind of blockchain can meet the requirements. So blockchain testing is a very important task. This testing mechanism can be categorized into two parts: standardization and testing. In the standardization phase, all the required criteria should be developed and verified. In the testing phase, blockchain testing is performed according to the predefined criteria. An online retail business user should care about the data transfer of the blockchain technology so the testing must verify the average time of one transaction as well as the capacity of a block.

Blockchain applications: Blockchain is mostly used in financial systems to make transactions secure and protect them from attackers. But now blockchain can be applied on traditional industries to enhance the system process. For example, user information can be stored in blockchain and at the same time it can be used for improving the performance of another specific domain. Arcade city is ridesharing startup platform where riders can communicate directly with the drivers by using blockchain technology. Smart contract is a computerized transaction protocol that executes the contract terms. The design and proposed method of this scheme was very long but now it is ready for implementing with blockchain. The role of smart contract in blockchain is a code fragment that is used to execute the miners automatically.

Stop the tendency to centralization: Blockchain is a decentralized system. But there is a trend that miners become centralized in the pool. Till now, the 5 top-rated mining pools own more than fifty percent of the total hash power in the blockchain network. Selfish mining strategy defines that pool with more than twenty percent of the computing power gaining more than their fair share of revenue. Blockchain does not fit well for some organization and some methods to solve this problem can be proposed.

Big data analysis: Blockchain can be integrated with big data very easily. The combination of these powerful technologies can be categorized into two different types: data management and information analysis. In data management, blockchain is used for important data storage. The originality of the data is also ensured by blockchain. For example, in health care, if blockchain is used for medical purposes then it is responsible for the originality of the patient health information. That information should not be altered or modified. In the data analysis approach, transactions are used to analyze the information in blockchain. A trading analysis can be done to extract or predict the trading behavior.

10.11 CONCLUSION

The different advantages of blockchain like decentralization, anonymity, auditability, and persistency make it very popular in enhancing the traditional industries. Blockchain-based technology is the proof of the transparency and usability of blockchain. Some insecurity issues faced by blockchain can be solved after setting a proper and secure blockchain platform. The blockchain architecture is a cyberattack resilient database that is supported by immutability, consensus platforms, and cryptography. These ingredients proved to be very effective tools for the implementation of a security system for information. In this chapter, we explained blockchain security aspects in many different application areas. We also explained challenges and problems faced at the time of designing the blockchain architecture for specific applications. We can say that the value of blockchain would increase in other fields in the future. And this technology will be more secure compared with today. In the future, blockchain will also move in two different directions. The first direction is those applications that follow the decentralized approach and secure network design. The other direction will be a power tool that advances artificial intelligence. The integration of artificial intelligence with blockchain technology will take it in the next stage. The advancement can be big data security, democratized ownership, and decentralized approach. Blockchain security can be used at the time of storing personal and sensitive information.

REFERENCES

1. www.investopedia.com/terms/b/blockchain.asp
2. www.researchgate.net/figure/Blockchain-Structure_fig1_325173502

3. https://hyperledger-fabric.readthedocs.io/en/release-2.2/network/network.html#:~:text=A%20blockchain%20network%20is%20a,(chaincode)%20services%20to%20applications.
4. https://blockgeeks.com/guides/what-is-blockchain-technology/
5. www.ibm.com/in-en/blockchain/what-is-blockchain
6. https://geek-university.com/ccna-security/confidentiality-integrity-and-availability-cia-triad/
7. www.dartexon.com/DARTexon-Blockchain-CyberSecurity
8. Ali, A., and M. M. Afzal, "Confidentiality in Blockchain", *International Journal of Engineering Science Invention (IJESI)*, 7(1), January 2018, 50–52. ISSN [Online]:2319–6734, ISSN (Print): 2319–6726. www.ijesi.org
9. www.csoonline.com/article/3519908/the-cia-triad-definition-components-and-examples.html
10. https://learning.oreilly.com/library/view/hands-on-cybersecurity-with/9781788990189/1d4a00e6-84d4-4bb8-9ec1-3253a1858952.xhtml
11. www.cisco.com/c/en_in/products/security/common-cyberattacks.html#~types-of-cyber-attacks
12. www.spheregen.com/blockchain-technology-basics/)
13. Kesavarapu, K. R. and V. P. Venkatesan, "Security Attacks on Blockchain", *International Journal of Computer Applications (0975–8887)*, 178(16), June 2019.
14. www.seba.swiss/research/are-blockchains-safe-how-to-attack-them-and-prevent-attacks
15. https://en.wikipedia.org/wiki/Mt._Gox
16. www.proofpoint.com/us/threat-reference/endpoint-delivered-threats
17. https://academy.binance.com/en/articles/sybil-attacks-explained
18. www.radixdlt.com/post/what-is-an-eclipse-attack/
19. www.usenix.org/conference/usenixsecurity15/technical-sessions/presentation/heilman
20. https://academy.binance.com/en/articles/what-is-an-eclipse-attack
21. https://btchijack.ethz.ch/#:~:text=An%20attacker%20can%20use%20routing,the%20creation%20of%20parallel%20blockchains
22. www.cm-alliance.com/cybersecurity-blog/the-future-use-cases-of-blockchain-for-cybersecurity
23. https://www2.deloitte.com/tr/en/pages/technology-media-and-telecommunications/articles/blockchain-and-cyber.html
24. https://builtin.com/blockchain/blockchain-cybersecurity-uses
25. Butcher, J. R., Steptoe & Johnson LLP, and C. M. Blakey, Paul Hastings LLP, *Cybersecurity Tech Basics: Blockchain Technology Cyber Risks and Issues: Overview*, 2019.
26. www.esecurityplanet.com/network-security/cybersecurity-risk management.html#:~:text=Cybersecurity%20risk%20management%20takes%20the,your%20organization%20is%20adequately%20protected
27. https://securityscorecard.com/blog/10-considerations-for-cybersecurity-risk-management
28. Feng, S., W. Wang, and Z. Xiong, "On Cyber Risk Management of Blockchain Networks: A Game Theoretic Approach", arXiv preprint arXiv:1804.10412, 1–15, 2018.
29. Khan, M. A., and K. Salah, "IoT Security: Review, Blockchain Solutions and Open Challenges", *Future Generation Computer System*, 82, 2018.
30. Ali, S., G. Wang, B. White, and R. L. Cottrell, "A Blockchain-based Decentralized Data Storage and Access Framework for PingER", 2018.
31. Nyamtiga, B. W., J. C. S. Sicato, S. Rathore, Y. Sung, and J. H. Park, "Blockchain-Based Secure Storage Management with Computing for IoT", 2019.
32. *Cyber Threat Alliance Joint Analysis: Securing Edge Devices*, 2019.
33. Khan, M. A., and K. Salah, "IoT Security: Review, Blockchain Solutions and Open Challenges", *Future Generation Computer System*, 82, 2018.

34. Antonopoulos, A. M., *Mastering Bitcoin: Unlocking Digital Cryptocurrencies,"* O'Reilly Media, Inc., Newton, MA, 2014.
35. Wibowo, S., and T. Sandikapura, "Improving Data Security, Interoperability and Veracity Using Blockchain for One Data Governance, Case study of Local Tax Big Data", in International Conference on ICT for Smart Society, 2019.
36. Zheng, Z., S. Xie, H. Dai, X. Chen, and H. Wang, "An Overview of Blockchain Technology: Architecture, Consensus, and Future Trends", in IEEE 6th International Congress on Big Data, 2017, DOI: 10.1109/BigDataCongress.2017.85
37. Bruce, J., "The Mini-Blockchain Scheme," July 2014. [Online]. http://cryptonite.info/files/mbc-scheme-rev3.pdf
38. van den Hooff, J., M. F. Kaashoek, and N. Zeldovich, "Versum: Verifiable Computations Over Large Public Logs," in Proceedings of the 2014 ACM SIGSAC Conference on Computer and Communications Security, New York, NY, USA, 2014, pp. 1304–1316.
39. Eyal, I., A. E. Gencer, E. G. Sirer, and R. Van Renesse, "Bitcoinng: A Scalable Blockchain Protocol," in Proceedings of 13th USENIX Symposium on Networked Systems Design and Implementation (NSDI 16), Santa Clara, CA, USA, 2016, pp. 45–59.
40. Meiklejohn, S., M. Pomarole, G. Jordan, K. Levchenko, D. McCoy, G. M. Voelker, and S. Savage, "A Fistful of Bitcoins: Characterizing Payments Among Men with no Names," in Proceedings of the 2013 Conference on Internet Measurement Conference (IMC'13), New York, NY, USA, 2013.
41. Barcelo, J., "User Privacy in he Public Bitcoin Blockchain", 2014. *http://www. dtic. upf. edu/jbarcelo/papers/20140704 User Privacy in the Public Bitcoin Blockc hain/paper. pdf* (Accessed 09/05/2016).
42. Moser, M., "Anonymity of bitcoin transactions: An analysis of mixing ¨ services," in Proceedings of Munster Bitcoin Conference ¨, Munster, ¨ Germany, 2013, pp. 17–18.
43. Bonneau, J., A. Narayanan, A. Miller, J. Clark, J. A. Kroll, and E. W. Felten, "Mixcoin: Anonymity for bitcoin with accountable mixes," in Proceedings of International Conference on Financial Cryptography and Data Security, Berlin, Heidelberg, 2014, pp. 486–504.
44. Maxwell, G., "Coinjoin: Bitcoin Privacy for the Real World," in Post on Bitcoin Forum, 2013. https://bitcointalk.org/index.php?topic=279249.0
45. Ruffing, T., P. Moreno-Sanchez, and A. Kate, "Coinshuffle: Practical Decentralized Coin Mixing for Bitcoin," in Proceedings of European Symposium on Research in Computer Security, Cham, 2014, pp. 345–364.
46. Miers, I., C. Garman, M. Green, and A. D. Rubin, "Zerocoin: Anonymous Distributed e-Cash from Bitcoin," in Proceedings of IEEE Symposium Security and Privacy (SP), Berkeley, CA, USA, 2013, pp. 397–411.
47. Sasson, E. B., A. Chiesa, C. Garman, M. Green, I. Miers, E. Tromer, and M. Virza, "Zerocash: Decentralized Anonymous Payments from Bitcoin," in Proceedings of 2014 IEEE Symposium on Security and Privacy (SP), San Jose, CA, USA, 2014, pp. 459–474.
48. Solat, S., and M. Potop-Butucaru, "ZeroBlock: Timestamp-Free Prevention of Block-Withholding Attack in Bitcoin," Sorbonne Universites, UPMC University of Paris 6, Technical Report, May 2016. [Online]. https://hal.archives-ouvertes.fr/hal-01310088
49. "Crypto-Currency Market Capitalizations," 2017. [Online]. https://coinmarketcap.com
50. "The Biggest Mining Pools," [Online]. https: //bitcoinworldwide.com/mining/pools/
51. Szabo, N., "The Idea of Smart Contracts," *Nick Szabo's Papers and Concise Tutorials*, 6(1), 1997.

11 The Implementation of Blockchain Technology to Enhance Online Education

A Modern Strategy

Apeksha Singh and Abhilasha Sisodia

CONTENTS

11.1 INTRODUCTION

"Distributed ledger technology (DLT) and the narrower concept blockchain is the subject of significant curiosity, boosterism, criticism, investment, and genuine, fast-moving innovation." (USAID)

Online education has reached a stage of explosive growth since the turn of the 20th century owing to the rapid expansion of Internet technology. Online education, also known as distance education or online learning, is a web-based teaching tool for the flow of information and easy learning using information technology and Internet technology. With the Internet as the platform, online teaching transcends the boundaries of place, setting, time, and teachers and provides quality teaching activities to students worldwide. The different forms of online education can be categorized into vocational training, test and examination preparation, development of professional skills, language education, early childhood education, K–12 education (kindergarten to 12th grade), etc. We are seeing the advancement of artificial intelligence, smart schools, and distance learning with the help of new technology today. Chances are, in the years ahead, blockchain will become an intrinsic part of college. The education tech market is rapidly growing and is expected to hit $93.76 billion globally by 2020. (Ayers, 2019). Blockchain is undoubtedly an evolutionary quirk. It's a platform that began as a foundation for virtual currency but it's rapidly becoming obvious that blockchain is more than just a cryptocurrency. Time will reveal if blockchain is a much faster and more efficient online asset sharing protocol or "the next big thing" as some would like to call it (Sharma, 2019).

The implementation of blockchain-based software could be split into three major phases: Blockchain 1.0, 2.0 and 3.0. Blockchain 1.0 was used by cryptocurrency and based on enabling easy cash transfers. (Gatteschi et al., 2018). Subsequently, Blockchain 2.0 was launched for properties and smart contracts. Such smart contracts enforced specific requirements and standards that must be fulfilled before they are registered in the blockchain. Registration shall be effective without the involvement of a third party. Several implementations have been built in Blockchain 3.0 in various fields, such as government, education, health, and technology.

Real-world blockchain technologies containing these multidisciplinary fields are shown in Figure 11.1 are often explored under the general term cryptoeconomics—described as, "a discipline concerned with the production, consumption and transfer of wealth using computer networks, cryptography, and game theory to enhance prosperity of groups in current and future digital market economies" (Lielacher, 2017).

On 31 October 2008, Satoshi Nakamoto released to a cryptography forum a short but pioneering article. In it, he presented a way to solve the double-spending situation—the problem that troubled earlier cryptocurrencies. Despite not specifically referencing blockchain, he defined its structure as a chain of hashed time stamps: "Each timestamp includes the previous timestamp in its hash, forming a chain, with each additional timestamp reinforcing the ones behind it" (Nakamoto, 2008). Hence, blockchain is the central infrastructure used to build the cryptocurrency, Bitcoin, by conserving irreversible distributed ledgers in thousands of nodes. (Ibid.). Although this approach was later modified for Bitcoin, the idea was laid down: a chain of blocks, each cryptographically linked to the previous one, using a hash index. From this we denote that a blockchain is nothing more than a sequence of records, each of which has been hashed and connected to the previous block. (Figure 11.2).

It has been perceived as a core component of the fourth industrial revolution following the advent of steam engines, electricity and information technology.(Chung and Kim, 2016; Schwab, 2017). This revolutionary technology would have a huge effect on national government, governmental structures, business processes, employment and our daily lives in the 21st century. It has the power to turn the modern Internet from the "Internet of information sharing" to the "Internet of value exchange."

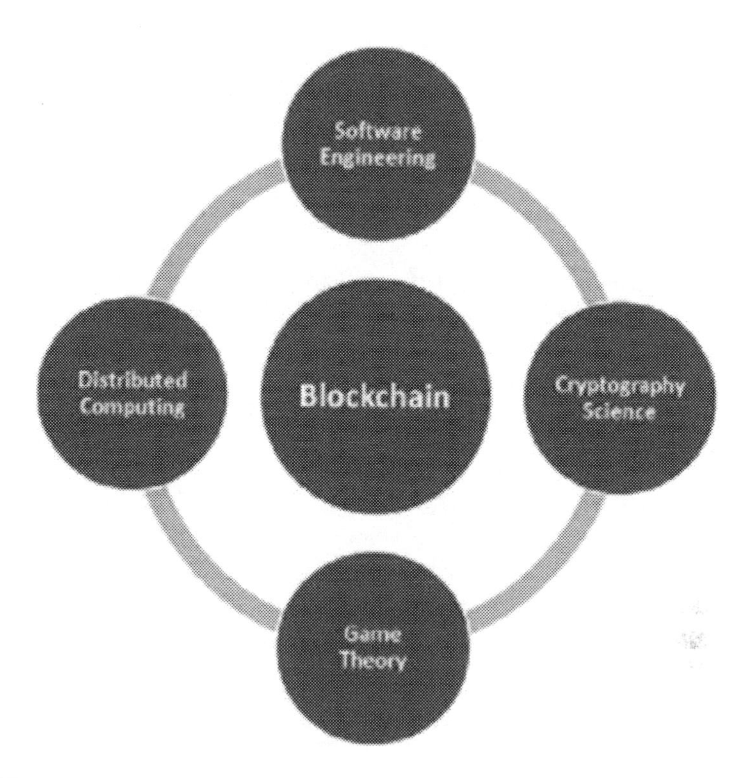

FIGURE 11.1 Multidisciplinary foundations of blockchain technology.

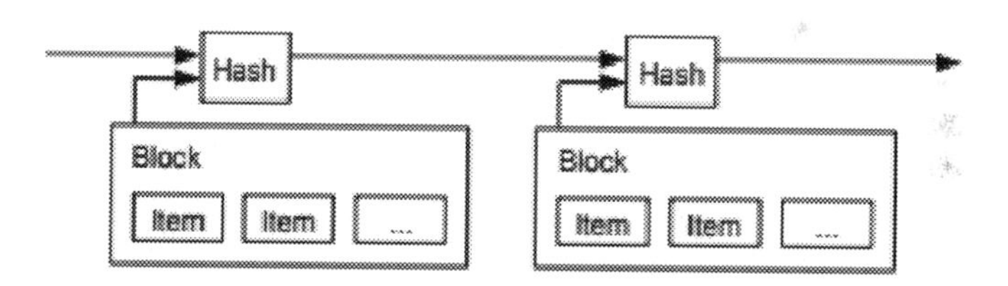

FIGURE 11.2 Nakamoto's blockchain proposal (Nakamoto, 2008).

Blockchain technology is projected to radically change the operational modes of finance, manufacturing and education, as well as to facilitate the exponential growth of the knowledge-based economy on a global scale. Owing to its irreversibility, openness and accountability for all transactions carried out in a blockchain network, this revolutionary platform has many future applications (Underwood, 2016). At the early stages of its existence, blockchain technology was unable to attract any interest. Nevertheless, as bitcoin proceeds to work successfully and consistently through the years, organizations have also become conscious of the immense potential of the fundamental technologies of this innovation in its implementation not just to bitcoin but also to several other fields (Collins, 2016). Blockchain technology has thus become a critical subject for growing numbers of governments, organizations, businesses and researchers.

Swan (2015) suggested that blockchain implementations could be broken down into three stages; blockchain 1.0, 2.0, and 3.0. Blockchain 1.0 is cryptocurrencies implemented as a peer-to-peer cash payment network. Blockchain 2.0 is the comprehensive application of blockchain other than for simple cash transactions, including investments, mutual funds, borrowings, smart property and smart contacts. Blockchain 3.0 is evolving uses for blockchain beyond money, banking and markets, such as infrastructure, health, technology, learning, entertainment and media. According to the preceding concept, the latest blockchain implementations are only in phases 1.0 and 2.0. Most people are unfamiliar with the term "blockchain," not to mention the potential uses of blockchain technology. Although researchers questioned the operational usage of blockchain; several researches have explored how blockchain technology can be used in education (Chen et al., 2018).

This chapter is a comprehensive analysis of research into blockchain-based educational applications. It works on three key topics: (i) educational technologies that have been incorporated using blockchain technology, (ii) advantages that blockchain technology may bring to academia, and (iii) barriers to the implementation of blockchain technology in teaching and learning. A thorough review of the outcomes of each concept was undertaken as well as an intense discussion focused on the conclusions. The study also gives insight into many aspects of education that may benefit from blockchain technology.

11.2 METHODOLOGY

This analysis was carried out in compliance with the recommendations issued by Okoli and Schabram (Okoli et al., 2010). The guidance lays out eight measures.

1. Defining the purpose and research problems of the review. This process is important to make the systematic analysis clear to readers. In addition, well-formulated research questions that improve the feasibility of the systematic review and minimize the costs and expense of obtaining the appropriate papers.
2. Creating a thorough evaluation protocol and preparing your reviewers on how to implement it. The protocol is a plan that outlines the basic measures and protocols to be taken throughout the analysis. This move is necessary to ensure that the reviewers are fully informed of the thorough protocol to be followed.
3. The search for the related papers. The primary means of literature research is currently electronic data. Nonetheless, it is important for reviewers to acknowledge the proper usage of electronic databases to scan such resources effectively.

4. Screening of articles for incorporation. In this stage, the reviewers determine which papers should be recommended for analysis and which should be discarded. They will need to mention the pragmatic reasons for eliminating each article.
5. An evaluation of the content of the papers. During this stage, the reviewers need to determine that the papers are of adequate quality to be included in the detailed review. This move has two purposes. Initially, in systematic reviews where there is a minimum quality requirement for incorporation, the quality assessment is used to disqualify papers that do not meet the standards of the reviewers. Secondly, there has to be some sort of quality assessment of all comprehensive evaluations, because the quality of the analysis relies to a large degree on the standard of the publications used.
6. Extricating the details from the papers. When all the documents that would be used in the analysis have been established, the reviewers need to carefully collect the necessary details from each article. These data would serve as a raw material for the process of synthesis. The quality of the data to be collected is decided on the basis of research questions encountered at the early stage of the analysis.
7. Analyzing the data collected. Also known as systematic review, this stage includes the collection, arrangement, analysis and interpretation of the information derived from the papers. The process used in this stage depends on whether the papers used are qualitative, quantitative or mixed. Qualitative, quantitative and mixed research may be qualitatively evaluated, whereas only quantitative studies can be quantitatively examined.
8. Writing a detailed analysis. Basic principles for writing research papers should be observed in this phase. The analysis should be stated in adequate depth such that the findings can be replicated separately.

11.2.1 Defining the Research Questions

Based on the objective of this report, the following research questions have been developed:

1. Which technologies have been developed for educational purposes using blockchain technology?
2. What advantages can blockchain technologies offer to education?
3. What are the advantages and disadvantages of the use of blockchain technologies in education?

11.3 LITERATURE REVIEW: WHAT IS BLOCKCHAIN?

Blockchain technology is often known as distributed ledger technology. Coinbase, the world's largest cryptocurrency exchange, defines blockchain as "a distributed, public ledger that contains the history of every bitcoin transaction" (Coinbase,

2017). The Oxford English Dictionary broadens the definition somewhat, defining blockchain as "a digital ledger in which transactions made in bitcoin or another cryptocurrency are recorded chronologically and publicly" (Oxford Dictionaries, 2018). A somewhat broader definition is offered by Webopedia where a blockchain is defined as "a type of data structure that enables identifying and tracking transactions digitally and sharing this information across a distributed network of computers, creating in a sense a distributed trust network. The distributed ledger technology offered by blockchain provides a transparent and secure means for tracking the ownership and transfer of assets" (Stroud, 2015). It helps participants to protect contract resolution, acquire contracts and transfer funds at a minimal cost (Tschorsch and Scheuermann, 2016). A summary flow of a blockchain transaction in cryptocurrencies can be seen as follows. User A initiates a connection to User B via a network of peer-to-peer blockchains. The network uses a cryptographic identification proof (a combination of public and private keys) to uniquely distinguish user A and user B. The transaction would then be sent to the blockchain network memory pool, requesting authentication and confirmation of the transaction. The new block is created by a certain number of accepted nodes; this is called consensus reaching. After consensus is achieved, the new block is created on the entire blockchain network, and each node restores its respective copy of the blockchain ledger. This block includes all of the transactions that took place during this period. By a digital signature, it is "connected" to the original block in the network (Yli-Huumo et al., 2016). The consensus level is achieved by using a consensus algorithm. The process is known as mining. Namely, the peer-to-peer network establishes consensus on the new public ledger status. Each node can vote by its computing power to approve valid blocks by taking extensions or by rejecting expansions. This consensus framework will enforce any rules and incentives necessary (Nakamoto, 2008; Kraft, 2016). Any transaction in a block is marked with a different time stamp. A time stamp also connects the two lines. The data on the ledger thus has a property of time, and the chain length is constantly increasing. It implies blockchain is a centralized version incorporating the time stamp function (Haber and Stornetta, 1991). Blockchain uses advanced hardware to develop a large cryptographic data chain, and the SHA-256 hash function is used to prohibit third-party consumer data from being tampered with (Tschorsch and Scheuermann, 2016). Any effort to change just a nugget of information destroys the current chains. In short, blockchain is a distributed public ledger that is transparent and secure. It makes use of collaborative strategies and consensus algorithms that all participants retain (Chen et al., 2018).

Blockchain is not only a modern form of centralized application-based Internet technology but also a different type of supply chain network. Blockchain is basically a distributed network of computers (nodes) used to manage the origins of the exchange of knowledge. Through maintaining a full collection of ledgers of past transactions, each node preserves the confidentiality and authenticity of the records. The programmer who generates a new block is the first to verify all the transactions in the block and tackle the calculation problem by creating a cryptographic signature for the block that satisfies a predefined principle using the hash function. The newly formed block will telecast to the entire blockchain network, enabling all nodes to keep the same full ledger (Tschorsch and Scheuermann, 2016).

The mechanism for consensus is accomplished through three main processes of verification. Bitcoin uses the Proof of Work (Nakamoto, 2008) authentication method. The miners are nodes that operate in a peer-to-peer network of blockchains. Their job is to verify all transactions used in one block and use a hash function to tackle the mathematical digital signature problem. The miners are interacting with each other, and the response will be exchanged with other mining nodes until someone solves the problem. The successful miner gets additional bitcoins as a reward. Many miners must recognize the job proof and connect the latest block to the blockchain network (Fanning and Centers, 2016). Cryptocurrency has four stages of development which include Frontier, Homestead, Metropolis and Serenity. The first three stages use the Proof of Work authentication method, and the fourth stage uses Proof of Stake. The Proof of Stake requires that the certifier demonstrates possession of any volume of cryptocurrencies (Sharples and Domingue, 2016)."Proof of Zero Information" is the collective process used in Zcash that can give its users more privacy. Proof of Zero Information has increased in reliability and performance relative to other authentication methods (Tschorsch and Scheuermann, 2016).

The idea of blockchain was created by a person called Satoshi Nakamoto in his thesis "Bitcoin: A Peer-to-Peer Electronic Cash System" published in 2008 (Nakamoto, 2016a). In this thesis, Nakamoto suggested the bitcoin system, but did not use the term blockchain. Alternatively, the block and chain are defined as the data structures that document the background of the bitcoin transaction database. The "block" corresponds to the distributed data, while the "chain" is the hierarchical set of blocks ordered by cryptographic methods. Together, the block and the chain are a continuous sequence of transactions. In general terms, blockchain technology often applies to centralized blockchain-based accounting technologies, including mutual consensus, privacy and security, peer-to-peer (P2P) networking, network protocols and smart contracts. Blockchain infrastructure combines three core concepts: transaction, block and chain. The transaction is a ledger process such as entering or withdrawing an object, which often corresponds to a change in the ledger status; the block tracks the effects of all transaction details over a period of time, and the chain is a sequential block string that represents all of the ledger's state changes.

The following methodology is applied with blockchain technologies. First, there has to be a centralized ledger within the network that only allows new data to be introduced. In other words, no data should be omitted from the record, thus guaranteeing the data is tamperproof. As seen in Figure 11.3, the blocks are in sequential order connected into a chain, with each block maintaining the previous block's hash value. When a new ledger transaction takes place, the entire network must track the transaction data block and connect it cryptographically to the chain using the elliptic curve digital signature algorithm (ECDSA) (Johnson et al., 2001).

The blockchain platform therefore has a de-centralized, de-trusted, open data management system. Using cryptographic methods, this system means that no transaction data can be changed and can be backtracked and checked. The hierarchical

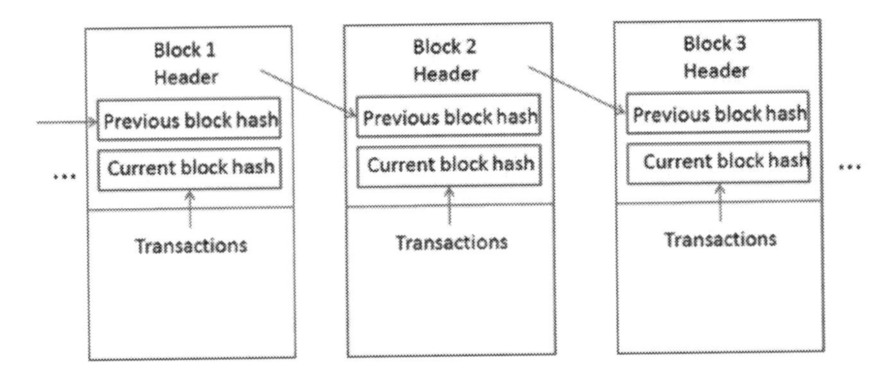

FIGURE 11.3 A simple blockchain structure.

and autonomous design of the network is covered by the distributed data management and mutual maintenance. The blockchain system reduces the risk of data failure on the whole network triggered by an attack on a single node relative to the conventional centralized databases.

The blockchain infrastructure is inherently autonomous, decentralized, de-trusted, reliable, voluntarily managed, and secure to privacy.

- **Decentralized:** Focused on a decentralized, distributed P2P network (Waterhouse et al., 2002), blockchain technology does not need a central node to test and track transaction records. The network nodes will share data directly, based on the trust scheme, which increases data exchange performance. The disruption of a single node, meanwhile, does not affect the entire network's results.
- **De-trusted:** Blockchain technology generates a connection block centered on the cryptographic value of the hash and, on the other hand, guarantees transaction protection by utilizing the digital signature created by asymmetric cryptography. Hence, nodes may render transactions securely without third-party oversight.
- **Reliable:** The blockchain platform uses distributed storage, that is, each node can receive a copy of all transaction records. This mode of storage ensures the quality and reliability of the records. In addition, all transaction data is reported on the basis of time stamps and identifiable source, implying that the data is non-modifiable.
- **Collaboratively managed:** Blockchain data is managed by all nodes in the network. Because no node is exempt from servicing, the fault of a single node has little effect on the data of the entire network.
- **Confidentiality secure:** Owing to the digital signature algorithm, the information is transmitted using a public key and a private key without revealing the identity of the node. In the transmitting process, the consumer becomes absolutely invisible (Sun et al., 2018).

With a high degree of integrity and reliability, blockchain technology offers a perfect solution to the issue of online education. Specifically, the blockchain will provide accurate and uneditable learning documents for online education without the need for third-party oversight to ensure equal approval of course credits. Blockchains and other forms of distributed ledger technology (DLT) are the backbone behind other high-profile innovations, such as cryptocurrencies (e.g., bitcoin), which aim to transform how data is handled when exchanged across a wide spectrum behind sectors, like education. Vendors, business groups and a range of education partners are working to utilize blockchain technologies to tackle the daunting challenges of data access and accountability posed by organizations, schools and students, whether by the integration of current data items and processes or through completely new applications.

11.4 DISCUSSION

11.4.1 Features of Blockchain Technology

Blockchain is a term that you may use often. Yet most of us don't know precisely what blockchain is or how to explain it to others. Many people think blockchain is bitcoin and vice versa. Yet this is not the case. In reality, bitcoin is a digital currency or cryptocurrency that operates with blockchain technology. The word blockchain was first defined in 1991. A team of researchers decided to construct a digital time stamp record resource so that it could not be backdated or updated. The technique was later developed and redefined by Satoshi Nakamoto. In 2008, Nakamoto developed the first cryptocurrency, a blockchain-based project called bitcoin. As the name suggests, blockchain is a chain of blocks containing information. Each block is made up of a number of transactions, and each transaction is documented in hash form. Hash is a unique address allocated to each block during its development, so any further alteration in the block will result in its hash changing.

A block has three main parts as shown in Figure 11.4.

1. Data/Information part—this contains the information of the transaction created.
2. Hash—This is the unique ID of the block.
3. Previous Hash—This is the hash of the previous block.

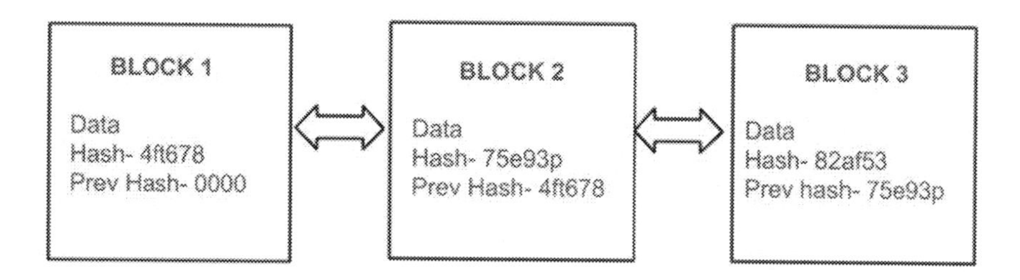

FIGURE 11.4 Parts of a blockchain.

Because each block has the hash of the previous block in the blockchain, if someone wants to interfere with the data of another block then the block hash will be modified. Therefore, they are going to have to change the "previous hash" of the next line. In doing so, the current hash of the next block would also alter.

From the technological point of view, blockchain technology has the following four attributes: decentralization, traceability, immutability and currency properties.

- **Decentralization**

Decentralization refers to the data authentication, preservation, management and blockchain sharing mechanisms that are based on a distributed network framework. The blockchain is a series of blocks containing personal information (database) but clustered together in a network (peer-to-peer) in a safe and true manner. In other words, blockchain is a blend of computers that are connected to each other rather than a central server, suggesting the whole network is decentralized. Under this system, trust between distributed nodes is created through mathematical methods rather than by centralized organizations. Blockchain is a decentralized and open ledger. The ledger is the database of transactions made and is thus considered an open ledger, because it is available to all. No entity or any agency is responsible for the transactions. Any link inside the blockchain network has the same copy of the ledger.

Blockchain is a decentralized network as shown in Figure 11.5, that is, it has no central authority to regulate the network because it operates in the client server model.

- **Traceability**

Traceability means that transactions in blockchain are organized in sequential order, and the cryptographic hash function binds a block with two neighboring blocks. Thus, each transaction can be traced by analyzing the block information connected by hash keys. Hashing is very complicated, so it

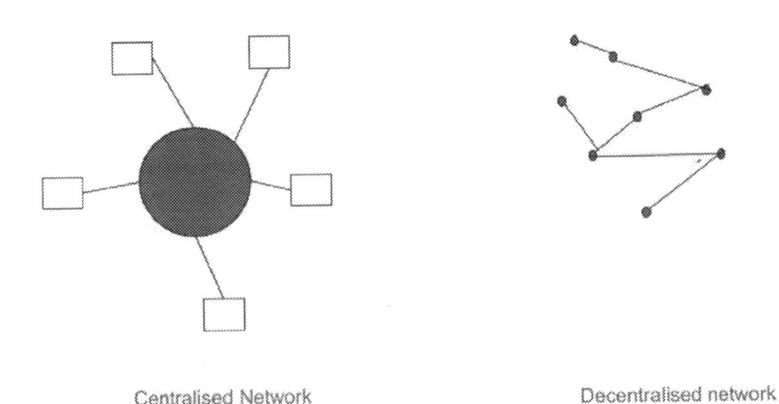

FIGURE 11.5 Centralized network and decentralized network structure.

can't be modified or reversed. Only one will come up with a public key and the private key. A simple adjustment in the input may still lead to an entirely different ID, but minor improvements in the system are inconvenient.

- **Immutability**

A Blockchain is a permanent and tamperproof record of transactions. There are two reasons why blockchain technology is immutable. On the one hand, with one hash key connecting from the preceding block and one hash key leading to the next block, all transactions are stored in blocks. Manipulating any transaction would lead to separate hash values and thus would be identified by all the other nodes running precisely the same validation algorithm. On the other hand, blockchain is a decentralized public database distributed on thousands of nodes, in which all the ledgers tend to converge in real time. Successful manipulation would entail a modification to more than 51% of the ledgers stored in the network (Tschorsch and Scheuermann, 2016).

- **Currency Properties**

Blockchain technology and cryptocurrency are indistinguishable, that is, every blockchain network has a type of cryptocurrency ownership. Blockchain offers a peer-to-peer network. This blockchain feature requires only two parties, the sender and the recipient, to be interested in the transactions. Therefore, it eliminates the "third-party authorization" condition because everyone in the network will approve the transactions themselves. Digital currency distribution based on blockchain technologies is fixed. In bitcoin in particular, the currency foundation is set at 21 million caps and digital currency is generated using a different mining algorithm and is bounded by a predetermined formula. Therefore, the problem of inflation, recession and so forth, does not occur. In implementations for Blockchain 2.0 and 3.0, the mixture of other activities such as political activities, educational activities and financial activities will make such non-financial activities have the property of currency.

11.5 ADVANTAGES OF BLOCKCHAIN TECHNOLOGY

Based on the four technological features listed previously, some of the advantages of their implementation using blockchain technology are defined as follows:

- **Reliability**

The decentralized design of the blockchain network switches the storage of all the transaction information from closed and centralized ledgers held by only a few approved organizations to open public ledgers managed by hundreds of thousands of nodes. The loss of a single node does not affect the function of the whole network. It eliminates the common failure point and guarantees the high reliability of systems based on the blockchain technologies.

- **Trust**

The network of blockchains makes trust decentralized too. Like the centralized trust we take for granted, such as central currency issuing governments and commercial banks, the blockchain network functions as new confidence-bearers of open ledgers. Those ledgers are exchanged by a tamperproof node network (Underwood, 2016).

- **Security**

The blockchain program uses a one-way hash function, which is a mathematical function that takes a variable-length input sequence and transforms it to a fixed-length binary sequence. The output does not bear any apparent relation to the data. The cycle is difficult to reverse because, provided only the output, the value cannot be calculated (Yli-Huumo et al., 2016). In addition, the newly created block follows strictly the linear sequence of time.

- **Efficiency**

All data is automatically processed via preset procedures. Thus, blockchain technologies can not only greatly minimize labor costs but also increase productivity. For the digital money in Blockchain 1.0, the management of the public ledger is essentially the management of settlement. By reducing the number of intermediaries involved and making the negotiation process quicker and more functioning, blockchain technology could speed up the processing and settlement of some financial transactions (Wang et al., 2016).

11.6 IMPLICATIONS OF BLOCKCHAIN TECHNOLOGY

Several colleges and institutes also utilize blockchain technology in education; most of them use it to help with student degree management and summative learning result assessment (Sharples and Domingue, 2016; Skiba, 2017).

- **Student Reports and Credentials**

When considering the standardized curriculum that involves instructional material and results, student success and academic recognition, it could be suggested that the entire text might be developed through blockchain technologies. Subsequently, knowledge on work experience, expertise, online learning experience as well as specific preferences is included in the informal learning sense. This data may be processed and viewed anonymously on a blockchain network in the appropriate manner. The University of Nicosia is the first school to use blockchain technologies to handle student credentials earned from MOOC systems (Sharples and Domingue, 2016). Sony Global Education has used blockchain technologies to build a decentralized appraisal framework for the delivery of degree knowledge collection and management services (Hoy, 2017). In fact, the Massachusetts Institute of Technology (MIT) and the Learning System organization collaborated together to develop a digital blockchain-based online learning credential. Students who have worked in the MIT Technology Lab programs and completed the test will receive a credential that will be held on a blockchain network (Skiba, 2017). Moreover, Holberton School is the first organization to utilize blockchain technology to store degrees and has announced that this knowledge would be exchanged from 2017 onwards. The blockchain ledger will connect the specific ID of the consumer with all sorts of educational knowledge. It involves in-class learning behavior, micro-academic project practice and experience in macro schooling, etc.

Some organizations see blockchain as a flawless technology for storing, tracking and using credentials from the students. Blockchain courses will allow learners to quickly and conveniently access their documents and communicate directly with prospective employers. As such, employers won't have to contact universities and colleges to get information on student achievements. Having a direct connection from an applicant will save time and will provide a complete overview of the skills and history of a candidate. The greatest thing is that the records won't be locked up but pertain to the learners. We can use these records freely at any time whenever it is needed.

- **Framework for Collaboration**

Blockchain acts not only as a transparent repository of information but also as a whole network for sharing and communication. Universities and colleges want to make sure that students have a productive academic experience. Interactions between students and professors are extremely important: when students receive sufficient advice and guidance, they can study without cost for the assignment. What's more, the platform based on blockchain enables departments to release information about notable events and lectures. This ensures long-term and effective alumni–faculty relations. Thus possible obstacles will easily be lowered.

- **The Defense of Copyrights and Online Privacy**

Plagiarism is a significant concern in academia. In fact, the blockchain platform aims to reduce degree manipulation. There have been several instances of theft of degrees in the past. However, it now can be avoided by using blockchain to confer and manage the degree of the student. The data matched to the user's ID and stored in blockchain is checked, validated and maintained by miners from around the world. The global blockchain database is transparent and secure. Consequently, both transparency and authority are assured and would effectively minimize degree fraud. It is possible to use blockchain systems to monitor the propagation of copyrighted content over the Internet. The technology's core purpose is the safe protection of the information stored within a chain. As such, data inside the chain cannot be manually changed because it is secured by the advanced cryptographic steps. This will make academic materials readily available yet secure and unalterable. Every use of the source material is recorded in the chain, and the owner can control the admin rights easily. The usage can be tracked online, and ownership is empirically verified.

- **Network for Innovative Technology Learning**

Exploration and discovering platforms are one of the best applications of blockchain in education. For example, the Education Ecosystem Platform is one of the first projects to connect academics, students, developers and content producers using blockchain. Colleges and schools need to use the same framework to build unique environments, enabling learners to access materials from research and share their experiments and proposals. Inbuilt tokens can be used to download books, order materials, and submit feedback and recommendations. These tokens are acquired when users make contributions, invite new entrants and watch videos. Web creators receive bonuses for their work, and they are given new tokens when people interact with their contents. As such, the more you study and practice, the more you get to continue learning content. Such an engagement degree will bring clear benefits to the academic system. Blockchain may also be viewed as a "capacity-currency transfer system." Blockchain learning ledger explicitly documents comprehensive details regarding the learning process of users and tracks the growth of their expertise and skills. Both of them, according to a set of rigorous criteria, may be converted into a kind of digital money and deposited on a blockchain. Students will receive rewards through their study efforts, which are called "learning is earning" (Sharples and Domingue, 2016). With this idea, some schools have begun the proposal, for example, Sharples and Domingue (2016) reported a kind of Education Reputation Currency called Kudos. It can be used and placed in a virtual wallet to calculate learning outcomes.

- **Human Resources: Regulatory Affairs**

Blockchain can make the job application process easier for both sides—Students and employers. As we have already stated, the technology allows students to store and share information on their background. At the same time, sophisticated authentication protocols are protecting this data from any forgery. As such, especially employees and HR managers get accurate and relevant information. It is not unusual for work applicants to lie about their degrees, qualifications and the colleges they attended. Those claims are hard to check in the current systems. The blockchain holds the appropriate information about the competence of an applicant by storing it in a protected network.

Hence, a blockchain is a historical database of transactions, almost like a conventional monetary database. New series of transactions or "blocks" are registered and connected to the previous record cryptographically, creating a chain. The defining features of a blockchain usually involves, (i) integrity—documents are connected cryptographically, making them very difficult to alter; (ii) transparency—because each network user has its own blockchain copy, notifications are circulated and transparent; (iii) Democracy—blockchains allow peer-to-peer transactions to be checked without a centrally controlled mediator.

Despite the enormous notoriety, the present form and systems of online education have many flaws in the face of the increasingly open and digital Internet. For instance, the education process and the outcomes of MOOCs are deprived of notoriety and official certification; the confidentiality of the learners is at risk, because the programs and data protection rely heavily on the centralized online education platform; personal data of students cannot be maintained effectively owing to the diversity of the Internet and data tampering; and there is no mature boundary-platform course sharing framework to fully divulge teaching resources. In order to make the learning process and the outcomes reliable, a centralized and trustworthy data storage system must be created to monitor the learning process of the students, reveal all learning information to the public and guarantee data confidentiality and non-tamperability.

Blockchain technology is a valuable method for addressing online education challenges including inadequate registration, lack of acceptance and data vulnerability. Nowadays, this technique is primarily used in areas such as banking, the Internet and the Internet of Things (IoT). Standard monetary implementations include digital currencies, trading and payment systems. Smart contracts, ranging from shares and inventories to bank loans, can be automatically executed without human intervention (Kosba et al., 2016). In terms of the IoT, the blockchain platform helps computers to autonomously connect and detect errors. (Emilie, 2016). Preliminary use of this technology was also rendered in the field of education. For example, Mike Sharples suggested the use of blockchain to recognize decentralized storing of educational data, creating the so-called information currency (Sharples, 2016). Several researchers proposed applying blockchain to authentication of credit cards, secure data protection and centralized data storage (Devaney, 2016). The MIT Media Lab has developed a network of digital learning certificates using blockchain technologies and an accessible certificate from Mozilla (Redman, 2016). Moreover, the blockchain platform has been implemented in the manufacturing industry for drug production. Under the Sony Corporation of Japan, a blockchain technology software network will exchange education courses and publish data freely and safely without revealing this knowledge to the Education Management Agency, thus making education fair and digitized. University College London uses blockchain technology to help credit risk management postgraduates test authenticity (University College London, 2018).

Considering the previously mentioned online education issues, this chapter discusses the essence and features of blockchain technology and applies this technology to online education with the goal of building a stable, accessible and trustworthy online educational platform. Study results pave the way for the establishment of decentralized online education.

11.7 FUTURE CREATIVE LEARNING SYSTEMS EMPLOYING BLOCKCHAIN TECHNOLOGY

Blockchain technology can be applied in many innovative ways to education beyond mere diploma management and assessment of achievements. Blockchain infrastructure has tremendous potential for broader development opportunities on formative assessment, design and execution of learning experiences for both teachers and students and keeping track of the whole learning processes. Some groundbreaking implementations for the usage of blockchain technology in the area of education are suggested as follows.

A smart contract operating on the Ethereum blockchain network is basically a code protocol that replicates a specific contract (e.g., economic transfers, jobs, etc.) (Kosba et al., 2016). This will promote the settlement of contracts, clarify the terms of contracts, enforce the implementation of contracts and validate the status of successful contracts. It labels the distinctive and correct identities of the parties in an agreement (contract subjects) in a digital manner and explicitly states by code the privileges and responsibilities of both sides (contract terms). The smart contract not only eliminates "third party expenses" in conventional contracts but also significantly improves the confidentiality and stability of the transaction. For instance, in the case of car payments, the customer negotiates directly with the vendor rather than with bank loans, avoiding any extra payment costs. The code will be enforced if the user violates the law, and the smart contract will be cancelled. The smart contract significantly improves corporate control and justice in comparison with conventional methods. Therefore, if professors and students carry out smart contract-based teaching and learning programs, some of the academic problems will be solved.

There are also certain detrimental psychological or quantitative influences from the student's viewpoint that affect bad learning results, such as loss of encouragement and financial burden. Because of the currency property feature, blockchain may be used to enforce "learning is earning" to inspire students (Sharples and Domingue, 2016).

The smart teachers–students' relationship may be extended to the educational situation. The teachers will offer real-time awards to students with only a few easy clicks. Depending on the smart contract, students will receive a certain number of digital currencies as rewards. This kind of money can be kept, used as tuition, or traded with actual currencies in the educational wallet (Chen et al., 2018).

Assessment is often an issue in the academic sector. Formative evaluation has long been championed and yet it is not yet ready as it is not easy to monitor any teaching and learning information. Applying blockchain and smart contracts should come up with this task. In particular, blockchain's immutability, traceability and transparency ensures the data stored in blockchain is more reliable, accurate and theft-proof. Take for example "collaborative study," which is deemed an outstanding way to carry out teaching on constructivism and to develop the willingness of students to collaborate with others. However, it is also followed by a question of free-riding that hinders equal assessment. Blockchain technologies may minimize this occurrence. A student publishes his/her work to the learning platform from his/her specific account; the smart contract operating on the platform will check the student's output and the results will be reported in blocks. All traits during collaboration will also be stored in blocks as evidence for evaluation. In comparison, the decentralized ledger is defined by decentralization. This implies that the public ledger guarantees the integrity of certain nodes. Therefore, as nodes in the blockchain network, the views of students should be taken into account when judging them. In this setting, blockchain guarantees the integrity of the evaluation.

From a teacher's point of view, teaching is complex and imaginative in such a way that it is impossible to determine. The conventional approach, focused on student input, appears to be one-sided, ignores subjectivity and is hardly beneficial for teacher development. A modern appraisal framework may be developed on the basis of a blockchain network and a smart contract.

In the first position, teachers ought to apply preplanned instructional events to the institutions as a smart contract. Any teaching events will be registered in the blockchain network during the instructional process. The smart contract will check the accuracy of the nature and implementation of learning and will be a significant predictor of instruction assessment. Furthermore, it is possible to verify and replenish one another with a smart contract between teachers and schools as well as the one between teachers and students. Teachers who meet the requirements would be rewarded with digital currencies. It acts as both an acknowledgment and support for the teaching skills of teachers.

In terms of career success, the instructor or college adviser is primarily responsible for coordinating the student's curriculum. We have the duty to support the subject in the preparation of study programs and keep aware of the learning efforts and success of the project. Such problems are not reviewed and monitored in practice, though, and it would be difficult to distinguish the obligations if anything bad occurs in the future. This circumstance would change if there is usage of smart contract and blockchain technologies in this field. All data should be tracked and registered in the blockchain ledger via the smart contract framework. Such as how many times during the previous semester did the teacher negotiate with students? How many times did the supervisor study the thesis in draft as well as final form? Did they offer the students sufficient feedback in the collection and study design of courses? Taking into consideration the traceability and immutability of blockchain technologies, the actions of both students and instructors can be documented in the blockchain ledger. The creative technology will protect all parties' preferences.

Broadly speaking, blockchain may be used to create a framework for the calculation of learning processes and performance. It is a credible and fair evidence of everybody's value. Theoretically, because of its high decentralization and immutability, blockchain can address knowledge asymmetry and confidence challenges among newcomers. It guarantees validity as the knowledge and meaning is publicly published and preserved. This offers a safe path for the expenditure of talent. The consumer with more digital currency awareness has plenty of chances for gaining recognition and investment. The blockchain database records everything you've ever experienced. Employers will use this knowledge to provide you with the work that suits your abilities. The subscriber who wants an outstanding employee, on the other hand, may also accede to the blockchain ledger. This would significantly reduce the possibility of prejudice and loss of investments. In a word, blockchain maximizes all parties' rights.

11.8 PROSPECTIVE CHALLENGES AND APPLICATION OF BLOCKCHAIN TECHNOLOGIES IN ONLINE EDUCATION

The possible risks to using blockchain technologies in education are evident.

In a dynamic method, the teachers must subjectively evaluate certain learning habits and learning results, such as articles and curriculum presentations. Without human involvement, it is very challenging to test this kind of learning practice via the preprogrammed smart contract.

If an educational blockchain program is introduced in classrooms, the educational details of all students will be merged into the blockchain ledgers. The blockchain technology's immutability functionality will serve as a double-edged sword. It removes the opportunity for any applicants to change their school reports for acceptable purposes.

Additionally, other technological problems or hurdles for the blockchain to be used in education are not discussed. For example, the classic method of agreement Proof of Research consumes resources and has low efficiency in terms of amount of transactions per second (Vukolić, 2015), which will incur an unnecessary expenditure and obstruct its use in schools.

As described previously, the technological features of the blockchain encourage a range of good solutions to the problems of online education. This chapter aims to apply blockchain technologies in the following fields of online education.

- **Complete learning trajectory record**

The blockchain stores data in a centralized archive, which tracks time stamp data blocks in sequential order. No new data blocks can be removed. The cryptographic algorithm is used to avoid data manipulation, leading to the complexity of deception. Currently, most online training sites are open, providing courses of varying consistency. Worse, the outcomes of research lack general recognition because of the absence of a centralized certification program. Predictably, online education has not been successful. The historical data storage of the blockchain is a reliable way to document the learning results of online education.

Student learning records, including class time, course files and test scores, can be documented on the blockchain in chronological order, and each student's record can be labelled with a time stamp. The integrity of the data is secured by a cryptographic recording system that avoids risks such as interference or deletion. Huge kudos to the decentralization, the centralized ledger and the ongoing management of the blockchain, every educational network or institution would be able to monitor the learning trajectories of students through regions and time. This would increase network performance and increase hardware costs.

In addition to completely preserving student learning data, the blockchain-based learning database avoids tampering and destruction and gives a good guarantee for the integrity of the student learning data. At the same time, the learning data, the confidentiality of which is guaranteed by the encryption technology, can be distributed across the network and conveniently accessed by the employer. From blockchain-based data, the employer can learn more about the learning progress of students and check their knowledge. Thus, blockchain technology will easily prevent document theft, bogus college degrees and other wrongdoing in higher education and set up a transparent network for graduates, teaching systems and employers.

- **Trustworthy and reliable certification in learning results**

Despite the overwhelming success of online training sites, after taking a few courses the students are not passionate because the learning outcomes are not widely accepted or officially accredited. It is because of the difficulty in moving forward the evaluation of learning outcomes. Currently, third-party organizations are inefficiently performing inspection for online education. This style can't satisfy the potential needs of the online education boom. When a student seeks work, his/her credentials are archived on the educational website or at the workplace, which the employer must check. If he/she loses a certificate, to get another copy of the certificate from the website or course, the student needs to go through a difficult and wasteful method. Furthermore, blockchain technology offers an easy, efficient solution for certifying the learning outcomes, especially academic certification.

Student credentials can be quickly checked even though they are misplaced. The blockchain pursues an asymmetric cryptographic encryption algorithm to ensure the data is safe and trustworthy. On this basis, a series of credential programs for learning outcomes can be developed. First, the electronic training site or the authorizing institution stores student learning data based on blockchain technologies, including basic information, course statistics, course grades, date of question, etc., and encrypts the data using the network or organization's private key. Then, the authenticated coded certificates inside the network are given to the students and other recipients. In this case, the contractor may use the public key of the network or company to conduct hash verification of the cryptographic certificates.

Blockchain records are tamperproof and cryptographic. The blockchain platform will offer a reliable mechanism for certifying the learning outcomes. The learners do not have to worry about the lack of credentials with this approach, the platform or company can streamline the process of obtaining credentials and the employer spends less on testing the learning outcome.

- **Decentralized Education Provision**

There are already several online training platforms available, providing a rich selection with varying courses. However, the courses are not distributed around the sites because of limitations such as instructional style, copyright, etc. The user experience for those taking different types of courses is indeed very bad, as they have to sign in to different courses. Similarly, graduates of higher learning have a very challenging time exercising the expertise of another school or specialty because of the absence of organized and effective usage, often basic course materials are missed. With the growing social economy (e.g., shared bikes), society is demanding greater use of resources. Sharing of resources represents the potential direction of development in the education field. Blockchain technology helps students to consider the exchange of information digitally (Ibid, 256–257).

As a standard implementation of blockchain technologies, the smart contract is a software structure built for the cryptographic security process. Complex transfers should take place without intervention by humans. The system also allows for automated deployment and authentication. Smart contract technology can streamline the transaction process, provide secure, automatic and transparent transactions and increase the security of transactions (Morrison, 2016).

Through decentralizing blockchain storage and shared control, students can access the strength of several networks by signing into a single node in the blockchain network. Furthermore, data from educational institutions cannot be replaced if particular nodes become disrupted in attacks, which provides a tremendous data protection assurance. Furthermore, existing information networks such as Wikipedia, research institutions, academic journals and other educational data are connected to the blockchain network that leverages blockchain technology, creating a wide knowledge base. Nodes in any blockchain network can have access to these information services. It dramatically increases learning performance and enhances the learning processes.

11.9 IN-RETROSPECT: TOWARD CONCLUSION

With the growth of Internet technologies, online and virtual education has become a modern way for people to learn. However, online learning often poses issues such as lack of validation of performance, loss of anonymity and loss of a networking system. As an evolving computing platform, blockchain has been extensively adopted in a number of fields owing to its transparent, trusted and secure functionality.

The deployment of blockchain technologies to the educational sector is in its adolescence. To the authors' knowledge, this is the first analysis on this topic. The analysis covered 31 papers that discussed three major themes: technologies, advantages and challenges. It gives rise to various observations. First, it stated that blockchain technology is mostly used to issue and validate academic credentials, share student expertise and learning accomplishments, and assess their technical skills. However, a broad variety of other technologies are growing rapidly. Second, it shows that blockchain will offer enormous benefits to education, including offering a safe platform for student data sharing, reducing costs and increasing transparency and accountability. Third, it demonstrates the usage of blockchain technology is not without hurdles. Managers and decision-makers must understand the related confidentiality, privacy, cost, accessibility and functionality barriers prior to implementation of the technology. Lastly, it demonstrates that the educational fields in which blockchain technology has been deployed are still minimal. The possibilities for blockchain are therefore still untapped.

To sum up, there are several ways by which blockchain will boost the education sector. This chapter explores the potential use of blockchain technology in education but focuses specifically on possible changes in the positions of both individuals and organizations participating in the teaching-learning process that this groundbreaking technology will bring to bear. Built on a peer-to-peer (P2P) topology, blockchain is a distributed ledger (DLT) platform that enables data to be stored on thousands of servers worldwide—allowing anyone on the network to track their inputs in near real time. That makes it complicated for a single person to take control over a network, or a game. The technology is perfect for secure information storage, sharing and networking. Many operations will become quicker, simpler and better with the aid of this advanced platform. It fills the gap in credentialing, copyright protection and efficient communication. These standard processes will surely benefit from blockchain very soon.

Technologies such as artificial intelligence and virtual reality are already entering the education sector. As we know that new technologies are entering our lives, we should be smartly applying them to allow progress to move in the right direction.

Online education, though, still poses problems including lack of validation of results, loss of privacy and the failure of a networking infrastructure. As an evolving computing technology, blockchain has been widely used in a number of fields because of its transparent, reliable and secure functionality.

By understanding "learning is earning," for example, blockchain technology will promote the learning drive of the students. It will store a full, reliable collection of records of instructional events in both structured and informal learning settings including the procedures and outcomes. It may also monitor the teaching activities and results of students, thereby offering a guide for evaluating teaching. Blockchain has significant possible applications for both learners and teachers in educational architecture, behavior tracking and interpretation, as well as seminal evaluation. At the same time, it poses threats and incentives to scholars, designers and instructors.

For scholars, blockchain has an outstanding ability to be commonly utilized in education platforms. However, very few research studies have been conducted. It will be difficult to research more closely on issues such as, what prospects would the education revolution offer? How to allow the effective use of digital currency properties to improve learning opportunities and accomplishments?

For designers, these innovative concepts are only the first moves to introduce blockchain to school. It is also an essential aspect of designing educational systems and applications that presents challenges for developers. How to create a blockchain network that meets users' customized needs? How can I integrate technology with blockchain to create an ecosystem for data collection and recording? How can you manage a large volume of educational transaction details and incorporate blockchain technology into current educational platforms and systems?

For instructors, some of the advantages of implementing blockchain technologies in the design of smart contact-based learning practices may be that they are verifiable, transparent and traceable. The confidentiality element offers great safeguards for teachers who have done an incredible job. In addition, school administration in the evaluation of teaching results will also be updated in order to incorporate this modern technology.

The affirmation of protection, anonymity, confidence and equality in the 21st century can be enforced through blockchain technology. Safety relates to the security of sensitive assets and knowledge. Nowadays, certain individuals have properties but are reluctant to assert possession, such as intellectual property lawsuits. This might contribute to dispute with others. Blockchain technologies may be used to verify the property by testing the documents in the ledger. A great deal of confidential knowledge, such as concept plans and strategic strategy, can be intercepted by industrial spies. Blockchain infrastructure may be used to secure this important enterprise by storing data on a blockchain network. The blockchain infrastructure safeguards the instructional architecture of teachers from expropriation, thereby enhancing the quality of intellectual property rights.

Security ensures that each and every node saves the entire ledger, including all details except the actual identity. For the purpose of protection, identification numbers are always given to consumers. It ensures that blockchain technology preserves the identity of the dealer, because no one else can have a secret key. In the educational framework, all details regarding the training opportunity documented on blockchain can only be accessed from the private key of its particular person. But most are not accessible, which ensures that the confidentiality of blockchain users may be assured.

However, for sanguinity, blockchain technologies will switch people's forms of rebuilding relationships from fostering it through a third-party entity to building it through technologies. Teacher and student activities are also registered and tracked while the hash value and blockchain are implemented. Security amongst respondents is focused on the technology itself, not on a third party.

Equality applies to the fair privileges and resources that everybody has in the blockchain network. The transparencies, limitlessness and unrestrained existence of blockchain technologies will allow anyone fair exposure to the platform and the network that is created around it. Anybody may register for an authentication server on a blockchain network. Blockchain infrastructure does not impose any limitations on consumers. Almost all universities, instructors and learners should implement it on a day-to-day basis, eliminating accountability-bias.

As a consequence, this chapter blends blockchain technologies with online education to address these challenges, creating a secure, open and accessible education distribution network. The results of the study reflect the emergence of online and virtual education.

REFERENCES

Ayers, R. (2019). How Will Blockchain Transform the Education System? Retrieved from https://dataconomy.com/2019/01/how-will-blockchain-transform-the-education-system/ Accessed on 30 April 2020.

Chen, G., Xu, B., Lu, M. et al. (2018). Exploring Blockchain Technology and Its Potential Applications for Education. *Smart Learning Environments*, 5: 1. Available at https://doi.org/10.1186/s40561-017-0050-x Accessed on 18 April, 2020.

Chung, M., and Kim, J. (2016, March 30). The Internet Information and Technology Research Directions Based on the Fourth Industrial Revolution. *KSII Transactions on Internet and Information Systems*, 10(3): 1311–1320. Retrieved from www.itiis.org/digital-library/manuscript/1286. Accessed on 10 June, 2020.

Coinbase. (2017). What is the Bitcoin Blockchain? Retrieved from https://help.coinbase.com/en/coinbase/getting-started/general-crypto-education/what-is-the-bitcoin-blockchain# Accessed on 12 June 2020.

Collins, R. (2016). Blockchain: A New Architecture for Digital Content. *EContent*, 39(8): 22–23. Retrieved from www.econtentmag.com/Articles/Editorial/Commentary/Blockchain-A-New-Architecture-for-Digital-Content-114161.htm. Accessed on 11 June 2020.

Devaney, L. (2016). 5 Things to Know About Blockchain Technology, 12: 11. Retrieved from www.pcworld.com/article/3053946/5-things-you-should-know-about-the-blockchain.html. Accessed on 25 April 2020.

Emilie, H. (2016). Investigating the Potential of Blockchains, 12: 11. Retrieved from http://www3.weforum.org/docs/WEF_Realizing_Potential_Blockchain.pdf. Accessed on 15 June 2020.

Fanning, K., and Centers, D. P. (2016). Blockchain and Its Coming Impact on Financial Services. *Journal of Corporate Accounting & Finance*, 27(5): 53–57. Retrieved from https://doi.org/10.1002/jcaf.22179. Accessed on 27 April 2020.

Gatteschi, V., Lamberti, F., Demartini, C., Pranteda, C., and Santamaría, V. (2018). Blockchain and Smart Contracts for Insurance: Is the Technology Mature Enough? *Future Internet*, 10: 20.

Haber, S., & Stornetta, W. S. (1991). How to Time-Stamp a Digital Document. *Journal of Cryptology*, 3(2): 99–111. Retrieved from https://doi.org/10.1007/BF00196791. Accessed on 10 June 2020.

Hoy, M. B. (2017). An Introduction to the Blockchain and Its Implications for Libraries and Medicine. *Medical Reference Services Quarterly*, 36(3): 273–279. Retrieved from https://doi.org/10.1080/02763869.2017.1332261. Accessed on 10 June 2020.

Johnson, D., Menezes, A., and Vanstone, S. (2001). The Elliptic Curve Digital Signature Algorithm (ECDSA). *International Journal of Information Security*, 1(1): 36–63. Retrieved from https://doi.org/10.1007/s102070100002. Accessed on 12 June 2020.

Kosba, A., Miller, A., Shi, E., Wen, Z., and Papamanthou, C. (2016). Hawk: The Blockchain Model of Cryptography and Privacy-Preserving Smart Contracts. *Security & Privacy*, 839–858.

Kraft, D. (2016). Difficulty Control for Blockchain-Based Consensus Systems. *Peer-to-Peer Networking and Applications*, 9(2): 397–413. Retrieved from https://doi.org/10.1007/s12083-015-0347-x. Accessed on 22 April 2020.

Lielacher, A. (2017). An Introduction to Cryptoeconomics. *BTCMANAGER*. Retrieved from https://btcmanager.com/an-introduction-to-cryptoeconomics/. Accessed on 15 June 2020.

Morrison, A. (2016). Blockchain and Smart Contract Automation: How Smart Contracts Automate Digital Business, 1: 7. Retrieved from https://declara.com/content/YamDM4e5. Accessed on 10 June 2020.

Nakamoto, S. (2008). *Bitcoin: A Peer-to-Peer Electronic Cash System*. Retrieved from https://bitcoin.org/bitcoin.pdf. Accessed on 12 June 2020.

Nakamoto, S. (2016a). Bitcoin: A Peer-to-Peer Electronic Cash System, 11: 29.

Nakamoto, S. (2016b). Sony Global Education Develops Technology Using Blockchain for Open Sharing of Academic Proficiency and Progress Records, 2: 16.

Okoli, C., and Schabram, K. (2010). A Guide to Conducting a Systematic Literature Review of Information Systems Research. *Sprouts: Working Papers on Information Systems*, 10: 1–51. Retrieved from http://sprouts.aisnet.org/10-26. Accessed on 12 June 2020.

Oxford Dictionaries. (2018). Blockchain | Definition of Blockchain in English by Oxford Dictionaries. Retrieved from www.lexico.com/definition/blockchain. Accessed on 15 June 2020.

Redman, J. (2016). MIT Media Lab Uses the Bitcoin Blockchain for Digital Certificates, 12: 11. Retrieved from www.newsbtc.com/2016/06/05/mit-uses-bitcoin-blockchain-certificates/. Accessed on 18 June 2020.

Schwab, K. (2017). *The Fourth Industrial Revolution*, The Crown Publishing Group, New York City, NY.

Sharma, R. C., Yildirim, H., and Kurubacak, G. (2019). *Blockchain Technology Applications in Education*. Preface, IGI Global. Retrieved from https://doi.org/10.4018/978-1-5225-9478-9.ch013

Sharples, M., and Domingue, J. (2016). The Blockchain and Kudos: A Distributed System for Educational Record, Reputation and Reward. In K. Verbert, M. Sharples & T. Klobučar (Eds.) *Adaptive and Adaptable Learning: Proceedings of 11th European Conference on Technology Enhanced Learning (EC-TEL 2015)*, Lyon, France, 13–16 September 2016. Switzerland: Springer International Publishing, pp. 490–496.

Skiba, D. J. (2017). The Potential of Blockchain in Education and Health Care. *Nursing Education Perspectives*, 38(4): 220–221 https://doi.org/10.1097/01.NEP.0000000000000190

Stroud, F. (2015). Blockchain: Webopedia Definition. Retrieved from www.webopedia.com/TERM/B/blockchain.html. Accessed on 7 May, 2020.

Sun, H., Wang, X., and Wang, X. (2018). Application of Blockchain Technology in Online Education, *IJET* 13(10): 252–259. Retrieved from https://doi.org/10.3991/ijet.v13i10.9455. Accessed on 30 April 2020.

Swan, M. (2015). *Blockchain: Blueprint for a New Economy*, 1st edn. O'Reilly Media, Sebastopol, CA.

Tschorsch, F., and Scheuermann, B. (2016). Bitcoin and Beyond: A Technical Survey on Decentralized Digital Currencies. *IEEE Communications Surveys and Tutorials*, 18(3): 2084–2123. Retrieved from https://doi.org/10.1109/COMST.2016.2535718. Accessed on 25 April 2020.

Underwood, S. (2016). Blockchain Beyond Bitcoin. *Communications of the ACM*, 59(11): 15–17. Retrieved from https://doi.org/10.1145/2994581. Accessed on 10 April 2020.

United States Agency for International Development (USAID), Primer on Blockchain. Available at: www.usaid.gov/sites/default/files/documents/15396/USAID-Primer-Blockchain.pdf. More information on blockchain technology is available from the National Institute of Standards and Technology. Retrieved from https://nvlpubs.nist.gov/nistpubs/ir/2018/NIST.IR.8202.pdf. Accessed on 7 April 2020.

University College London using Bitcoin Verification to Overcome cv fraud, 1: 29 (2018). Retrieved from http://blockchain.cs.ucl.ac.uk/press-release/. Accessed on 10 June, 2020.

Vukolić, M. (2015). *The Quest for Scalable Blockchain Fabric: Proof-of-Work vs. BFT Replication*. Open Problems in Network Security, Springer, Cham, pp. 112–125. Retrieved from https://doi.org/10.1007/978-3-319-39028-4_9. Accessed on 18 April 2020.

Wang, H., Chen, K., and Xu, D. (2016). A Maturity Model for Blockchain Adoption. *Financial Innovation*, 2(1): 12. Retrieved from https://doi.org/10.1186/s40854-016-0031-z. Accessed on 17 June 2020.

Waterhouse, S., Doolin, D. M., Kan, G., and Faybishenko, Y. (2002). Distributed Search in P2PNetworks. *IEEE Internet Computing*, 6(1): 68–72. Retrieved from https://doi.org/10.1109/4236.978371. Accessed on 25 June 2020.

Yli-Huumo, J., Ko, D., Choi, S., Park, S., and Smolander, K. (2016). Where is Current Research on Blockchain Technology?—A Systematic Review. *PLoS One*, 11(10): e0163477. Retrieved from https://doi.org/10.1371/journal.pone.0163477. Accessed on 12 June 2020.

Index